COMPUTATIONAL METHODS AND TESTING FOR ENGINEERING INTEGRITY

FIRST INTERNATIONAL CONFERENCE ON COMPUTATIONAL METHODS AND TESTING FOR ENGINEERING INTEGRITY

CMT 96

CONFERENCE DIRECTORS

T.V. Duggan

Wessex Institute of Technology, UK

C.J. Flavell

Dartec Limited, UK

INTERNATIONAL SCIENTIFIC ADVISORY COMMITTEE

Organised by:

Wessex Institute of Technology, UK

Co-Sponsored by:

Dartec Limited, UK

COMPUTATIONAL METHODS AND TESTING FOR ENGINEERING INTEGRITY

EDITOR:

T.V. Duggan
Wessex Institute of Technology, UK

Computational Mechanics Publications
Southampton Boston

1996

T.V. Duggan
Consultant
Wessex Institute of Technology
Ashurst Lodge
Ashurst
Southampton SO40 7AA, UK

Published by

Computational Mechanics Publications
Ashurst Lodge, Ashurst, Southampton SO40 7AA, UK
Tel: 44 (0)1703 293223; Fax: 44 (0)1703 292853
E-mail: cmp@cmp.co.uk

For USA, Canada and Mexico

Computational Mechanics Inc
25 Bridge Street, Billerica, MA 01821, USA
Tel: 508 667 5841; Fax: 508 667 7582
E-mail: cmina@ix.netcom.com.

British Library Cataloguing-in-Publication Data

A Catalogue record for this book is available
from the British Library

ISBN: 1 85312 412 5 Computational Mechanics Publications, Southampton

Library of Congress Catalog Card Number 96-83304

The texts of the various papers in this volume were set individually by the authors or under their supervision

Printed and bound in Great Britain by Antony Rowe Ltd.

PREFACE

This volume represents the papers presented at the First International Conference on Computational Methods and Testing for Engineering Integrity, which took place on 19-21 March 1996, in Kuala Lumpur, Malaysia. The conference was organised by the Wessex Institute of Technology and co-sponsored by Dartec Limited.

Assessing the engineering integrity of materials, components, structures and systems is a fundamental requirement of engineering design and analysis. The ubiquity of the computer and the development of sophisticated software and testing facilities have led to simulation models which can reliably predict behaviour and performance of systems of all kinds with a high level of confidence. As new materials and simulation models are developed further, it becomes even more important to test for reliability and improve confidence levels. The increasing success in simulating a wide range of engineering problems requires the effective validation of these models against experiment. The next important advance is likely to be the ability to use real time processing during actual testing, and apply appropriate simulation models to assess integrity. Such an approach will enable concurrent engineering to be more effectively achieved and improve confidence levels in assessing the performance and behaviour of systems of all kinds. This applies just as much to the testing of software for life prediction models as it does, for example, to obtaining experimental data for materials, components and structural behaviour.

The book, which includes papers by internationally recognised specialists, covers a wide range of topics associated with the application of computational methods and testing in order to assess risk and performance behaviour, particularly in the areas of the mechanical behaviour of materials, experimental and computational simulation, and in design and model simulation. These will enhance the knowledge and appreciation of engineering integrity and its importance in different disciplines and parts of the world, and will consequently provide a useful reference which will contribute to a better understanding of engineering integrity and its relevance to economic and strategic issues.

The organisers wish to express their appreciation to the invited speakers and all the delegates for their contributions and active participation in the conference. In particular, acknowledgement is due to the members of the International Scientific Advisory Committee for their strong encouragement and support throughout the planning and implementation of this event, and to the co-sponsors Dartec Limited.

The Editors
March 1996

CONTENTS

Section 1: Mechanical Behaviour of Materials

Section 2: Experimental and Computational Simulation

SECTION 1: MECHANICAL BEHAVIOUR OF MATERIALS

Crack propagation in pipes under cyclic axial loading

Andrea Carpinteri, R. Brighenti, A. Spagnoli
Dipartimento di Ingegneria Civile, Università di Parma, Viale delle Scienze, 43100 Parma, Italy

Abstract

The propagation of a circumferential external surface flaw in a metallic round pipe under constant amplitude cyclic axial loading is examined through a two-parameter theoretical model. The surface flaw is assumed to present an elliptical-arc shape with aspect ratio $\alpha = a_{el}/b_{el}$. The relative depth ξ of the deepest point on the crack front is equal to the ratio between the maximum crack depth a_{el} and the pipe wall thickness t. The edge flaw being considered is shown to follow preferred propagation patterns in the diagram of α against ξ.

1 Introduction

Several K_I-solutions for edge flaws in plates and round bars have been deduced by using different methods [1-4], but only few authors have examined the behaviour of surface flaws in pipes [5-9].

In the present study, a circumferential external surface flaw is assumed to exist in the centre-line cross section of a metallic round pipe subjected to a constant amplitude cyclic axial loading. The defect presents an elliptical-arc shape with aspect ratio $\alpha = a_{el}/b_{el}$ (Fig.1), while the relative depth ξ of the deepest point A on the crack front is equal to the ratio between the maximum crack depth a_{el} and the wall thickness t of the pipe. The aspect ratio α is made to vary from 0.0 to 1.2 ($\alpha = 0.0$ corresponds to the straight front, and $\alpha = 1.0$ represents the circular-arc front), whereas the relative crack depth ξ ranges from 0.05 to 0.8. Moreover, the ratio R/t between the internal radius of the pipe and the wall thickness is assumed to be equal to 1 or 10.

The stress-intensity factor variation along the crack front is de-

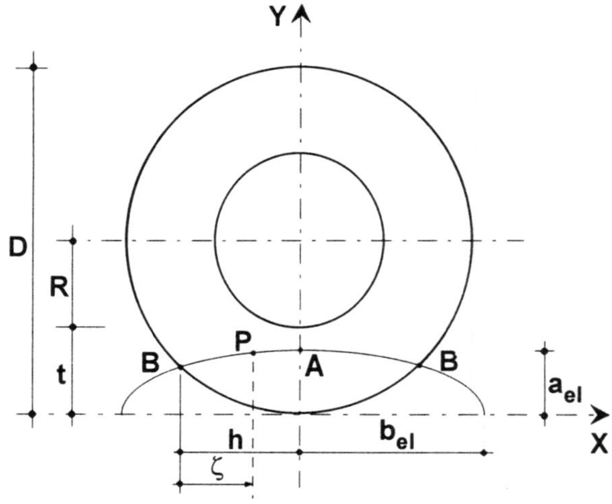

Fig. 1. Circumferential external surface flaw in a round pipe.

termined through a finite element analysis carried out with 20-node brick elements. Then the fatigue crack propagation is analyzed by means of a two-parameter theoretical model. The problem is complex since the flaw is an elliptical arc during the whole propagation, but the aspect ratio changes. The edge flaw considered is shown to follow preferred propagation patterns in the diagram of α against ξ.

2 Finite element analysis

The stress field for the part-through cracked pipe under tension loading is determined by means of a finite element analysis. Because of symmetry, only one-fourth of the flawed structural component is modelled by 20-node isoparametric solid elements. A total of 3731 nodes and 720 finite elements are employed (Fig.2). The pipe sizes are assumed to be equal to D = external diameter = 50 mm and l = half-length = 200 mm.

Since the stress square-root singularity near the flaw is obtained by using quarter-point finite elements, reliable results cannot be deduced in a boundary layer near the intersection point B between crack front and external surface (Fig.1), but this effect is confined in a small region near point B [1]. Therefore, only the results for ζ/h from 0.10 (point C) to 1.00 (point A) are considered in the following.

The dimensionless stress-intensity factor $\widetilde{K}_{I,F}$ along the elliptical-arc crack front is calculated as follows :

$$\tilde{K}_{I,F} = \frac{K_{I,F}}{\sigma_F \left(\pi\ a_{el}\right)^{1/2}} \tag{1}$$

where

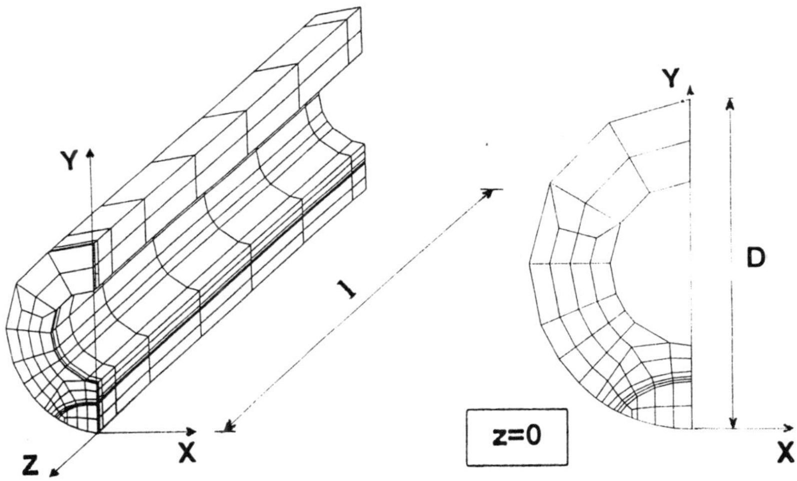

Fig. 2. Finite element mesh.

$K_{I,F}$ = stress-intensity factor for tension = $\sigma_{Z,F}\,(2\pi r)^{1/2}$, with $r \longrightarrow 0$,

$\sigma_{Z,F}$ = opening stress (perpendicular to the crack plane) at the distance r from the generic point P on the crack front, with r evaluated perpendicular to the front (Fig.1),

$\sigma_F = F/[\pi(D^2 - 4R^2)/4]$ = applied uniform tensile stress.

The stress-intensity factor $\widetilde{K}_{I,F}$ along the crack front is plotted in Fig.3 for different values of ξ, in the case of $R/t = 1$. For circular-arc front ($\alpha = 1.0$), the maximum stress-intensity factor is attained near the external surface.

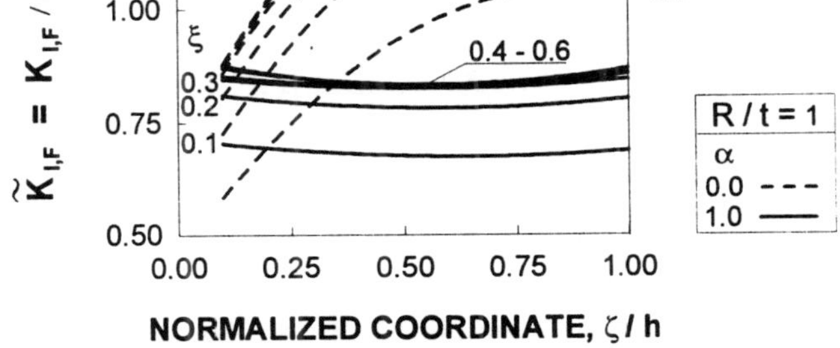

Fig. 3. Stress-intensity factor variation under tension, for $R/t = 1$.

On the contrary, all the dashed curves in Fig.3, which are related to the case of straight-fronted crack ($\alpha = 0.0$), increase by increasing the parameter ζ/h and, therefore, present the maximum value at point A.

The dashed curves for R/t = 10 (Fig.4) are qualitatively analogous to those for R/t = 1 (Fig.3). The continuous curves for thin-walled pipes increase for $\xi > 0.1$ and decrease for $\xi \leq 0.1$.

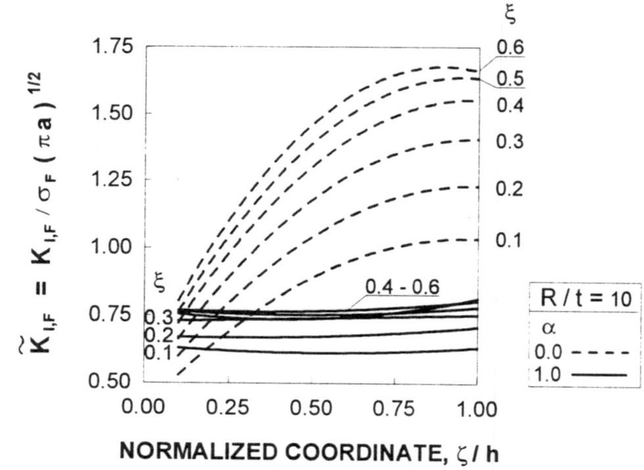

Fig. 4. Stress-intensity factor variation under tension, for R / t = 10.

Figure 5 shows the transition from $\alpha = 0.0$ to $\alpha = 1.2$ for $\xi = 0.3$, in the case of R / t = 1. The slope of the curve $\widetilde{K}_{I,F}$ becomes negative for α greater than or equal to about 1.0, that is to say, the maximum stress-intensity factor is attained at point A for $\alpha \leq 1.0$ and near point B for $\alpha \geq 1.0$. This transition phenomenon can also be observed for other values of relative crack depth, but the change of slope sign

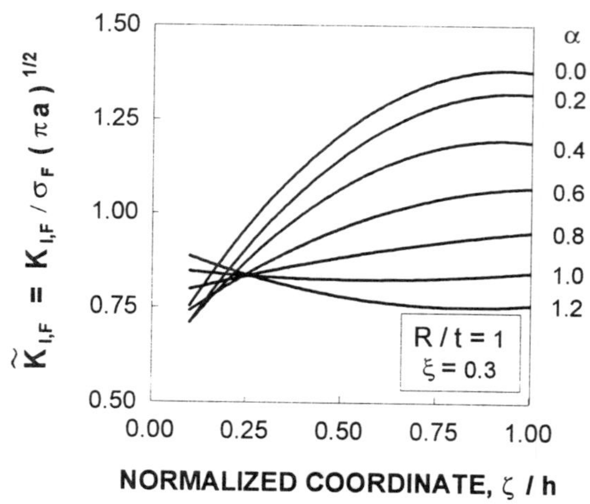

Fig.5. Influence of the crack aspect ratio α, in the case of R / t = 1.

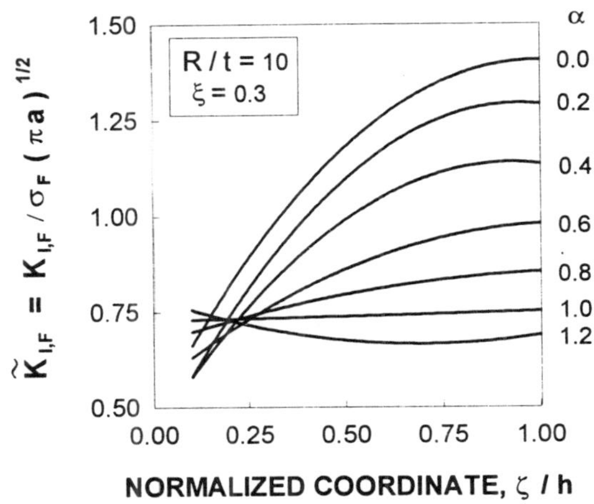

Fig.6. Influence of the crack aspect ratio α, in the case of $R/t = 10$.

occurs in correspondence to a value of aspect ratio which decreases by increasing ξ. The same phenomenon is displayed in Fig.6 for $\xi = 0.3$ and $R/t = 10$.

The results discussed in the present Section are quite close to those of other Authors [7,9].

3 Fatigue propagation model

Several authors have shown that the surface flaws in flat plates tend to follow preferred fatigue propagation paths, that is, the crack aspect ratio is a function of the relative depth (crack depth / plate thickness) for both axial and bending cyclic loading [1,10-13]. An analogous conclusion has also been drawn for surface flaws in round bars [14,15].

The fatigue growth of circumferential external surface flaws in pipes is theoretically examined. Since the ellipse centre is assumed to lie on the external perimeter of the pipe cross-section during the whole propagation (Fig.7), the generic crack front after one cyclic loading step can be expressed by the following equation:

$$\frac{x^2}{\left(b_{el}^{*}\right)^2} + \frac{y^2}{\left(a_{el}^{*}\right)^2} = 1 \qquad (2)$$

where the two unknowns $a_{el}{}^{*}$ and $b_{el}{}^{*}$ can be obtained from the condition that the coordinates of the points A* and C*, deduced from the Paris-Erdogan law [16], must satisfy eqn(2).

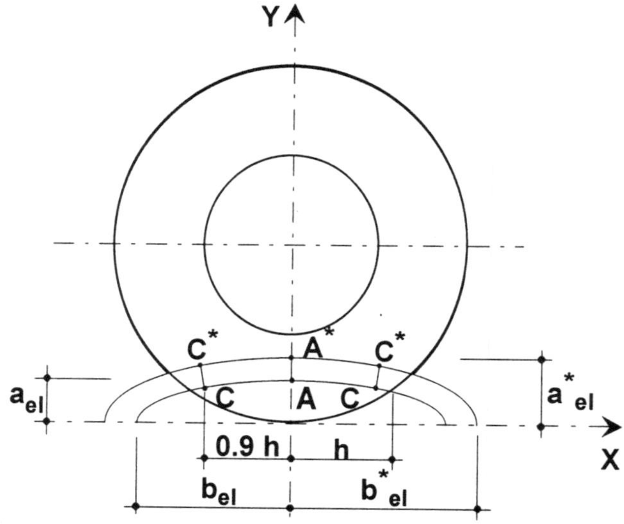

Fig. 7. Three-parameter model for fatigue crack propagation.

As an example, the propagation patterns for surface flaws under cyclic tension with stress range $\Delta\sigma_F = 100$ N mm^{-2} are plotted in Fig.8 for $R/t = 1$ and in Fig.9 for $R/t = 10$, respectively. The material constants in Paris-Erdogan equation are assumed to be $m = 2$ and $A = 1.64\times10^{-10}$, for da/dN in mm cycle^{-1} and $\Delta K_{I,F}$ in N mm$^{-3/2}$. The initial flaws examined have relative depth ξ_0 equal to 0.05, 0.10 or 0.20 and crack aspect ratio α_0 equal to 0.001 (straight front), 0.3, 0.5 or 1.0 (circular front). For each initial crack configu-

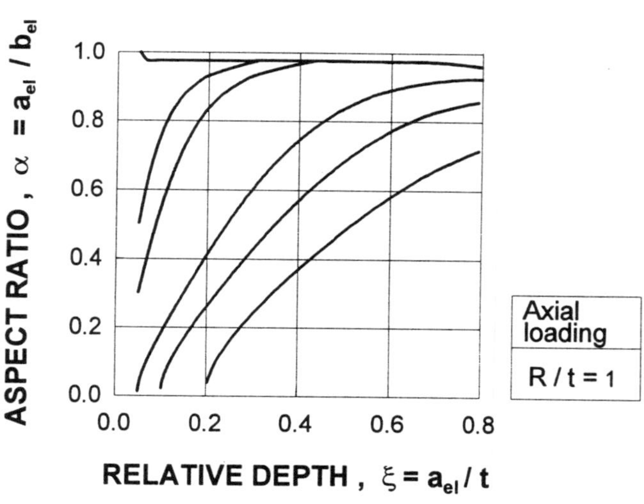

Fig.8.Crack propagation paths under cyclic axial loading, for R /t = 1.

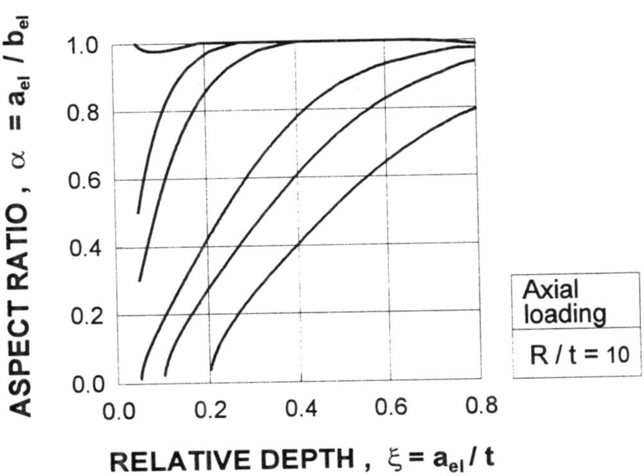

Fig.9.Crack propagation paths under cyclic axial loading, for R /t = 10.

ration analyzed (ξ_0, α_0), the propagation path tends to converge to a horizontal asymptote, with α included in the range 0.7 ~ 1.0 for $\xi = 0.8$.

4 Conclusions

The behaviour of circumferential external surface flaws in round pipes subjected to cyclic axial loading with constant amplitude has been examined. The following conclusions can be drawn :

(1) For both thin and thick wall thickness of the pipe, the stress-intensity factor variation along the crack front is more pronounced for high values of the relative crack depth $\xi = a_{el}/t$. For a given value of the parameter ξ, the variation is remarkably dependent on the crack aspect ratio $\alpha = a_{el}/b_{el}$.

(2) Fatigue crack propagation patterns have been obtained by means of a two-parameter theoretical model. It can be remarked that, for each initial flaw configuration, the crack growth path converges to a horizontal asymptote in the diagram of α against ξ.

Acknowledgements

The authors gratefully acknowledge the research support for this work provided by the Italian Ministry for University and Technological and Scientific Research (MURST) and the Italian National Research Council (CNR).

References

1. A. Carpinteri (Editor), *Handbook of Fatigue Crack Propagation in Metallic Structures*, Elsevier Science Publishers B.V., Amsterdam, The Netherlands, 1994.
2. A. Carpinteri, Stress-intensity factors for semi-elliptical surface cracks under tension or bending. *Engng Fracture Mech.* **38**, 327-334, 1991.
3. A. Carpinteri, Elliptical-arc surface cracks in round bars. *Fatigue Fracture Engng Mater. Struct.* **15**, 1141-1153, 1992.
4. A. Carpinteri, Surface flaw stress intensity factor computation with quarter-point elements. *J. Strain Anal. for Engng Design* **28**, 117-123, 1993.
5. F. Delale and F. Erdogan, Application of the line-spring model to a cylindrical shell containing a circumferential or axial part-through crack. *J.Appl.Mech.* **49**, 97-102, 1982.
6. R.G. Forman and V. Shivakumar, Growth behavior of surface cracks in the circumferential plane of solid and hollow cylinders. In *Fracture Mechanics : Seventeenth Volume*, ASTM STP 905, pp. 59-74, 1986.
7. I.S. Raju and J.C. Newman, Stress intensity factors for circumferential surface cracks in pipes and rods. In *Fracture Mechanics : Seventeenth Volume*, ASTM STP 905, pp. 789-805, 1986.
8. H. Gao and G. Herrmann, On estimates of stress intensity factors for cracked beams and pipes. *Engng Fracture Mech.* **41**, 695-706, 1992.
9. M. Bergman, Stress intensity factors for circumferential surface cracks in pipes. *Fatigue Fracture Engng Mater. Struct.* **18**, 1155-1172, 1995.
10. L.H. Burch, Fatigue growth of surface cracks in bending. *Engng Fracture Mech.* **9**, 389-395, 1977.
11. M.A. Mahmoud, Surface fatigue crack growth under combined tension and bending loading. *Engng Fracture Mech.* **36**, 389-395, 1990.
12. A. Carpinteri, Crack propagation under cyclic loading. *Fatigue Fracture Engng Mater. Struct.* **15**, 365-376, 1992.
13. A. Carpinteri, An automated procedure for fatigue crack growth analysis. *Int. J. Computers & Struct.* **44**, 1317-1338, 1992.
14. A. Carpinteri, Shape change of surface cracks in round bars under cyclic axial loading. *Int. J. Fatigue* **15**, 21-26, 1993.
15. A. Carpinteri and R. Brighenti, Part-through cracks in round bars under cyclic combined axial and bending loading. *Int. J. Fatigue*, to be published, 1995.
16. P.C. Paris and F. Erdogan, A critical analysis of crack propagation laws. *J. Basic Engng Trans. ASME* **85D**, 528-534, 1963.

Fatigue crack growth in welded joints

W.D. Dover,[a] C.C. Monahan[b]
[a]*NDE Centre, University College London, Torrington Place, London WC1E 7JE, UK*
[b]*Memorial University, Newfoundland, Canada*

Abstract

Recent developments in crack measurement technique ACPD/ACFM and parametric equations for weld toe non-linear stress distributions has allowed the validation of early fatigue crack growth models for tubular welded connections. A series of fatigue tests on X and multi-planar nodes have been conducted and these have provided early crack growth data. This information has been utilised to validate crack growth models for tubular welded connections.

1 Introduction

Welded joints, when subjected to even modest levels of cyclic loading, can be susceptible to fatigue damage. For this reason inspection and fatigue analysis of welded connections is very important. The nature of welded connections is such that crack initiation may not be a significant fraction of the total life leading to the studies, over many years, of fatigue crack growth in welded joints. At UCL, studies of fatigue crack growth have been made on tubular connections, butt welds and tee butt welds. This work has included constant amplitude, simulated service loading (e.g. WASH), and studies of the effect of cathodic protection on fatigue crack growth . From this work has emerged software for fatigue crack growth analysis (FACTS), techniques for crack detection, monitoring and sizing (ACPD and ACFM), information on inspection reliability (Probability of Detection (POD) curves) and inspection scheduling software (RISC). Much of this work has contributed to the revised HSE Fatigue Guidance Notes for Offshore Steel Structures (to be published shortly [1]).

A major outcome from these studies has been the development of crack detection and sizing techniques, now available commercially [2], based on the principle of alternating current field measurement. These techniques, used in either laboratories or in service, have reached the level where it is now possible to detect and monitor small cracks thus enabling further fatigue studies on early crack growth. This area has often been neglected partly because of the non linear stress gradients in this near surface region and also because of the difficulties in validation. The work reported here uses newly developed through thickness stress distributions in an attempt to predict early crack growth.

2 ACPD/ACFM Crack Detection & Sizing

Alternating current potential difference (ACPD) is a non-destructive testing (NDT) technique for the measurement of surface voltages adjacent to a crack. The raw data is theoretically interpreted to give an estimate of the crack size. The use of ACPD for measuring crack depth and monitoring crack extension in a wide variety of fatigue crack growth specimens and welded components has been possible for some time [3]. Crack sizing with ACPD has been made possible through the development of highly quality electronic instruments capable of measuring small surface voltages, and the development of a theoretical solution for interpreting the AC field distribution.

In order to achieve the most accurate results using ACPD, special attention must be directed towards producing a uniform electric field on the surface of the metal (in the anticipated region of the crack), and positioning the probe and field wires properly during measurement. The design of the ACPD probe is also important.

Some of the difficulties associated with ACPD have been overcome by the introduction of alternating current field measurement (ACFM). ACFM is a technique for measuring the magnetic field perturbations associated with electric field perturbations [4] [5]. The probe deployed in ACFM incorporates one large induction coil for generating the surface current and two small pick-up coils for measuring the magnetic field components in the vicinity of the crack. The inversion of magnetic field measurements to obtain surface voltages (using theory) effectively provides a non-contacting ACPD system. An added advantage of ACFM is that it can be used for crack detection as well as sizing. ACFM is now used widely in offshore, process, power-generating and construction industries for detecting and sizing surface breaking cracks.

The ACPD and ACFM technique measure the absolute values of the electric surface voltage and magnetic field strength, respectively, and use a theoretical interpretation of these measurements to give the crack size directly. This approach is in sharp contrast to other, more conventional techniques (e.g. eddy current systems) where only relative measurements are taken and calibration (often on a notched block) is necessary. ACPD and ACFM both have theoretical solutions incorporated into laboratory and in-service equipment and, because they do not require calibration, are quite versatile in use.

The techniques can both be used for crack sizing, either by periodic inspection, or by continuous monitoring. For ACPD, monitoring can be achieved with just a few spring loaded contacts or spot-welded connections as in compact tension fatigue crack growth tests, or with multiple connections (often spot-welded) as in tubular joint corrosion fatigue tests conducted for the offshore industry [6]. Various probe sizes are available for ACPD and the most appropriate has to be chosen for a particular application. ACFM can be carried out by moving a single sensor probe over the cracked region, or by positioning an array of sensor coils near the crack site. An interesting application of a linear flexible array has been described elsewhere [7], where it was used to monitor and interrupt a fatigue test on a welded joint, at the end of a crack initiation period. This avoided the need for periodic surveillance near the beginning of the test and allowed for continuous running of the fatigue test machine. Subsequent monitoring with the ACFM array probe demonstrated that the fatigue crack growth curve could be generated in crack depth increments as small as 0.1mm.

An alternative to induced field arrangements for ACFM is to use pick-up coils with an injected field; an arrangement which is closer to ACPD. For improved measurement precision, the pick-up coils could be mounted on an X-T (distance-time) tracking system. In this way, readings could be made along, say, a weld toe where a crack might be expected to initiate.

The basic principles of ACPD measurement can be illustrated by considering the case of a flat plate containing a shallow edge crack of depth a (see Figure 1). A uniform high frequency AC field is injected into the surface of the cracked component by means of shielded field input connections. The high frequency associated with the injected field causes the current to flow in a thin skin (δ_s) close to the metal surface. If the voltage drop measured in front of the crack (V_1) using a fixed probe spacing (Δ_p) is compared with the voltage drop measured across the crack (V_2) using the same probe spacing, it will be found that $V_2 > V_1$. This is because the crack increases the surface distance between the probe contacts by an amount equal to twice the crack depth (d_1). This change in surface distance ($2d_1$) is related to the increase in voltage drop by the equation

$$\frac{V_2}{V_1} = \frac{\Delta_p + 2d_1}{\Delta_p} \tag{1}$$

which, when solved for d_1 becomes

$$d_1 = \frac{\Delta}{2}\left(\frac{V_2}{V_1} - 1\right) \tag{2}$$

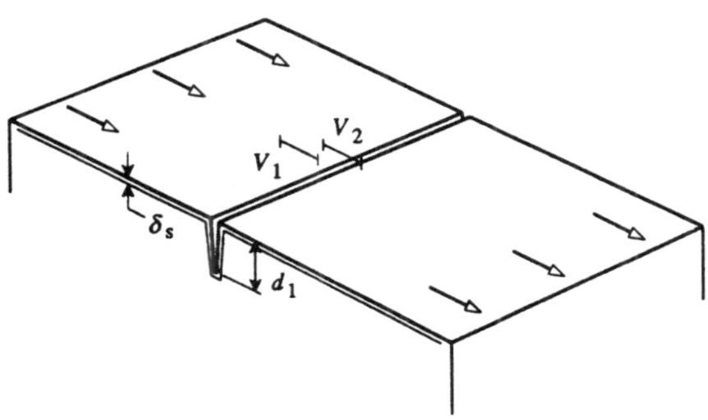

Figure 1 Prediction of crack depth using ACPD.

Equation (2) is referred to as the one-dimensional interpretation of ACPD crack depth. It is reasonably accurate for surface cracks when the crack length (2c) is large compared with the ACPD probe spacing, and the crack aspect ratio (a/2c) is less than about 0.10. When either of these conditions is not met, the one-dimensional estimate needs to be corrected using a theoretically derived modification factor called a "multiplier" [8].

ACFM in contrast can be illustrated by Figure 2 which shows some ACFM data collected above a well-defined fatigue crack in a submerged arc-welded flat plate. Figure 3 illustrates the geometry for the case being considered. The two magnetic field components of interest are the ones which are parallel to the crack (B_X) and perpendicular to the surface of the metal (B_Z). The magnitudes of these two parameters are recorded as a function of time as the ACFM probe is traversed along the toe of the weld. The probe deployed in ACFM incorporates one large induction coil for generating the surface current and two small pick-up coils for measuring B_X and B_Z in the vicinity of the crack. The introduction of a uniform high frequency surface current causes a B_X magnetic field component to be present everywhere above the surface of the plate. The presence of a surface discontinuity such as a fatigue crack results in less current flowing towards the defective region and a noticeable drop occurs in the strength of the B_X signal. As the current is diverted around the ends of the defect, positive and negative peaks appear in the B_Z signal and these are used to size the crack for length. Interpretation of the crack depth is based on the remote and minimum B_X levels and on the B_Z estimate of crack length (9). A semi-elliptical crack shape is assumed in the calculations.

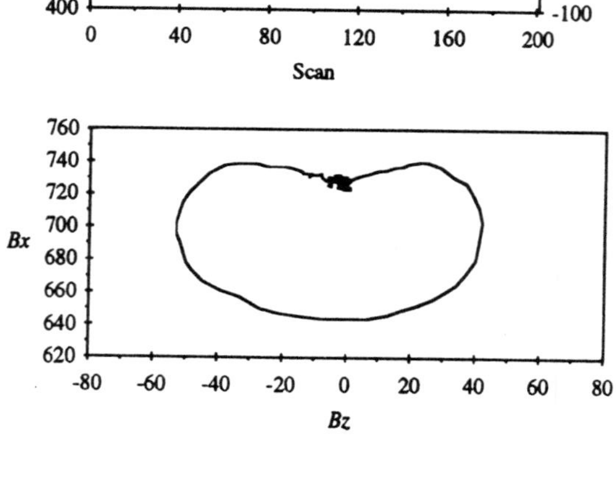

Figure 2 ACFM signals for a surface crack in a flat plate (a=4.5mm, 2c=39mm).

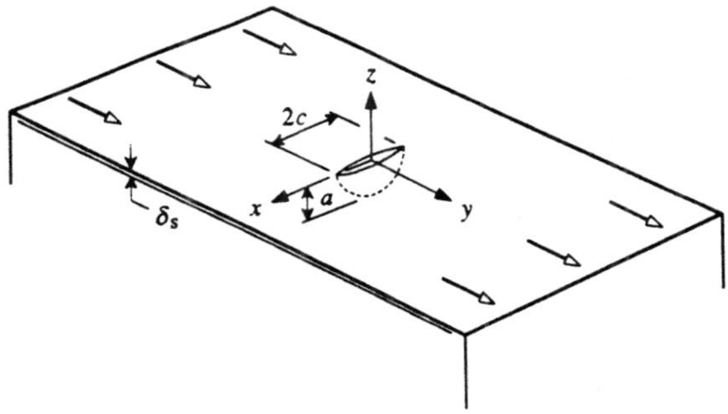

Figure 3. Geometry for the case of a surface crack interrogated by ACFM.

Plotting B_X against B_Z effectively removes the time base from the measurement and results in what is called a "butterfly plot" (see Figure 2). The advantage of the butterfly plot becomes apparent when the technique is used for crack detection. Changes in magnetic permeability near the toe of a welded joint are not uncommon and can lead to variability in the B_X and B_Z scans. In such cases, the butterfly plot enables the operator to clearly distinguish between spurious indications and true crack signals.

There are several obvious advantages of ACFM over the conventional ACPD method for monitoring the growth of small isolated fatigue cracks. Defects can be spotted at a very early stage in their development and by the time they reach 0.5 mm in depth, an indisputable crack signal is usually present. Measurements of crack depth are stable and repeatable, even when the sensing coils are deployed manually on unpolished welds, and interference from stray electrical noise is seldom a problem. With ACPD, early crack detection is not normally possible. Also, the successful application of ACPD requires a shiny metal surface for good electrical contact, and careful positioning of the field wires and probe pins. However, the ACPD technique becomes an indispensable tool when closely situated defects interact to produce long shallow fatigue cracks that are not semi-elliptical in shape.

3 Fatigue tests on tubular joints

Two multi-brace nodes (M-nodes) and one X-node, all fabricated from BS 4360 50D structural steel, were used for stress analysis and fatigue testing. The two different node geometries are depicted in Figures 4 and 5. All three nodes were fabricated in accordance with standard offshore welding practices and inspected using ultrasonics prior to leaving the supplier. Steel rings were later welded to the open ends of each chord and brace for the purpose of clamping the specimens in a test rig.

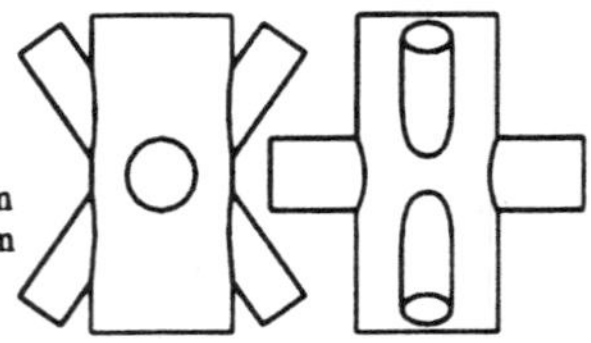

Figure 4 Multi-Brace (M-) Node geometry.

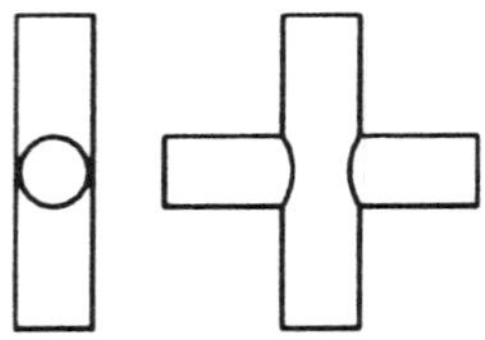

Figure 5 X-Node geometry.

The tests were carried out using in-plane bending (IPB) and out-of-plane bending (OPB). Four of the Y-braces were tested in OPB and one was tested in IPB. One of the larger T-braces was tested under IPB and two were tested in OPB. One of the smaller T-braces was tested in OPB.

Prior to fatigue testing strain gauge rosettes were used to measure the principle surface strains in the sites of maximum stress, the hot spot stress regions.

The choice of which side of the weld to place the gauges (i.e. brace toe or chord toe) depended on where the fatigue cracks were expected to grow. The Hellier-Connolly-Dover (HCD) parametric equations [10], were used to predict the most likely locations for fatigue cracks to develop.

Strain measurements from each gauge were recorded while the specimen was held static at a number of different load levels. Caution was exercised to ensure that the peak load used for stress analysis was always less than that which was to be used for fatigue testing. The strain measurements and load settings were used to calculate the maximum principle geometric surface stress per unit load (i.e. MPa/kN) at each gauge station. These results were then extrapolated to the weld toe to obtain estimates of the maximum principle geometric surface stress per unit load in that region.

Once the experimental stress analysis was complete, fatigue tests were carried out on each of the strain-gauged joints. A constant amplitude, sinusoidal load cycle and a test frequency of 1.0 Hz were used in all cases. Crack growth was monitored using a combination of the ACPD and ACFM techniques. The environment was room temperature (15-20°C) air. The T-OPB experiments involving the X-node and four of the Y-OPB experiments with the M-nodes were conducted under fully-reversed loading (R=-1). The remaining experiments involving the M-nodes were performed using a load ratio of R=0.05. The hot-spot stress range ($\Delta\sigma_h$) used in each test (confirmed from the experimental strain measurements) is given in Table 1.

Table 1 also gives details about other experimental parameters relevant to each test.

Brace Code	Node	Weld Toe	R	$\Delta\sigma_h$ (MPa)
T1-OPB	M	chord	0.05	191
T2-OPB	M	chord	0.05	200
T4a-OPB	X	chord	-1	250
T4b-OPB	X	chord	-1	230
Y1a-OPB	M	chord	-1	195
Y1b-OPB	M	chord	-1	172
Y2a-OPB	M	chord	-1	167
Y2b-OPB	M	chord	-1	200
Y1c-OPB	M	chord	0.05	200
Y2c-OPB	M	chord	0.05	199
T1-IPB	M	chord	0.05	200
Y1a-IPB	M	brace	0.05	141

Table 1. Fatigue test parameters

4 Results

Figures 6 and 7 show examples of the fatigue crack growth behaviour of the chord-toe cracks studied. Crack growth and shape changes are plotted in terms of the dimensionless parameters a/t and a/2c, where a is crack depth, t is chord wall thickness and 2c is crack surface length. In some cases a sudden acceleration in crack growth, following an abrupt decrease in aspect ratio (a/2c), can be seen. This effect was noticeable but less pronounced for other cracks. Figures 6 and 7 show examples of the sharp change in aspect ratio thought to be due to the merging of separate cracks. It appears that aspect ratio could exert an influence on the fatigue crack growth behaviour of welded tubular joints. The results from these tests were used for new models for early crack growth.

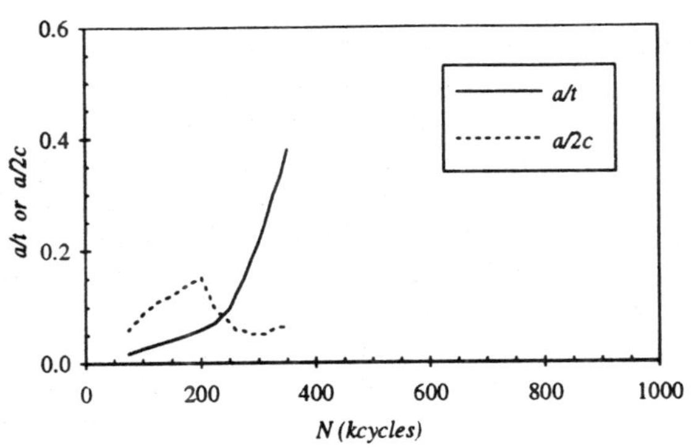

Figure 6 Fatigue crack growth curves for T2-OPB.

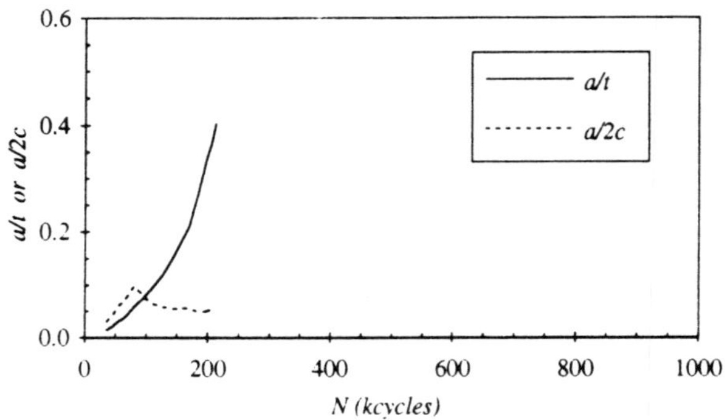

Figure 7 Fatigue crack growth curves for T1-IPB.

5 Discussion

The Mode I stress intensity factor (SIF) for a surface crack in a welded structure can be written as follows:

$$K_I = Y_s Y_w Y_e Y_g Y_k Y_m \ \sigma_{no} \sqrt{\pi \ a} \tag{3}$$

where σ_{no} is the nominal surface stress; a is the crack depth; Y_s corrects for a free front surface; Y_w accounts for finite plate width; Y_e corrects for crack shape; Y_g accounts for a non-uniform stress field; Y_k corrects for the presence of a geometrical discontinuity; and Y_m accounts for changes in structural restraint.

Several different SIF Solutions for surface cracks in flat plates exist e.g. [11] and these have been incorporated into FACTS [12]. In most cases they can be adapted to fit welded joints by using a non uniform stress correction. Ref [13] has shown that it is possible to incorporate the stress distribution using the following equation.

$$Y_g = \frac{2}{\pi} \sum_{i=1}^{n} \frac{\sigma_{x_i}}{\sigma} \left(sin^{-1}\left(\frac{x_{i+1}}{a}\right) - sin^{-1}\left(\frac{x_i}{a}\right) \right) \tag{4}$$

The stress fields surrounding the weld toe of a tubular joint are considerably more complex than those associated with a simple T-plate. For a T-plate tested under load control, the reactions at the supports (i.e. the boundary conditions) remain constant for the duration of the test and their influence on the stress intensity factor can be predicted without too much difficulty. For a complete local structure such as a tubular joint, the situation is different. To begin with, the local boundary conditions are not nearly as well defined, and to confound matters further, they vary with crack size. In addition to this, the through-thickness stress distribution will include a tensile component as well as a bending component, and the ratio of the two will depend on the overall joint geometry. As a result of these complications and others, an analytical SIF solution for weld toe cracks in tubular joints does not yet exist.

It is possible to model the fatigue behaviour of tubular joints using a flat plate SIF solution, if a force/moment release function is used to account for the stress release (load shedding) which accompanies crack growth in redundant structures. Finite element studies have shown that both the bending moment and membrane force decrease as the crack grows in tubular joints but that the bending moment effect is much greater than that of the membrane force. The assumption of linear moment release (LMR) for bending, and the retention of membrane stresses, is a conservative approximation to the real behaviour.

The LMR postulate states that as the crack grows through the chord wall, the membrane force acting within the cracked section remains constant, but the bending moment decreases to zero in a linear fashion. Thus, no correction is necessary for the component of membrane (tensile) stress, but the bending stress component needs to be multiplied by the release function

$$R(a/t) = 1 - \frac{a}{t} \tag{5}$$

where t is the chord wall thickness.

The fracture mechanics analysis of welded tubular joints requires knowledge of the stresses acting perpendicular to the prospective plane of the crack. The relative magnitude of these stresses will vary in the through-thickness direction and the shape of the distribution will depend on the local geometry of the weld toe and the global geometry of the tubular joint. Ref [14] has shown that finite-element (FE) calculated weld toe stress distributions in tubular joints are well-represented by combining the results of T-plate models loaded in pure tension and pure bending, provided the weld geometries are the same in both cases. This finding warrants the use of simple, two-dimensional FE models to calculate through-thickness stress distributions in welded tubular joints.

In subsequent studies [15], several different finite element models were used to calculate through-thickness stress distributions in T-plates subjected to either pure tension or pure bending. Numerous cases of weld geometry were investigated with values of α ranging from 30 to 60 degrees, and weld toe radius to plate thickness ratios (p/t) ranging from 1/15 to 1/50. These finite element results were used to derive parametric equations for predicting weld toe stress distributions for pure tension and pure bending as a function of the local weld geometry (α, p/t).

More recent work [16] extended the range of parameters to include the attachment size giving both updated stress distribution equations and parametric equations for SIF incorporating weld angle, weld toe radius, attachment size and aspect ratio. These are also now incorporated into FACTS.

In order to validate crack growth predictions, using say Newman & Raju (N&R) with non-uniform stress corrections (NSC) and linear moment release (LMR), it is necessary to have experimental data reduced to a convenient form. Typically the experimental crack growth needs to be analysed to give an empirical Y versus a/t relationship.

For the case of a welded tubular joint, it is possible to define an experimental stress intensity calibration factor,

$$Y_{EXP} = \frac{\Delta K_{eff}}{\Delta \sigma_{co}\sqrt{\pi a}} \tag{6}$$

where ΔK_{eff} is the effective range of the crack tip stress intensity factor, $\Delta\sigma_{co}$ is the crack-site surface stress range, and a is the depth of the crack. Values of Y_{EXP} may be calculated from fatigue crack growth rate data, if the relationship between ΔK_{eff} and the crack growth rate (da/dN) is known. It has been shown elsewhere [17] that an equation of the form

$$\frac{da}{dN} = C(\Delta K_{eff} - \Delta K_{th})^m \tag{7}$$

provides a reasonably good fit to surface fatigue crack growth data collected for 50D steel. In the above expression, ΔK_{th} is the threshold stress intensity factor range, and C and m are empirical crack growth constants. If Equation (7) is re-arranged to get

$$\Delta K_{eff} = \Delta K_{th} + \left(\frac{1}{c}\cdot\frac{da}{dN}\right)^{1/m} \tag{8}$$

and the right hand side of this expression is substituted in place of = ΔK_{eff} in Equation (6), then

$$Y_{EXP} = \frac{\Delta K_{th} + \left(\frac{1}{C}\cdot\frac{da}{dN}\right)^{1/m}}{\Delta\sigma_{c_o}\sqrt{\pi a}} \tag{9}$$

Calculation of Y_{EXP} thus requires knowledge of the empirical crack growth "constants" = ΔK_{th}, C and m. Equation (16) has shown that all three of these values are sensitive to mean stress at low and intermediate levels of = ΔK_{eff}, and that the following equations may be used to characterise the mean stress effect:

$$\Delta K_{th} = 8.6 - 9.5R_{eff} \tag{10}$$

$$C = 2.8\times10^{-11}\,(1 + R_{eff} - 3.16R^2_{eff}) \tag{11}$$

$$m = 2.76 - 0.65R_{eff} + 1.75R^2_{eff} \tag{12}$$

$$\text{for } 0.2 \le R_{eff} \le = 0.6$$

where R_{eff} is the effective stress ratio. At this stage, it becomes important to distinguish between the applied stress ratio (R) and the effective stress ratio (R_{eff}). If there are no residual stresses to consider and material response is entirely elastic, R and R_{eff} will be the same. However, in cases where residual stresses are known to exist and/or weld toe plasticity is present, the two stress ratios will be different. If we assume elastic-perfectly plastic material behaviour with the flow stress (σ_f) acting as a yield criterion, it is possible to calculate Reff from a simple set of rules. In cases where the weld toe stress range (σ_w) is less than $2\sigma_f$ and the sum of the maximum weld toe stress (σ_w^{max}) and the residual stress (σ_r) is less than σ_f, R_{eff} is given by

$$R_{eff} = \frac{\sigma_w^{max} + \sigma_r}{\sigma_w^{max} + \sigma_r} \tag{13}$$

where = σ_w^{max} is the minimum weld toe stress.

$$\text{If} = \Delta\sigma_w < 2\sigma_f, \text{ but} = \sigma_w^{max} \geq \sigma_f$$

$$R_{eff} = \frac{\sigma_f - \Delta\sigma_w}{\sigma_f} \tag{14}$$

Finally, if = $\Delta\sigma_w \geq 2\sigma_f$,

$$R_{eff} = -1 \tag{15}$$

If the weld toe and residual through-thickness stress distributions, $\sigma_w(x/t)$ and $\sigma_r(x/t)$, are known for a tubular joint, Equations (13-15) may be used to predict the effective stress ratio as a function of depth.

A numerical model (16) based on these simple rules was used to estimate R_{eff} for each of the dominant fatigue cracks. In all cases, the residual stress distribution was assumed to take the shape

$$\sigma_r(a/t) = \sigma_y (1 - a/t) \tag{16}$$

where a is crack depth, t is chord wall thickness and σ_y is yield stress (σ_y=386MPa for BS 4360 50D steel).

Once R_{eff} was estimated, appropriate values of = ΔK_{th}, C and m were computed using Equations (10-12) and substituted into Equation (21) to calculate Y_{EXP}. In cases where R_{eff} was estimated to be greater than 0.6, the values of = ΔK_{th}, C and m corresponding to R_{eff}= 0.6 were used to calculate Y_{EXP}. In cases where R_{eff} is estimated to be less than 0.2, suitable values for = ΔK_{th}, C and m may be found by setting R_{eff}= 0.2 in Equations (10-12), but this was not necessary for the data considered here.

Y_{EXP} values for four of the eight cases considered are plotted in Figures 8 to 11.

Figures 8-11 also include predictions using $Y_{N\&R}$ (plate solution) and $Y_{N\&R+NSC+LMR}$. It can be seen that the $Y_{N\&R+NSC+LMR}$ are quite good predictions for Y_{EXP}. However, in search for even greater accuracy in prediction it was noted that the effect of crack shape may be more significant than calculated.

Monahan & Dover [18] showed that a further crack shape correction may be necessary. This was based on the observation ψ defined below varied with aspect ratio a/2c as shown in figure 12 and that when this empirical factor was included excellent predictions were obtained.

$$\psi = \frac{Y_{EXP}}{Y_{N\&R+NSC+LMR}} \tag{17}$$

The general trend is for ψ to decrease with increasing crack aspect ratio (a/2c). If ψ is assumed to equal unity for $a/2c \leq 0.05$, then the following equations provide a conservative approximation to this behaviour:

$$\psi = 1 \quad \text{for} \quad a/2c \leq 0.05, \tag{18a}$$

and

$$\psi = \frac{1}{1+0.7(a/2c-0.05)^{0.4}} \quad \text{for} \quad 0.05 < a/2c < 0.26 \tag{18b}$$

The physical reasons for the empirical corrections are not as yet understood but it is possible that adjacent cracks may be an influence. Including the empirical crack shape correction (CSC) with other modifiers (e.g. $Y_{N\&R+NSC+LMR+CSC}$) gave very good predictions for Y_{EXP} as can be seen from figures 8-11.

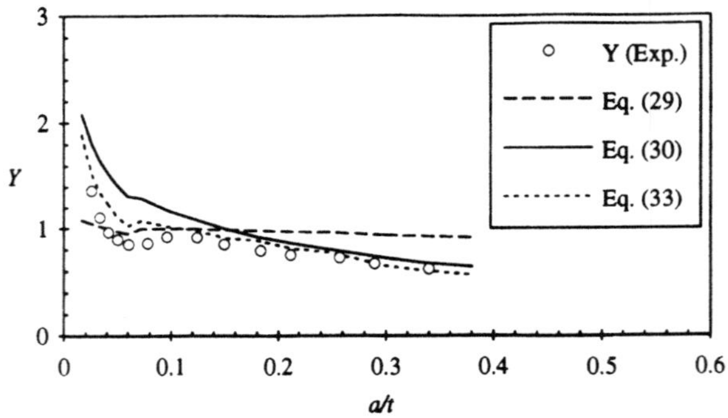

Figure 8 Comparison of Y_{EXP} with model predictions for T2-OPB.

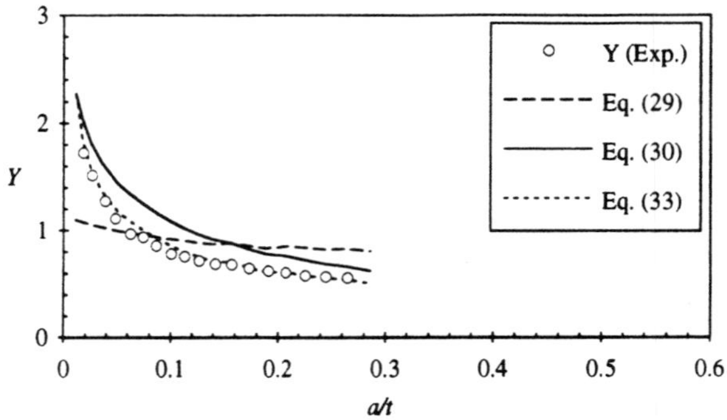

Figure 9 Comparison of Y_{EXP} with model predictions for YIC-OPB.

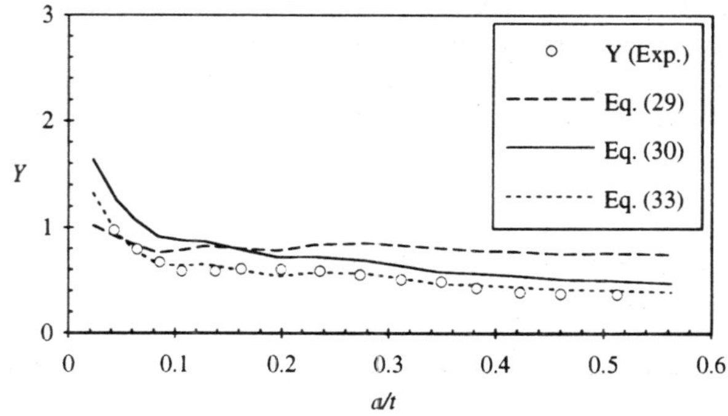

Figure 10 Comparison of Y_{EXP} with model predictions for T4b-OPB.

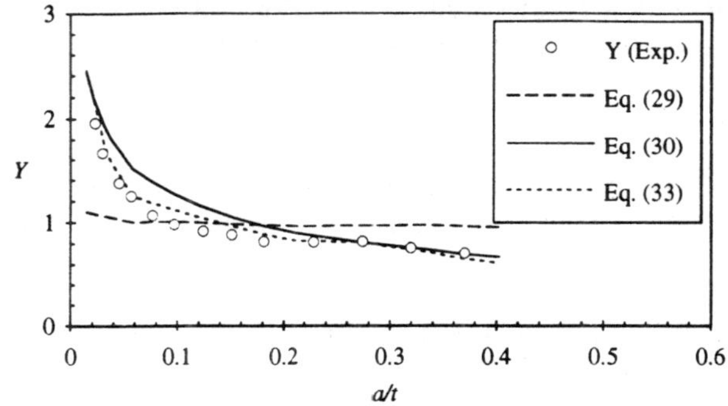

Figure 11 Comparison of Y_{EXP} with model predictions for T1-IPB.

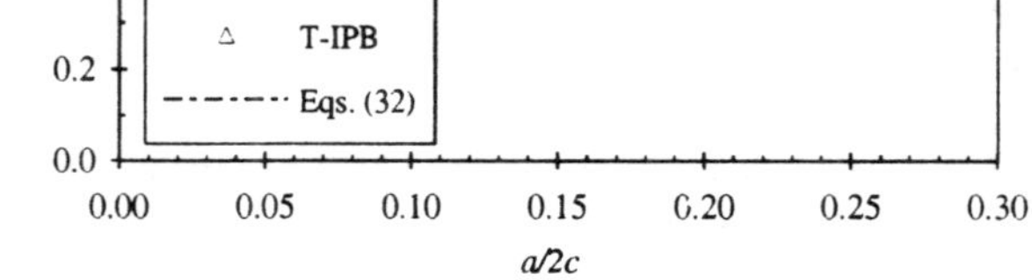

Figure 12 Empirical crack shape correction factor for tubular joints.

6 Conclusions

1. New developments in ACPD and ACFM now make it possible to detect and size cracks in welded connections. With careful instrumentation even small cracks can be measured and monitored.

2. Parametric equations are now available for predicting stress concentration factors and stress intensity factors in tubular joints and tee butt welds. These have been incorporated into a software package (FACTS) allowing rapid calculation of fatigue crack growth and remaining life.

3. The production of parametric equations describing the non-linear stress gradients in the near surface region and the advent of ACPD/ACFM equipment capable of measuring small cracks has enabled validation of early crack growth predictions for tubular joints. These improved predictions can be used for inspection scheduling.

References

[1]. "Fatigue Guidance Notes for Offshore Fixed Platforms", Health & Safety Executive, UK, to be published.

[2]. TSC Ltd., 6 Mill Square, Featherstone Road, Wolverton Mill, Milton Keynes MK12 5RB.

[3]. W.D.Dover, F.D.W.Charlesworth, K.A.Taylor, R.Collins and D.H.Michael (1981), "The use of AC Field Measurement to determine the size and shape of a crack in a metal", ASTM STP 722, G. Brinbaum and G. Free, Eds., ASTM Philadelphia, pp.401-427.

[4]. A.M.Lewis, D.H.Michael, M.C.Lugg and R.Collins (1983), "Thin skin electromagnetic fields around surface breaking cracks in metals", J.Appl. Phys. 65, pp.3777-3784.

[5]. W.D.Dover, R.Collins and D.H.Michael (1991), "Review of developments in ACPD and ACFM", Brit.J.NDT 44, pp.121-127.

[6]. W.D.Dover and J.A.Austin (1991), "Corrosion fatigue crack growth in offshore structures", J.Instit.Mech.Eng., Nov., pp.1-11.

[7]. W.D.Dover, J.R.Rudlin, C.C.Monahan and M.C.Lugg (1992), "ACFM inspection of offshore structures", Proc.6th Int.Conf.Behaviour of Offshore Structures (BOSS'92), London, pp.643-657.

[8]. R.Collins and D.H.Michael (1982), " Crack monitoring using AC field measurement - Multipliers for semi-elliptical cracks", Contract Report No. NCS 362/3007, London Centre for Marine Technology, University College London.

[9]. Lewis, A.M. (1991), "The Modelling of Electromagnetic Methods for the Nondestructive Testing of Fatigue Cracks," PhD. Thesis, University of London.

[10]. Hellier, A.K., Connolly, M.P., Kare, R.F. and Dover, W.D. (1990b), "Prediction of the Stress Distribution in Tubular Y- and T-joints," Int J of Fatigue, Vol 12, No 1, pp 25-33.

[11]. Newman, J.C., and Raju, I.S. (1981), "An Empirical Stress Intensity Factor Equation for the Surface Crack," Engrg Fract Mech, Vol 15, No 1-2, pp 185-192.

[12]. "Fatigue Crack Growth Software", FACTS, available from TSC Ltd.

[13]. Albrecht, P., and Yamada, K. (1977), "Rapid Calculation of Stress Intensity Factors," J of Struct Div, ASCE, Vol 103, pp 377-389.

[14]. Kare, R.F. (1989). Influence of Weld Profile on Fatigue Crack Growth in Tubular Welded Joints, PhD Thesis, The City University, London.

[15]. Hall, M.S., Topp, D.A. and Dover, W.D.(1990), "Parametric Equations for Stress Intensity Factors", Project Report: TSC/MSH/0244, TSC Ltd., Milton Keynes, U.K.

[16]. Brennan, F.P.B., Dover, W.D., Kare, R.F. and Hellier, A.K. (1995), "Parametric Equations for Weld Toe Stress Intensity Factors", Project Report: UCL/TSC Ltd. Report.

[17]. Monahan, C.C. (1995). Early Fatigue Crack Growth at Welds, Topics in Engineering, Vol 26, C.A. Brebbia and J.J. Connor, Eds., Computational Mechanics Publications, Southampton, U.K.

[18]. Monahan, C.C. and Dover W.D. (1995), "Stress Intensity Factors for Small Fatigue Cracks in Tubular Joints" (ISOPE'95), Colorado.

New mage matching technique for microgrid displacement detection

S. Sokol,[a] M.B. Rapaccini,[b] M. Samuelides,[a] J.M. Rubianes,[c] J.-P. Bernadou,[d] G.Sala[b]
[a]*Département de Mathématiques, Ecole Nationale Supérieure de l'Aéronautique et de l'Espace, 10 Avenue Eduoard Belin, 3 1055 Toulouse, France*
[b]*Dipartimento di Aerospaziale, Politecnico di Milano, via Golgi 40, 20133 Milano, Italia*
[c]*SOLLAC LEDEPP, 17 Avenue des Tilleuls, 57151 Florange, France*
[d]*Laboratoire de Métallurgie, Ecole Nationale Supérieure de l'Aéronatique et de l'Espace, 10 Avenue Eduoard Belin, 31055 Toulouse, France*

Abstract

The minimal strain distribution around a fatigue crack-tip initiation is evaluated by coupling a lithographically deposited grid onto the sample surface and one of two proposed displacement calculation methods. Low-carbon steel flat samples were submitted to fatigue conditions using a new piezoelectric test system that permits in-situ Scanning Electron Microscope (SEM) tests. The fine period of the grid ($2\,\mu$m) allows a description of the minimal strain field at the mesoscopic level. The local strain calculated from the data obtained using the image matching is compared with the results based on the acquirement of the grid lines intersection coordinates after the fatigue test. A discussion of the advantages and limitations of two proposed methods is presented.

1 Introduction

The study of the stress and strain fields around a crack is essential for the understanding of the strain mechanisms and to get acquainted of the parameters having an influence on the crack propagation. the knowledge of these parameters is important at both the macroscopic and microscopic levels. The size and the morphology of the plastic zone are directly related to the metallurgical structure (grain size, inclusions, grain boundaries, etc.).

In order to establish the distribution of the minimal local strains at the crack tip and thus determine the sample's degree of damage under fatigue conditions, extensometric microgrids were used.

Previous works have already dealt with the use of grids to quantify the local strains under fatigue and uniaxial tensiion conditions [1, 2, 3, 4] as well as with the velocity measurement of a propagating crack [5]. This work is aimed to compare two different methods to establishing the minimal strain. The two methods use a lithographically deposited network and allow the calculation of the local minimal strain. The first one is based on the displacement measure of the grid intersection points after the test and the second one use a new image matching technique. Since the grid period is extremely fine ($2\,\mu$m) both methods permit the determination to the local damage at the mesoscopic level.

In the case of constant amplitude fatigue the plastic zone around the crack can be divided in three different zones: a monotonic plastic zone, a cyclic plastic zone and a very small and highly strained fracture elaborating zone which is located at the crack tip and inside of which the decohesions are produced [6]. Our study is focused on this this later zone.

The minimal strain It is possible to calculated the minimal local strain by the application of the theory of the minimal strain. The transformation between an initial configuration C^o and a final configuration C^t is defined at each point by the linear tangent application tensor F:

$$F_{ij} = \frac{\partial x_i}{\partial X_j} \tag{1}$$

One particle at the point M_o and in an initial configuration C^o during it's transformation to the C^t configuration and to it's final position M is submitted not only to the displacement (translation and rotation) but also to the strain. In order to describe this strain the righthand Cauchy–Green tensor is introduced. It is defined by:

$$C = F^T F \tag{2}$$

The complete description of the solid body strain requires the use of the Green–Lagrange strain tensor, E, which is described from a lagrangian point of view by the following expression:

$$E = \frac{1}{2}(C - 1) \tag{3}$$

or

$$E_{ij} = \frac{1}{2}\left(\frac{\partial u_i}{\partial X_j} + \frac{\partial u_j}{\partial X_i} + \frac{\partial u_k}{\partial X_i}\frac{\partial u_k}{\partial X_j}\right) \tag{4}$$

The generalized strain is expressed as a function of C by the following expression [7]:

$$\bar{\varepsilon} = \int_0^t \sqrt{\frac{1}{6}\mathrm{tr}\,(C^{-1}\dot{C})^2}\,dt \tag{5}$$

This strain which is represented by a scalar, is a kind of quadratic average of the strain rate tensor components.

There are infinity of deformation path from the initial configuration of an material element to the final one. Dammame [8] prove the theorem of minimal strain announced as follows:

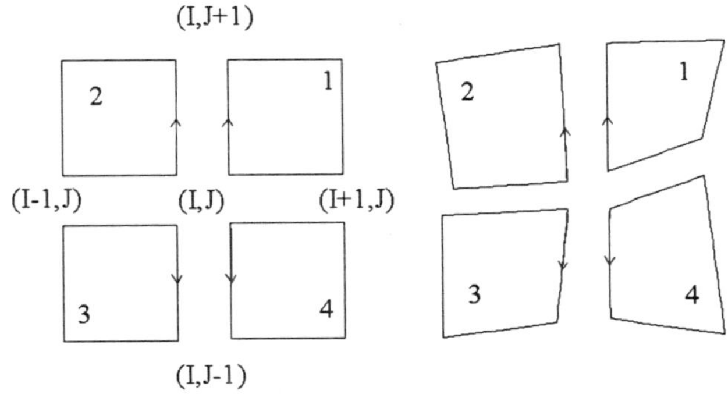

Figure 1: Node average

Theorem : *The path where the principal strain directions remain fixed and the logarithmic expansions follow this three directions at fixed ratios, minimize the generalized strain.*

In consequence the minimal strain is equal to:

$$\bar{\varepsilon} = \sqrt{\frac{1}{6}(\ln^2 \lambda + \ln^2 \mu + \ln^2 \nu)} \tag{6}$$

where λ, μ, ν represent the eigenvalues of Cauchy–Green C tensor of the final state.

2 Two methods for minimal strain computation

2.1 The node displacement method

This method use the node displacement measures of grid deposited on the sample surface and deformed by the fatigue test. The node average technique [9] is used to associate a minimal strain value to nodes of network.

This method considers the node $[I, J]$ belonging to four adjacent meshs (cf. fig. 1), which are deformed differently. The Cauchy–Green tensor C is a function of nodes $[I, J]$ et $[I+1, J]$, $[I, J-1]$, $[I-1, J]$, $[I, J+1]$. The node coordinates are picked up after strain thus the eigenvalues λ and μ can be calculated assuming that $\nu = 1$ which corresponds to plane deformation. Finally the arithmetic average of four minimal strains is attributed to node $[I, J]$.

2.2 The image matching method

Generally the aim of image matching is to establish the correspondence between the elements of two images. We consider the images as functions 2D taking the values between 0 and 255 above a rectangle $[0,\ N_x-1]\times[0,\ N_y-1]$ and 0 elsewhere. Here N_x and N_y are the numbers of pixels at the directions

x and y respectively. The elements to be matched are all pixels of the rectangle.

For each pixel (i, j) of the image f_f remaining fixed we shall match a correspondent point $(U(i, j),\ V(i, j))$ in the transformable image f_t. It is not imposed that $U(i, j),\ V(i, j))$ are entires so the bilinear interpolation is used to get a matched value of f_t at this point. This formulation of the matching problem allows to consider the problem also as transform coordinates one.

Having two digitalised SEM photographs: one of a deformed grid on the sample subjected to fatigue test and another one of the original unstrained grid. Once the matching is done the displacement field is immediately available. This permits to compute the tensor F (eqn (1)) at each pixel thus to calculate the Cauchy–Green tensor (eqn (2)) and his eigenvalues λ and μ. The value of ν is calculated as $1/(\lambda\mu)$ to satisfy as far as possible the condition of volume conservation. Three eigenvalues allow to have the minimal strains according to eqn (6).

Elastic model of image matching The coordinates transform is searched in the form:

$$U(i, j) = i + u(i, j) \tag{7}$$
$$V(i, j) = j + v(i, j) \tag{8}$$

This form correspond to the fact that a point $(U(i, j), V(i, j))$ of transformable image is translated by $(-u(i, j), -v(i, j))$ to find up a pixel of fixed image that has a the same grey level. It is clear that there may be few candidates having the searched grey level. Thus transform criterian is chosen in order to minimize the pixel retina distortiont.

More precisely, the proposed matching model searches a coordinates transform that modifys the image f_t so that an energy function E_{tot} (eqn (9) below) is minimized. This function takes into account the distortion between two images f_f and f_t (this is the first term E_d defined by eqn (10)) and penalizes the transforms that attribute very different displacements to neighbour pixels (this is the second term E_{el} defined by eqn (11)). The presence of the second term permits to speak of the elastic matching model [10]. This model admits an interpretation: the transformable image is drawn on an elastic material, an elastic retina and the stresses are retorted on this material to make this image as similar as possible to the fixed images with a minimum of elastic energy accumulated.

$$E_{tot} = E_d + E_{el} \tag{9}$$

The distortion energy is given by:

$$Ed = \sum_{ij} (f_t(U(i, j), V(i, j)) - f_f(i, j))^2 \tag{10}$$

and the elastic energy is expressed in terms of:

$$E_{el} = \beta \sum_{ij} [(u(i,j) - u(i+1,j))^2 + (u(i,j) - u(i,j+1))^2 +$$

$$(v(i,j) - v(i+1,j))^2 + (v(i,j) - v(i,j+1))^2] \qquad (11)$$

The constant β can be interpreted as pixel retina rigidity. The energy cost of a transform is higher as β increases. Thus to look for matching is equivalent to minimize the total energy E_{tot}.

Energy minimization The total energy depends on $2 \times N_x \times N_y$ variable. This number is excessively big to permit the use of stochastic algorithms like simulated annealing aimed to find up the global energie minimum [11] and requiring a lot of computations. The deterministic Newton type second order algorithms have important memory requirement.

The minimisation problem was resolved by the novel method that has adopted the multi-scale approach. This apprroch has proved its efficiency at numerous applications in imagery [12]. The matching problem was resolved according to following algorithm:

1) the images to be matched are smoothed by convolution with a low-pass filter;

2) the absence of fine details in the smoothed images allows the image undersampling without information loss. Each second pixel in each direction is omited. So the coarse detail scale is reached.

3) the images issued from precedent step contain 4 times less pixels than the original ones. It's possible to come down to even bigger scales or to carry out the matching at current scale with a method described below;

4) once matching is carryed out for a great scale, the variable vector $\{u_{ij}\}$ and $\{v_{ij}\}$ initial approximation for a smaller scale can be obtained by simple interpolation. So to go back to the small scales we have to fit the transformation just as a whole with the fine details.

A original iterative determinist algorithm is used to carry out the matching at given scale. According to this algorithm the pixels are visited at chess-board order. At each visit one of two components of displacement vector is updated, for instance u:

$$u_{ij}^{n+1} = u_{ij}^{n} + \delta_{ij} \qquad (12)$$

δ_{ij} is chosen to minimize the energy function. Since the other variables remain fixed the problem becomes 1D and an second order algorithm can be used.

A linear interpolation technique proposed in [13] is used for updating iterations. According to this technique three deformed image values are calculated using some (i.e. bilinear) interpolation:

$$f_0 = f_t(i + u_{ij}, j + v_{ij}) \tag{13}$$

$$f_1 = f_t(i + u_{ij} - d, j + v_{ij}) \tag{14}$$

$$f_2 = f_t(i + u_{ij} + d, j + v_{ij}) \tag{15}$$

Then the searched image value is presented as

$$\hat{f}(\delta_{ij}) = f_0 + \frac{\delta_{ij}}{2d}(f_2 - f_1) \tag{16}$$

Thus the distortion energy becomes a quadratic function of δ_{ij}.

$$E_d(\delta_{ij}) = (\hat{f}(\delta_{ij}) - f_f(i,j))^2 \tag{17}$$

The elastic energy function is also quadratic. This makes the energy minimum easy to find. After the simplifications the updating value δ_{ij} takes the form

$$\delta_{ij} = \frac{\Delta u_{ij} + (f_1 - f_2)(f_0 - f_f(i,j))/2\beta d}{4 + (f_2 - f_1)^2/4\beta d^2} \tag{18}$$

The equations for v evolution are identical.

This updating rule makes u_{ij} to tend to the the average of its four neighbours (because of the term Δu_{ij}) and to the difference minimum between $\hat{f}(\delta_{ij})$ and $f_f(i,j)$. The new value of u_{ij} is a compromise between this two tendencies.

3 Experimental equipment

A prototype of a test machine conceived to observe the surface damage of flat samples under mode I fatigue in vacuum was used. An original feature of this equipment is the fact of using the piezoelectric effect on the main actuator and thus the size of the test machine is small (215×60×42 mm). The dimension of the SEN type fatigue samples were dictated by the size of the testing device.

The material tested was a low alloyed steel used for deep drawing applications, with an average grain size between 10 and 20 μm. The extensometric microgrid was elaborated using a microlithographic technique and a SEM. The microgrid is formed by square elements having a period of 2 μm and a line thickness of 0.3 μm. Microgrids on notched samples were deposited from the end of the notch.

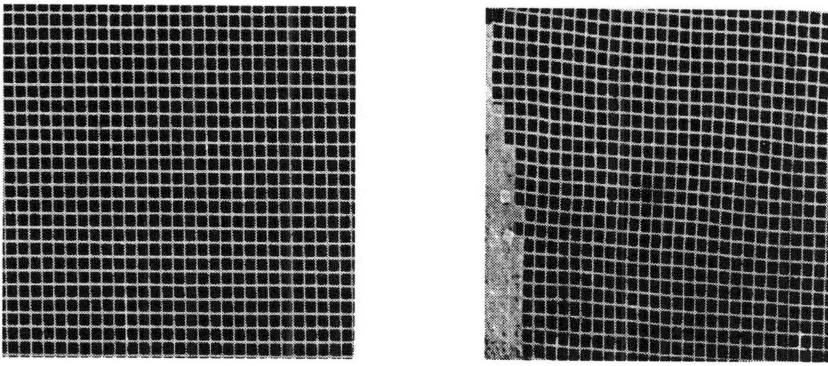

a) b)

Figure 2: a) Grid unstrained (the transformable image);
b) Grid after 500 000 cycles (the fixed image).

4 Results and discussion

A microgrid deposed on the sample's surface is shown in fig. 2.a. This image counts 512 $\times$ 512 pixels and was acquired by the SEM image analyser. The grid corresponding to 500 000 cycles under the fatigue conditions ($\Delta\sigma =$ 85 Mpa et $R = 0.42$) is presented in the fig. 2.b. The photograph is localised near the end of the notch lays on the left hand side from the picture and the crack propagates from left to right.

The matching is carryed out in two steps. First the ridge lines images are matched. Every sixth line is thicken out to produce the big scale details. The ridge lines images difference is presented in the fig. 3.a. It permits to have a better idea on the transformation undergone by the grid. The left part of image does not contain the ridge lines because of the resin undissolved at the grid elaboration stage. In this figure the negative values are black (the undeformed grid), the positives values are white (the deformed grid) and zeros are grey.

The initial matching step is helpful to increase the multi-scale approach efficiency and to avoid false matchings near the line intersection of two images.

At the second step the matching between the fig. 2.a and 2.b is carryed out. The result of this matching is presented in the fig. 3.b as the difference of the fig. 2.b and the fig. 2.a transformed according to the low $\{U(i,j), V(i,j)\}$ found during the matching procces. The matching seems to be correct. The difference is almost everywhere zero or near (the grey color in the figure) excepting the left part of image that was excluded of matching. This fact helps to interpret the minimal strain results presented

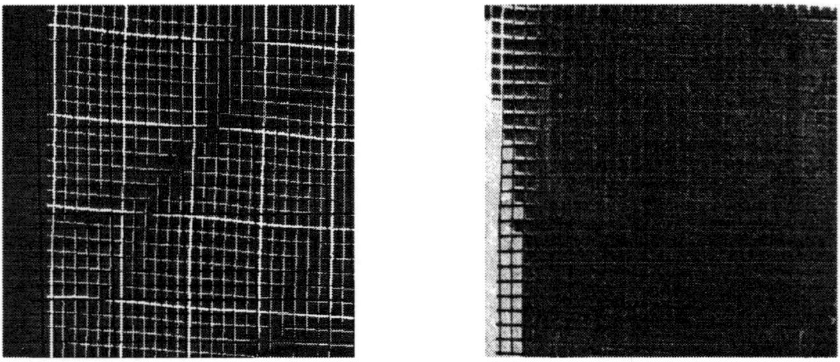

a) b)

Figure 3: a) Difference of ridge lines images;
b) Difference of the fig. 2.a and the transformed fig. 2.b

in the fig. 4.a. The great $\bar{\varepsilon}$ values at the left part seem to be due to the side effects. On the other hand an oblique band starting at bottom left corner corresponds probably to important plastification. In fact, in the further tests a micro-crack was initiated at this region.

The results issued from the matching method are compared with those obtained by the "node average" method proposed by Bernadou [9] (cf. fig. 4). To superpose the image 4.b and the 4.a the fig. 4 must be laid on the bottom and on the right hand side of the fig. 4.a.

A qualitative coherence can be noted. As to quantitative difference it can be explained by:

– acquisition system of the SEM that making appear a square grid as a rectangular or oblique one. The two methods of strain computation are influenced differently by this error. Beside the image acquisition defect the method "node average" is sensible to the node coordinates acquisition error.

– the resolution of the methods, i.e. the capacity to detect and localize a small strain, which is not the same for two methods. In fact, for "node average" the minimal strain is calculated at 6×6 points retorted in the image. To calculate $\bar{\varepsilon}$ at all pixels an interpolation has to be used and its choice influences the final results. In the case presented in the fig. 4.b the bilinear interpolation was used.

5 Conclusions

A novel pixel by pixel image matching technique is proposed. It was successfully used to the displacement field measurement on the sample's surface provided of a micro-grid deposited by lithography.

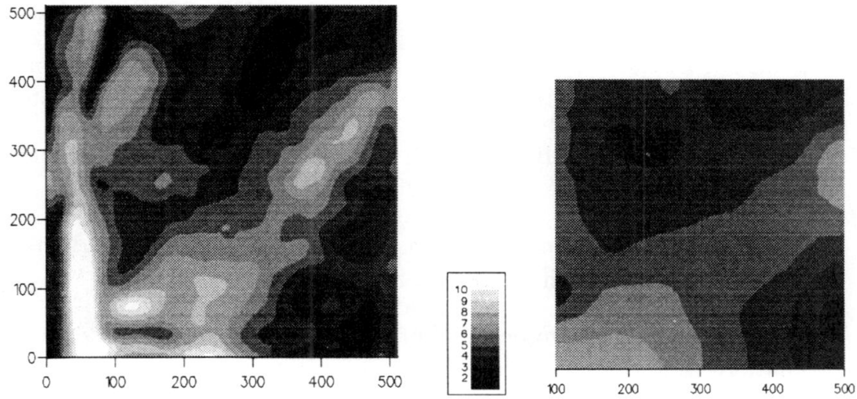

a) b)

Figure 4: a) minimal strains issued from image matching (%);
b) minimal strain by "node average" (%)

The matching is carryed out by an elastic model that leads to the energy function minimization. An iterative deterministic minimization algorithm sparing the computations is also proposed. It uses the multi-scale approach that accelerates the convergency considerably and makes it more robust .

The displacement field issued from matching is known at each pixel and permits to calculate the minimal strain according to the Dammame [8]. The results are in qualitative good accord with another method said "node average" proposed by Bernadou [9].

The new method seems to allow to fully profit of the resolution offered by the micro-grid technique and to carry out the measures at the mesoscopic scale. This improves the localization and quntification of the plastification zone that are important for the initiation and propagation crack studies.

References

[1] Mari, D., Marti, U. & Silva, P. A new photolithographic technique to detect the local deformation of materials: application to wc-co composites. *Materials Science and Eng.*, 1992, **A158**(2), 203-206.

[2] Karimi, A. Plastic flow using the microgrid technique. *Materials Science and Engineering*, 1984, **63**, 267-276.

[3] Mabille, P., Bernadou, J-P. & Thirion, I. Détermination des hétérogénéités locales de déformation par des essais de traction sous MEB. In *I.C.E.M. 13.*, pp. 43 to 44, Paris, 1994.

[4] Sirkis, J. & Lim, T. Displacements and strain measurement with automated grid methods. *Experimental Mechanics*, 1991, **31**(4), 382-388.

[5] Martin, D., Davido, K. & Scott, W. Slow crack growth measurement using an electrical grid. *Am. Ceram. Soc. Bull.*, 1986, **65**(7), 1052-1056.

[6] Guerre-Rosa, L., Moura Branco, C. & and Radon, J.C. Monotonic and cyclic crack tip plasticity. *Int. J. Fatigue*, 1984, **6**(1), 17-24.

[7] Sidoroff, F. *Formulations élasto-plastiques en grandes déformations*, Ecole d'été "Grandes Déformation", Sophia Antipolis, 1982/51.

[8] Dammame, G. Minimum de la déformation généralisée d'un élément de matière isolé pour des chemins de déformation passant d'un état initial à un état final. In *C.R. Acad. Sci.*, t.287, A-895, Paris, 1978.

[9] Bernadou, J-P. Déformation minimale en forçage axisymmetrique. Technical report, Centre de recherche matériaux atelier de construction de Tarbes, ATS, 1984.

[10] Kervann, C. & Heitz, F. A hierarcnical statistical framework for the segmentation of deformable objects in image sequence. Projet TEMIS 2133, INRIA, Programme 4 - Robotique, image et vision, Dec. 1993.

[11] Yuille, A. & Kosowsky, J. Statistical physics algorithmes that converge. *Neural Computation*, 1994, **6**, 341-356.

[12] Mallat, S. A theory for multiresolution signal decomposition: the wavelet representation. *IEEE Trans. Pattern Anal. Machine Intell.*, 1989, **2**(7), 674-693.

[13] Srinivasan, M. An image-interpolation technique for the computation of optic flow and egomotion. *Biol. Cybern.*, 1994, **71**, 401-415.

Mechanical behaviour of fly ash concrete

A. Surahman,[a] W. Merati[b]
[a]*Structural Mechanics Laboratory, and Department of Civil Engineering, Inter-University Center for Engineering Sciences, Institut Teknologi Bandung, Jl. Ganesha 15, Bandung 40132, Indonesia*
[b]*Center for Earthquake Engineering Studies, and Institute for Research, Institut Teknologi Bandung, Jl. Ganesha 15, Bandung 40132, Indonesia*

Abstract

A research on mechanical behaviour on fly ash concrete has been carried out. The aim of the research was to find out the behavioural changes of concrete mixed with fly ash. The materials have been obtained locally from various locations in the vicinity of Bandung area. Fly ash has been used as cement substitute to reduce the concrete cost and increase the concrete strength. Several mechanical properties have been investigated. The test results showed that fly ash significantly changes mechanical behaviour of concrete.

1 Introduction

The use of high strength concrete is getting more popular due to high competitiveness in construction business and scarcity of some construction materials. By using very high strength concrete, the material volume and space used will be reduced. Due to larger modulus of elasticity and tensile strength, the use of high strength and prestressing steel will be more feasible. One way to increase the concrete strength is by adding fly ash to the concrete mixture. Fly ash is a kind of material with its grains considered to be finer than commonly used cement. Grains with finer size are expected to fill the gaps left by much coarser grains of sand and cement. Due to the addition of the new material, some changes in mechanical behaviour are anticipated. This paper describes tests and researches conducted at the institute [6, 7, 8, and 9].

2 Details of the experiment

The laboratory experiments include evaluation of concrete ingredient materials. The cement used was Type I of *Tiga Roda* brand with a specific gravity of 3.15.

The sand is obtained from nearby *Galunggung* mountain area. Visual test showed that the grains were fine and round shaped. The sand material was tested for sieve analysis which was acceptable according to [2], bulk specific gravity (2.60), absorption (capacity 2.29%), total moisture content (3.12%), and fineness modulus (2.64). The coarse aggregates have also been tested with the following resulting data given in Table 1.

Table 1: Properties of coarse material

	diam. 5 - 10 mm	diam. 10 - 20 mm	diam. 20 - 40 mm
Bulk specific gravity	2.72	2.72	2.72
Absorption capacity	1.13 %	0.91 %	0.88 %
Total moisture content	2.00 %	1.07 %	0.89 %
Dry-rodded unit weight	1445 kg/m^3	1460 kg/m^3	1535 kg/m^3

With a water to cement ratio of 0.3 and entrapped air content of 1%, the material composition was as given in Table 2. For bond stress test specimens [6] somewhat different material composition was used (fly ash content of 15%) as shown in Table 3.

Table 2: Material composition for normal concrete

Cement	550.000 kg
Sand	624.235 kg
Gravel diam 5-10 mm	259.176 kg
Gravel diam 10-20 mm	345.569 kg
Gravel diam 20-40 mm:	501.075 kg
Water	165.000 lt

Table 3: Material composition for bond stress test specimens

Cement	467.500 kg
Sand	639.175 kg
Gravel diam 5-10 mm	490.896 kg
Gravel diam 10-20 mm	599.984 kg
Fly ash	82.500 kg
Water	159.695 lt

The test setup and procedures were based on standard manuals and specifications [1, 2, 3, 4, and 5]. Various amounts of fly ash were tried to obtain the optimum strength. The amounts of fly ash were varied according to weight proportions. The specimens were also tested according to the concrete ages (7, 14, 28, 56, and 90 days) to observe the effects of fly ash content on concrete curing [7].

3 Results and discussions

The experiments were carried out at the laboratory under supervision which is expected to be better than at ordinary construction site. Nevertheless, Fig. 1

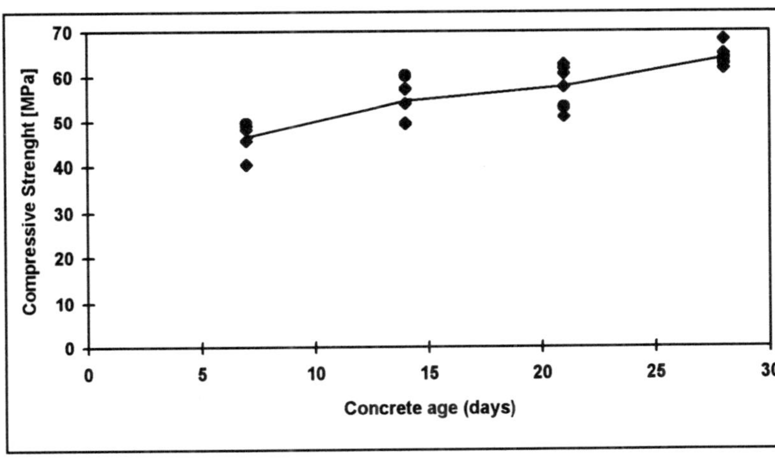

Figure 1: Scatter and average compressive stresses for different concrete ages

shows significant variations in the compressive strengths. The volume weights for these specimens varied between 2.38 to 2.452 kg/l. At the concrete age of 28 days, the variations tend to diminish. Similar behaviour has also been shown by fly ash concrete [7] with increase of strength at ages later than 28 days as shown in Fig. 2. The number of specimens for each case was six, and in the presentations to follow, the values representing the mechanical properties were taken as the average from six specimens.

The results of the compressive strength for different fly ash to cement percentages are given in Fig. 2. It is shown that the variations are relatively small; however the maximum value of compressive strength was achieved at the fly ash amount of 15%. The compressive strength of the concrete at this composition was approximately 76.3 MPa for a concrete age of 90 days. Comparing tensile strengths of 28 day concrete shown in Fig. 3, the stress for the optimum (15%) fly ash concrete was 4.15 MPa, which was smaller than that of normal concrete.

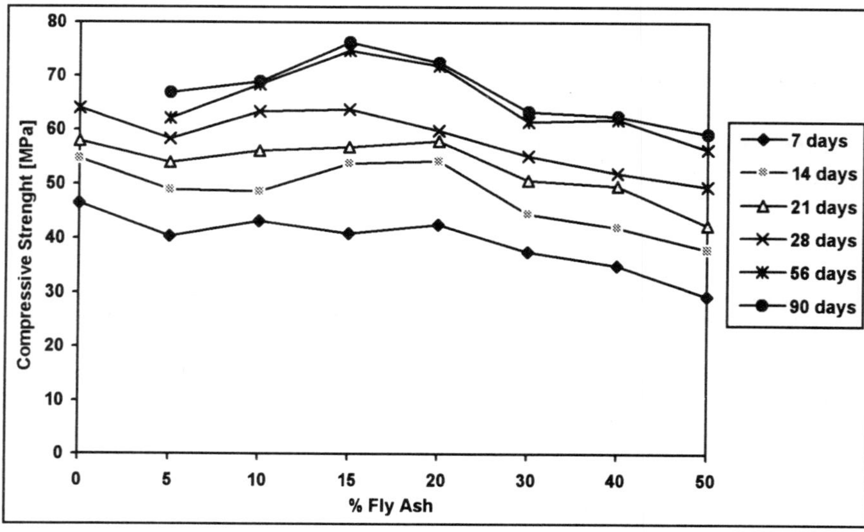

Figure 2: Compressive strength for different fly ash contents

The relationship between the tensile strength and compressive strength have been proposed by many researchers [7]:

ACI-318 (psi units):

$$f_{sn} = 6.50 \sqrt{f_c'} \quad (1.a)$$

Shah & Ahmad (in psi units):

$$f_{sp} = 6.50 (f_c')^{0.55} \quad (1.b)$$

Yamamoto & Kobayashi (in MPa units):

$$f_{sn} = 0.06 f_c' + 0.8 \quad (1.c)$$

where
f_{sn} = maximum tensile stress that can be carried by the concrete
f_c' = maximum compressive stress that can be carried by the concrete

Figure 3. shows that whereas for normal concrete the relationship basically matches the empirical formulas, the experimental tensile strengths for fly ash concrete tend to be slightly smaller than predicted.

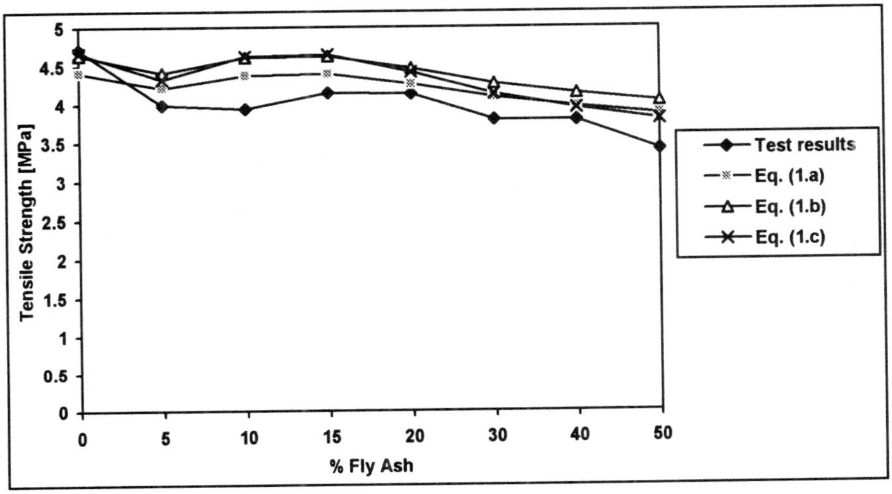

Figure 3: Tensile strength compared with several empirical formulas for different fly ash contents (concrete age of 28 days)

Using the values of concrete specific gravity and compressive strength, the modulus of elasticity has also been proposed by several researchers [7]:

ACI 318-83:

$$E_c = 33\ w_c \sqrt{f_c'} \quad (2.a)$$

ACI 318-83 for heavy concrete:

$$E_c = 57000\sqrt{f_c'}\ (w_c/145)^{1.5} \quad (2.b)$$

Carrasquillo, Nilson and Slate:

$$E_c = 40000\sqrt{f_c'} + 1.000.000 \quad (2.c)$$

Shah & Ahmad:

$$E_c = (w_c)^{2.5}(f_c')^{0.325} \quad (2.d)$$

where
E_c = modulus of elasticity [psi]
w_c = specific gravity [lbs/ft^3]
f_c' = compressive strength [psi]

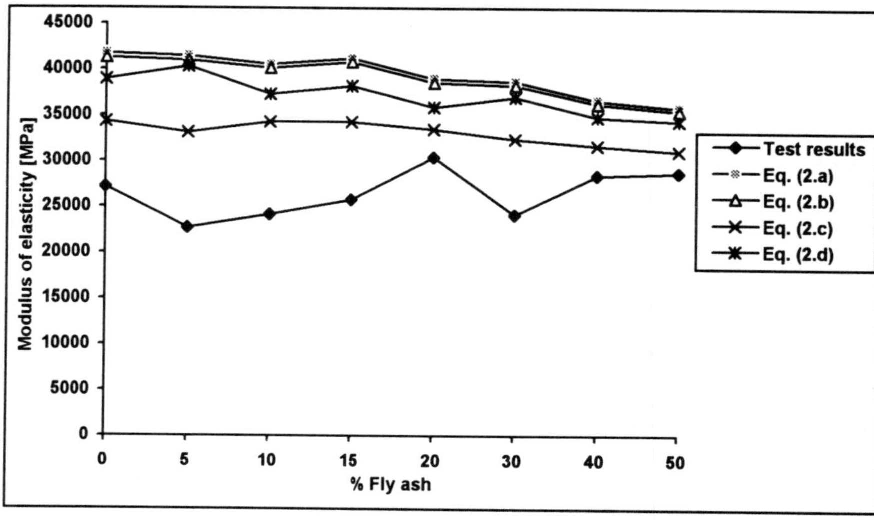

Figure 4: Modulus of elasticity for different fly ash contents compared with several empirical formulas (concrete age of 28 days)

Figure 4 shows that the experimental results show lower values for all the cases. This shows that the experimental conditions were not the same with those under which the empirical formulas were derived [7].

The bond stresses were determined by using pull-out tests of plain round bars of 12 and 16 mm. The embedded lengths were taken between 15 to 23 times the bar diameters. The results shown on Fig. 5 were taken from the fly ash composition of 15%.

It is shown that the bond stresses are larger than ordinary used in the design. The results on the larger diameter of 16 mm showed lower values of bond stresses since during experiments [6], some other failure modes occurred. Tests on deformed bars have also been carried out [6], and the results showed, as expected, higher values of bond stresses. This is obvious since the roughness of deformed bar produces interlocking effects.

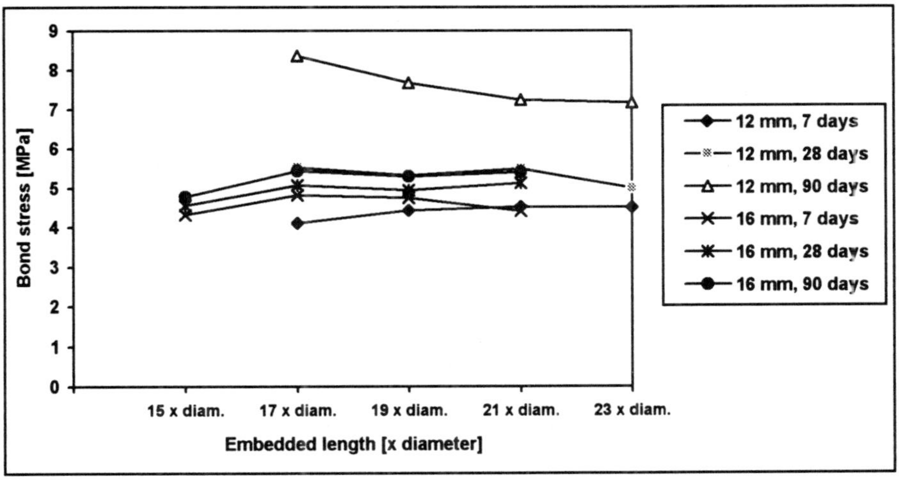

Figure 5: Bond Stresses for different bar diameters and embedded lengths

The above mentioned results are the kind of mechanical properties that of interest to structural engineers. Other tests such as permeability and shrinkage tests have also been carried out. Results of the tests showed no conclusive pattern can be deducted from permeability tests, whereas shrinkage tests showed that concrete specimens with larger amount of fly ash also exhibit larger amount of shrinkage [7].

4 Conclusions

For high strength concrete obtained by mixing with fly ash, the experiments showed that the optimum fly ash content is 15% with the compressive strength of 90 day concrete of 76.30 MPa. The strength of fly ash concrete was still increasing at ages much later than 28 days due to the slower rate of the curing process. The tensile strength of 28 day fly ash concrete was lower than that of normal concrete of the same age. Whereas the empirical formulas for normal concrete match the test results of normal concrete specimens, for fly ash concrete specimens the test results showed lower values of tensile strengths. Bond stresses between the plain reinforcing bars and the high strength concrete have been found to be significantly high such that failure of specimens with larger bar diameters occurred due to other reasons. No comparisons were made with normal concrete. More tests are needed to evaluate bond stresses of large diameter bars and to evaluate modulus of elasticity.

REFERENCES

1. ACI Committee 363, "*State -of-the-Art Report on High-Strength Concrete*", ACI Journal, July-August, 1984.

2. ASTM C 33-82, "*Standard Specification for Concrete Aggregate*", 1982.

3. ASTM C 39-83a, "*Standard Test Method for Compressive Strength of Cylindrical Concrete Specimen*", 1983.

4. ASTM C 33-82, "*Standard Method of Making and Curing Concrete Test Specimens in the Laboratory*", 1981.

5. ASTM "*1991 Annual Book of ASTM Standards*", Vol. 04.02, Concrete and Aggregates, ASTM, 1991.

6. Ginting, M., "*Kuat Lekat Beton Mutu Tinggi dengan Baja Tulangan*", Thesis submitted in partial fulfillment of the degree of Master of Science, Civil Engineering Department, Institut Teknologi Bandung, 1994.

7. Merati, W., Surahman, A., and Suprobo P.,"*Structural Behaviour of Very High Strength Concrete*", The Asahi Glass Foundation, 1994.

8. Udin, M., "*Optimasi Kadar Abu Terbang Untuk Beton Mutu Tinggi*", Thesis submitted in partial fulfillment of the degree of Master of Science, Civil Engineering Department, Institut Teknologi Bandung, 1994.

9. Wijaya, D., "*Penentuan Parameter k Untuk Lentur Balok Beton Mutu Tinggi*", Thesis submitted in partial fulfillment of the degree of Master of Science, Civil Engineering Department, Institut Teknologi Bandung, 1994.

Relation between existence of fatigue limit and non-propagating micro-cracks

S. Nishida,[a] N. Hattori,[a] T. Shimada[b]
[a]*Faculty of Science and Engineering, Saga University, Honjo-machi 1, Saga 840, Japan*
[b]*Yawata R&D Laboratory, Nippon Steel Corporation, Tobata 1-1, Kitakyushu 804, Japan*

Abstract

Fatigue tests had been performed by rotating bending fatigue testing machine about HM steel (0.2%C) specimen with comparing that of SUS304 one using successive-taken replica method. It is clarified that the fatigue strength of HM steel and SUS304 shows comparatively the same result, that is, the same fatigue limit and distinct knee point. In addition, some non-propagating micro-cracks are observed in the specimen's surface of HM steel subjected to the stress of fatigue limit by 10^7 cycles, while no micro-cracks exist in the specimen of SUS304 subjected to the same stress condition.

Following to the above tests, the authors have accomplished the same kinds of fatigue tests to investigate the effect of carbon content on the existence of non-propagating micro-cracks in the austenitic steels using the four kinds of high manganese steels with changing and decreasing the carbon content between 0.025%～0.100%. It is clarified that the non-propagating micro-cracks are not observed in the four kinds of HM steels (0.025%～0.100%C) under the stress of each fatigue limit by 10^7 cycles. From this fact, the limit of the existence of non-propagating micro-cracks is concluded to be between 0.1%～0.2% of carbon content, and the characteristics for crack initiation under the above stress conditions are founded to be predominantly influenced by its carbon content.

1. Introduction

It is said that high manganese non-magnetic austenitic steels (HM steels) has been expected to be a promising component material of the track for a linear induction motor train, nuclear fusion electric power systems etc., because it shows extra-stable non-magnetic properties, high tensile strength and ductility even in low temperature. This alloy steel contains much manganese, which exists in large quantities in the ocean depth, instead of the expensive nickel composition. Therefore, there would be no problems for its as being rich in natural resource.

As HM steel is a kind of new material and does not have come into wide use yet, there are few reports about its mechanical properties, still less its fatigue strength[1-3].

Then, fatigue behavior of HM steel (0.2% C) have been investigated in this paper, especially in view point of fatigue crack initiation by successive-taken replica method under a cyclic stress with comparing that of the representative austenitic stainless steel (SUS304), which is also well-known as a non-magnetic one. In addition, the effect of carbon content on the existence of non-propagating micro-cracks in the austenitic steels has been researched using four kinds of HM steels with changing and decreasing the carbon content only between 0.025%~0.100%.

2. Experimental procedure

First step ; the materials used in this test are HM steels (0.2%C) and typical austenitic steels (SUS304), which are both non-magnetic austenitic steels of rolled thick plates for a market. Tables 1 and 2 list their chemical compositions and mechanical properties, respectively.

Second step ; following to the above tests, the four kinds of HM steels with changing and decreasing carbon content were used in this test. Tables 3 and 4 list their chemical compositions and mechanical properties, respectively.

Figure 1 shows the dimensions of fatigue specimen, which was cut out by coinciding the rolling direction with the specimen's axis and by doing a partial shallow notch with its rolling surface side[4]. This shallow notch exists for limiting fatigue-damaged part and does not affect for its fatigue strength at all. All specimens were annealed in vacuum for one hour at temperature of 600℃ after polishing with fine emery paper to the longitudinal direction, and thereafter electro-polished to the depth of about 50 μm.

A rotating bending fatigue testing machine of uniform bending moment type (Ono-type fatigue tester, capacity; 98kN-m, frequency; 56.7Hz) had been used. The stress used in this paper is the nominal stress at the minimum defined by neglecting the existence of a partial notch; that is

$$\sigma = 32M/(\pi d^3) \qquad (1)$$

where M is the bending moment and d is the specimen diameter.

In the case of fatigue test of SUS304, testing part of each specimen was cooled by pouring fresh water for prevention of exothermic fracture. It is confirmed that there is no difference between the result under low frequency in air and that under high frequency with cooled by distilled water[5]. The observation of micro-crack initiation was adopted by successive-taken replica method in the circumferential direction at the specimen's surface and defined

Table 1 Chemical composition. (mass%)

	C	Si	Mn	P	S	Ni	Cr	Nb	N	Al
HM	0.178	0.23	23.8	0.024	0.001	0.23	2.05	0.04	0.043	0.003
SUS304	0.048	0.50	0.81	0.026	0.007	8.47	18.49	-	0.047	-

Table 2 Mechanical properties. (MPa, %)

	$\sigma_{0.2}$	σ_B	El
HM	251.9	751.7	56.0
SUS304	253.8	623.5	66.0

Table 3 Chemical composition. (mass%)

	C	Si	Mn	P	S	Ni	Cr	Nb	N	Al
HM-1	0.027	0.20	23.7	0.016	0.006	0.27	2.08	0.04	0.032	0.009
HM-2	0.047	0.20	23.4	0.017	0.006	0.32	2.19	0.04	0.031	0.010
HM-3	0.048	0.44	23.6	0.015	0.004	0.30	4.82	0.04	0.030	0.012
HM-4	0.107	0.44	23.3	0.016	0.004	0.30	4.82	0.04	0.029	0.010

Table 4 Mechanical properties. (MPa, %)

	$\sigma_{0.2}$	σ_B	El	RA
HM-1	196.5	783.0	41.6	62.6
HM-2	164.6	766.9	45.7	66.0
HM-3	181.8	746.3	46.5	68.8
HM-4	219.5	770.3	47.5	70.5

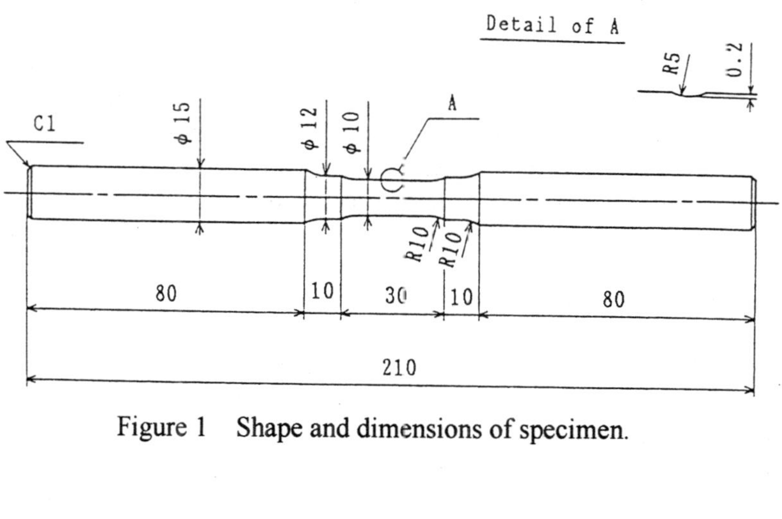

Figure 1 Shape and dimensions of specimen.

as the length of one grain or so.

3. Result and discussion

3.1 Comparison of HM steel (0.2%) and SUS304

3.1.1 Fatigue strength

Figure 2 shows the S-N curves of HM steel and SUS304. As shown in this figure, there exists the same fatigue limit (σ_{w0}=265MPa) about both materials, while the gradient of SUS304 becomes a little smaller than that of HM steel. When considering the relation between mechanical properties and its fatigue limit, it is generally said that the fatigue limit of steel specimen σ_{w0} is closely dependent on its tensile strength σ_B. As HM steel shows higher tensile strength and a little smaller proof stress than SUS304, that is, the former has smaller yield ratio than the latter, the fatigue limit of both materials happened to agree with each other and shows higher value than proof stress by about 5 %.

3.1.2 Fatigue crack initiation

Figure 3 shows the representative successive observation result of fatigue crack initiation of HM steel under cyclic stress amplitude σ_a =320 MPa. Fatigue cracks initiate in grain boundary or in its neighborhood and propagate mainly as a trans-crystalline crack. Figure 4 shows the representative successive observation result of fatigue crack initiation of SUS304 under cyclic stress amplitude σ_a =280 MPa. Fatigue cracks also initiate and propagate in the same manner as the above. Therefore, there is no difference about crack initiation between both materials.

On the other hand, Figure 5 show the surface state under the fatigue limit (σ_a =265 MPa) by 1×10^7 cycles about both materials. The non-propagating micro-cracks are observed in the specimen's surface of HM steel and its length becomes about 45 μm, while the non-propagating micro-cracks are not observed in the case of SUS304, though there appeared only slip bands. It is considered that the above phenomena would be caused by the difference of carbon content, corrosion resistance and yield ratio, etc.. In additon, the micro-Vickers hardness number, where slip bands are generated, becomes increase by about 30% than that of the grains of before test or the grains where slip bands are not generated under the above cyclic stress amplitude by 1.21×10^7 cycles.

3.2 HM steels(carbon content; 0.025～0.20%)

3.2.1 Fatigue strength

Figure 6 shows the S-N curves of HM-1 to HM-4 with decreasing carbon content than 0.2% C. When defined the fatigue limit as the fatigue strength

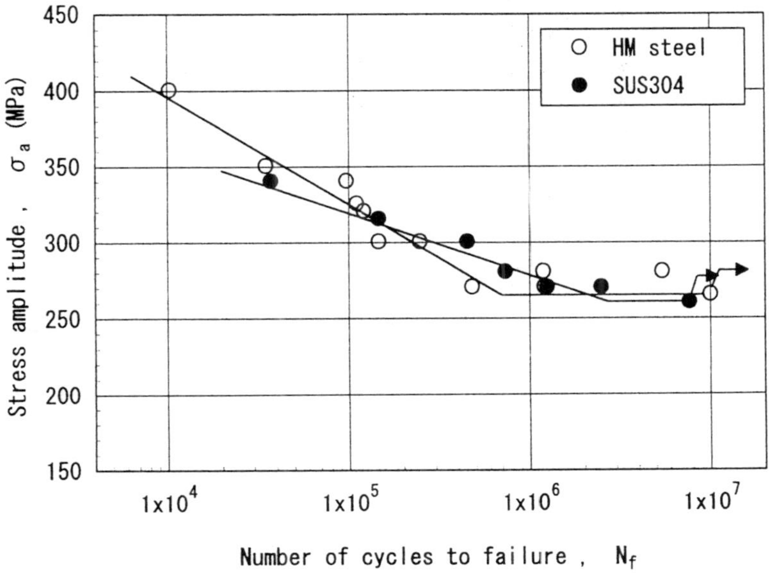

Figure 2 S-N curves.

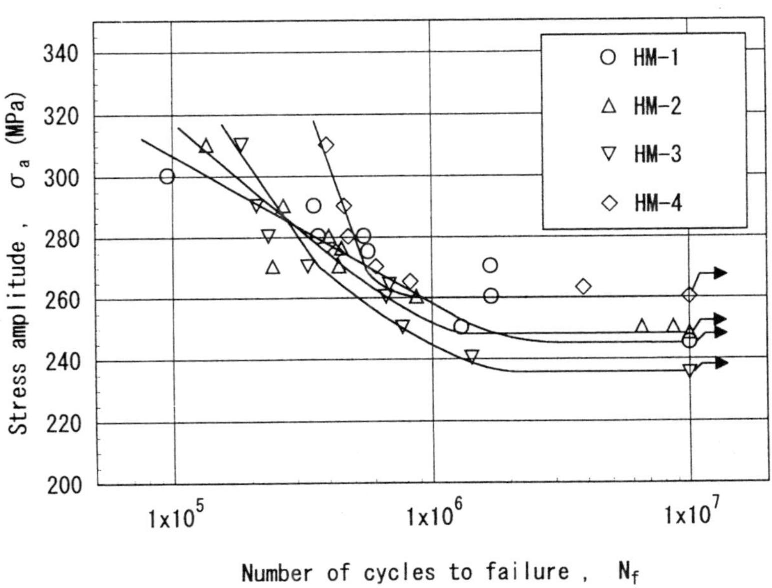

Figure 6 S-N curves.

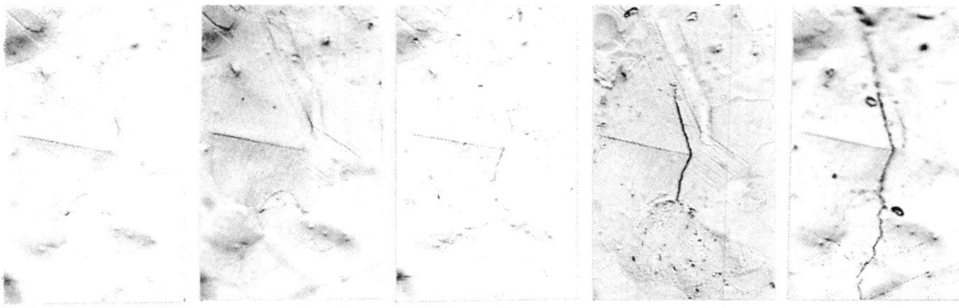

N=0 | 1.5×10^4 | 4.0×10^4 | 5.0×10^4 | 7.0×10^4 cycles

⟷ Axial direction | 20 μ m

Figure 3 Successive observation of fatigue crack initiation.
(HM steel, $\sigma_a = 320$ MPa, $N_f = 1.21\times10^5$ cycles)

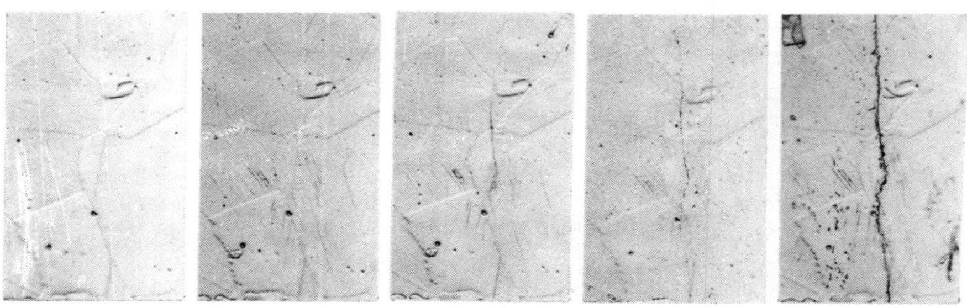

N=0 | 7.0×10^4 | 1.0×10^5 | 2.0×10^5 | 3.0×10^5 cycles

⟷ Axial direction | 20 μ m

Figure 4 Successive observation of fatigue crack initiation.
(SUS304, $\sigma_a = 280$ MPa, $N_f = 7.33\times10^5$ cycles)

(a) Non-propagating micro-crack of HM steel. | (b) Slip bands of SUS304.

⟷ Axial direction | 20 μ m

Figure 5 Specimen's surface state subjected to the stress amplitude of fatigue limit by 1×10^7 cycles.

of 10^7 cycles, the fatigue limits of HM=1, HM-2, HM-3 and HM-4 become 245, 248, 235 and 262 MPa, respectively. As shown in this figure, the four kinds of HM-1 to HM-4 steels do not show distinct knee point at all.

Figure 7 shows an illustration of S-N curves as shown in Figure 6. Each S-N curve in Figure 6 deviates from a linear line at A point and its gradient becomes smaller according to approaching the stress level (B point) of fatigue limit. Table 5 lists the value of stress level at A and B point.

3.2.2 Fatigue crack initiation

As an example, Figure 8 shows the representative successive observation result of fatigue crack initiation under cyclic stress amplitude between A point and B point (see in Figure7). Fatigue cracks initiate in grain and propagate accross the grain boundaty. Although not shown the results, it seemed that the fatigue crack behavior under cyclic stress amplitude at above A point of each material shown in Table 5 is different from the results under the above condition. Namely, the fatigue cracks propagation under the stress level at the above A point are affected grain structures, while those at between A point and B point are not so affected.

3.3 Effect of carbon content on fatigue properties

Figure 9 shows the specimen's surface state subjected to the stress amplitude of fatigue limit by by 1×10^7 cycles. As shown in Fig.5, the specimen of HM steel (0.20%) exists the non-propagating micro-cracks. On the other hand, it can be seen that only slip bands exist in the specimen's surface of HM-1 to HM-4, though the non-propagating micro-cracks are not observed in it. It is considered that this phenomena would be caused by the difference of carbon content.

Figure 10 shows the relation between carbon content and $\overline{BC}$ value (see in Figure 6). As shown in this figure, the carbon content becomes more, the value of $\overline{BC}$ becomes smaller, that is, the stress level of A point approaches the fatigue limit. Therefore, the S-N curve shows distinct knee point. The carbon content affect to the existence of non-propagating micro-cracks in the specimen's surface of the HM steel specimen subjected to the stress of fatigue limit by 10^7 cycles.

4. Conclusions

Fatigue tests had been accomplished by rotating bending fatigue tests about plain specimen of HM steel with a partial shallow notch with comparing that of SUS304. The main results obtained in this study are as follows;

(1) The fatigue strength of the HM steel and SUS304 shows comparatively the same result, e.g., the same fatigue limit and distinct knee point. In addition, the fatigue micro-cracks have initiated at one or two grains and propagated into

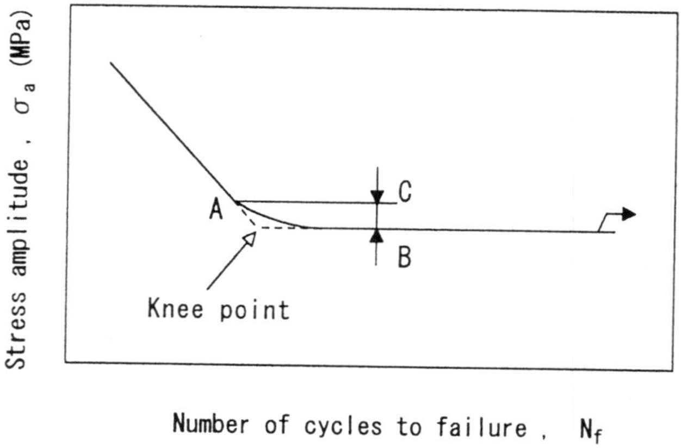

Figure 7 Typical S-N curve.

Table 5 The stress value at point A and B. (MPa)

	Stress of the A	Stress of the B	A-B
HM	265	265	0
HM-1	278	245	33
HM-2	272	248	24
HM-3	264	235	29
HM-4	273	262	11

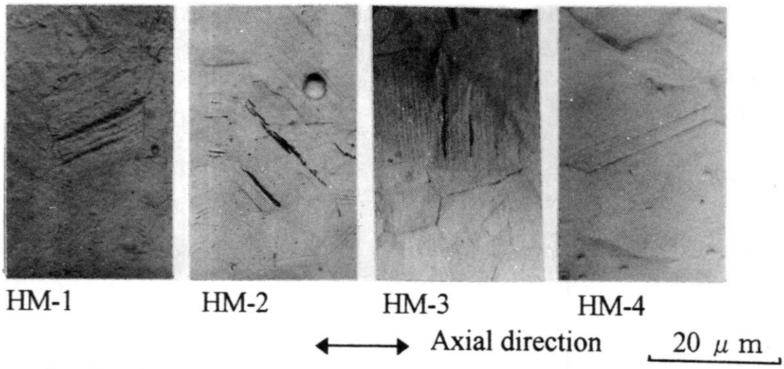

Figure 9 Specimen's surface state subjected to the stress amplitude of fatigue limit by 1×10^7 cycles.

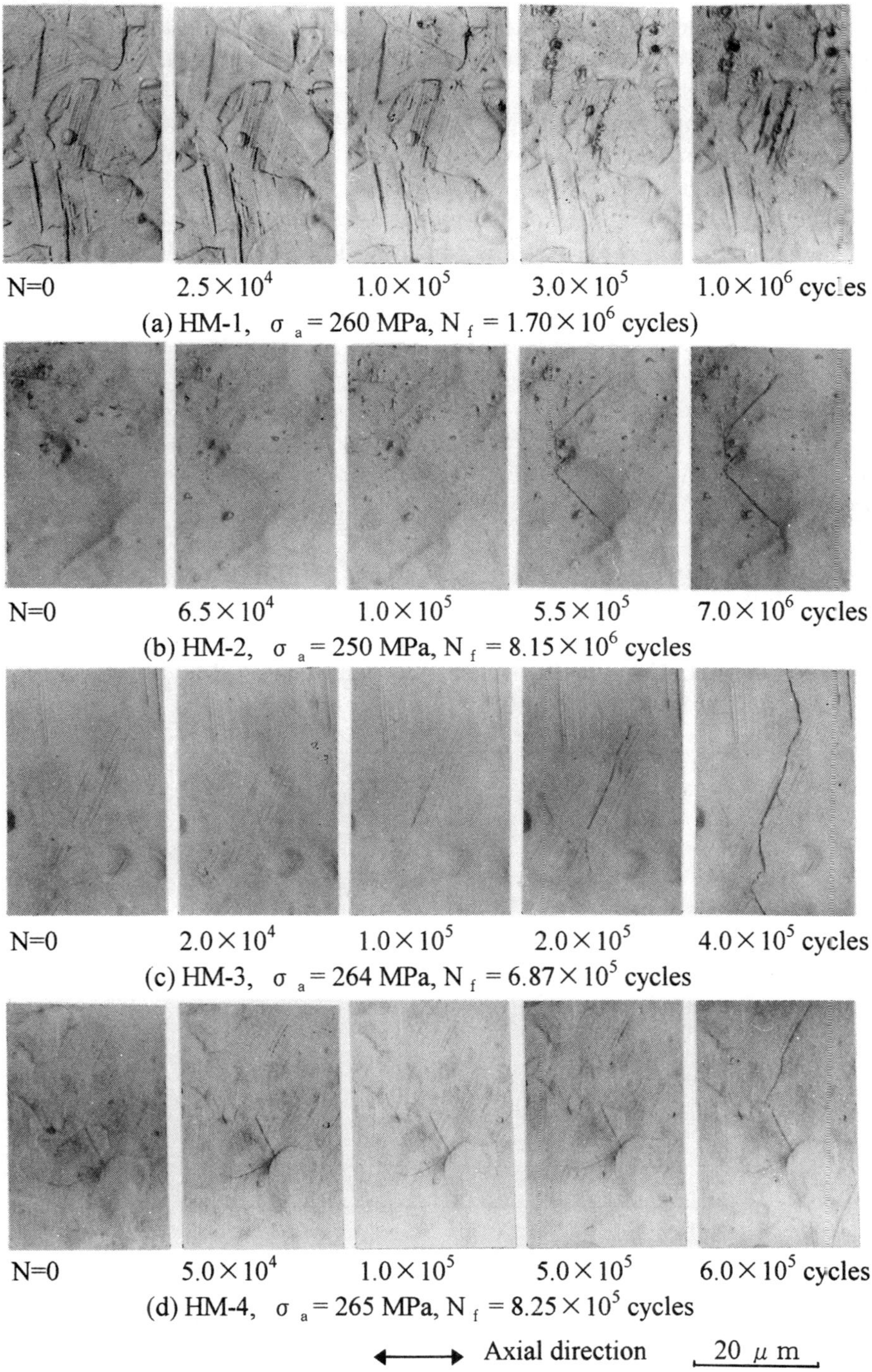

Figure 8 Successive observation of fatigue crack initiation.

final fracture in both cases.

(2) Some non-propagating micro-cracks (45 μm in length) are observed in the specimen's surface of the HM steel specimen subjected to the stress of fatigue limit by 10^7 cycles and no micro-cracks exist in the SUS304 specimen subjected to the same condition.

(3) From the above fact, the fatigue limit of the HM steel is based on the limit of propagation of micro-crack, while that of SUS304 is based on the limit of micro-crack initiation.

(4) The limit of the existence of non-propagating micro-cracks under the stress of fatigue limit by 1×10^7 cycles is concluded to be between 0.1%~0.2% of carbon content. In addition, the characteristics for the above crack initiation are predominantly influenced by its carbon content.

References

(1)I. Maekawa, Y. Tanabe, S. Nishida, T. Ashizawa, H. Ogawa, JSME (in Japanese), Vol.56. No.225, 1990, pp.1051-1057

(2)S. Nishida, N. Hattori, T. Shimada, Proceeding of Fatigue'93, Vol. I, 1993, pp.347-352

(3)S. Nishida, N. Hattori, T. Shimada, S. Iwasaki, JSME(in Japanese), Vol.61, No.582, 1995, pp.239-244

(4)e.g. H. Nisitani, Y. Hasuo, JSME (in Japanese), Vol.44, No.377, pp.1-7

(5)H.Nisitani, H. Noguchi, S. Nisida, K. Yamaguchi, JSME (in Japanese), Vol.56, No.525, 1990, pp.1067-1073

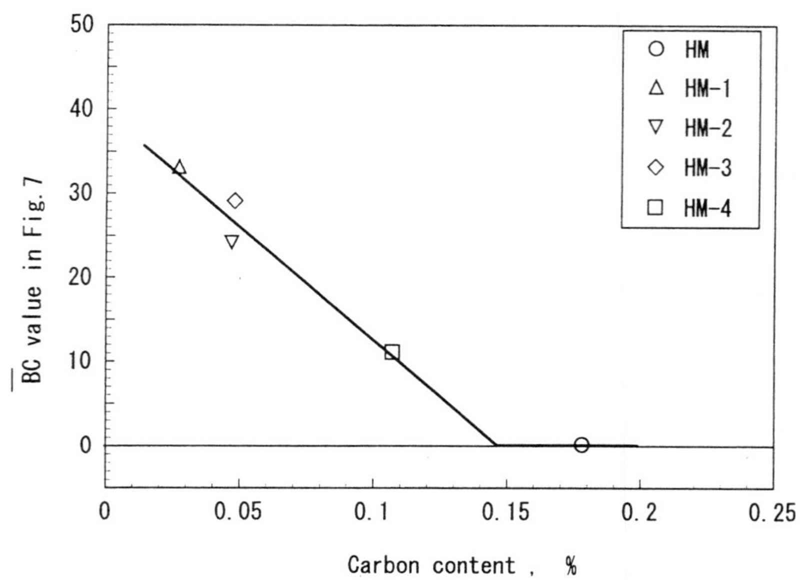

Figure 10 Relation between carbon content and $\overline{BC}$ value.

Comparison of the static and dynamic properties of RPV steels

P. Novosad
Division of Integrity and Materials, Nuclear Research Institute Rez plc., 250 68 Rez, Czech Republic

Abstract

Paper describes comparison of the static fracture toughness, dynamic fracture toughness and notch toughness of the nuclear reactor pressure vessels materials by testing the properties dependence on test temperature. Summarized are results of study of copper and phosphorus influence on radiation induced changes of Cr-Ni-Mo-V steel mechanical properties and influence of annealing on Cr-Mo-V steel.

1 Introduction

The effects of neutron irradiation upon the properties of reactor pressure vessel materials is an important factor which may, among others, limit the plant life.

In power reactor surveillance programes, based on tensile and Charpy-V type specimens, the limit plant life used to be derivec from the drop of the upper shelf energy of Charpy specimens and the shift of their transition temperature under influence of neutron irradiation. The evaluation of their irradiation embrittlement measured by fracture toughness is frequently based on the correlation with the notch toughness obtained from Charpy-V specimens. In this paper some examples of such correlations are described.

2 Materials and irradiation condition

Cr-Ni-Mo-V

For this experiments were used six heats doped with copper and phosphorus, made from original Cr-Ni-Mo-V Grade AA heat. All six experiment materials have received simulated heat treatment: 920°C/2 hours/air + 650°C/10 hours/air + 620°C/25 hours followed 650°C/20 hours/furnace and up to 300°C/air. The irradiation temperature was 288°C, the attained fluence is $2{\cdot}6E10^{23}$(E > 0.5 MeV) The irradiation was performed in experimental reactor LVR-15 in NRI Rez.

Chemical composition of each examined heats is given in Table 1.

	1	2	3	4	5	6
C	0.15	0.15	0.15	0.15	0.16	0.16
Mn	0.48	0.48	0.34	0.34	0.34	0.34
Si	0.17	0.17	0.21	0.21	0.20	0.20
P	0.012	0.018	0.015	0.021	0.014	0.021
S	0.010	0.010	0.009	0.009	0.009	0.009
Cr	2.06	2.06	1.89	1.89	2.14	2.14
Ni	1.28	1.28	1.28	1.28	1.27	1.27
Mo	0.56	0.56	0.55	0.55	0.58	0.58

Cu	0.30	0.30	0.52	0.52	0.08	0.08
V	0.10	0.10	0.09	0.09	0.10	0.10

Table 1: Chemical composition of examined steels Cr-Ni-Mo-V

Cr-Mo-V

For this experiments the base metal and weld metal of nominal composition (Table 2) were used, their heat treatment likewise complied with technical specifications.

Alloy	Cr-Mo-V-BM Th. Annealing	Cr-Mo-V-WM Th. Annealing
C	0.13	0.04 - 0.15
Mn	0.40	0.67 - 1.56
Si	0.25	0.30 - 0.45
P	0.009	0.012 - 0.039
S	0.012	0.011 - 0.017
Cr	2.77	1.40 - 2.48
Ni	0.16	0.20 - 0.24
Mo	0.62	0.43 - 0.61
V	0.30	0.19 - 0.30
Cu	0.13	0.15 - 0.25

Table 2: Chemical composition of examined steels Cr-Mo-V

Dynamic and static fracture toughness levels vs. test temperature are fitted (in transition region) by relationship:

$$K = A + B.\exp(C.T)$$

Transition temperatures and transition temperature shifts are established for 100 $MPam^{1/2}$ level.

The relations are in Figs.1 to 4.

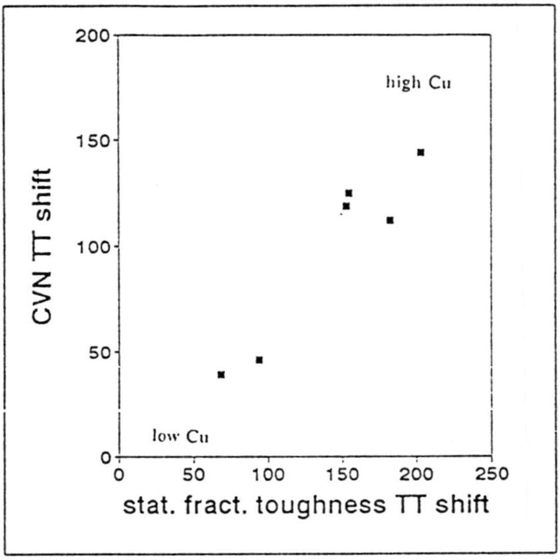

Fig. 1: Correlation between static fracture toughness transition temperature shifts and Charpy-V impact energy transition temperature shifts for heats with different Cu and P content Cr-Ni-Mo-V steel

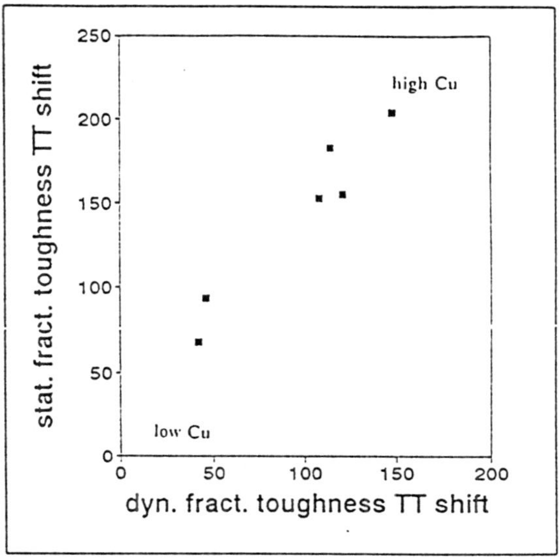

Fig. 2: Correlation between dynamic fracture toughness transition temperature shifts and static fracture toughness transition temperature shifts for heats with different Cu and P content Cr-Ni-Mo-V steel

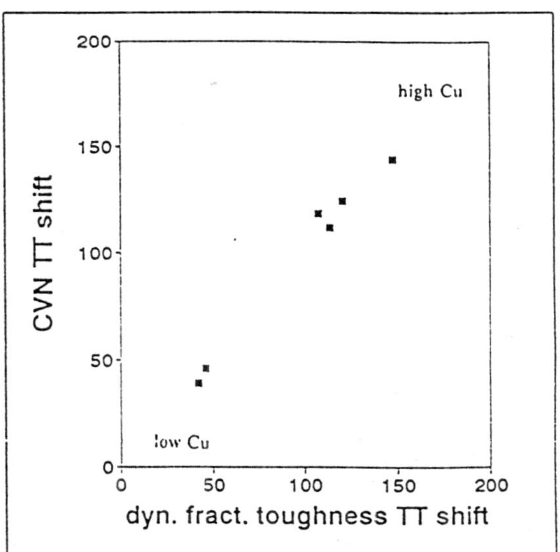

Fig. 3: Correlation between dynamic fracture toughness transition temperature shifts and Charpy-V impact energy transition temperature shifts for heats with different Cu and P content for Cr-Ni-Mo-V steel

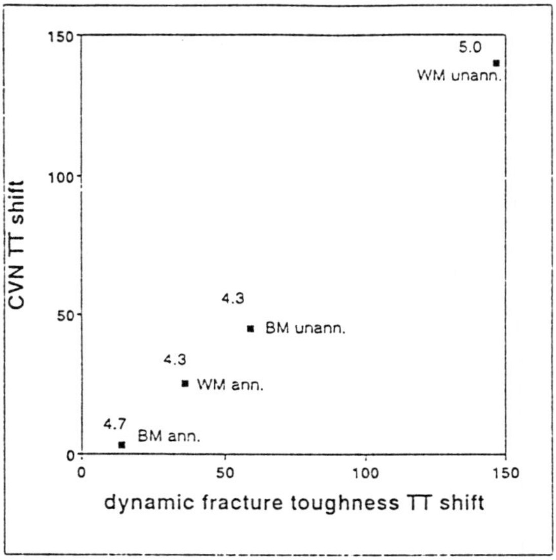

Fig. 4: Correlation between dynamic fracture toughness transition temperature shifts and Charpy-V impact energy transition temperature shifts for annealed and unannealed Cr-Mo-V steel

4 Discussion

Irradiation embrittlement measured by TT shift for static fracture toughness, dynamic fracture toughness and Charpy tests showed the same picture. Irradiation embrittlement increased with copper content, by negligible effect of phosphorus (for Cr-Ni-Mo-V). For Cr-Mo-V is irradiation embrittlement higher for the weld material than base metal The cross-correlations between static fracture toughness transition temperature shift, dynamic fracture toughness transition temperature shift and Charpy transition temperature shift appears to be good linear Fig.1 to 4. It is recognized, that usually quoted correspondence 1 to 1 is only between CVN (Charpy) and K_{ID} (dyn. fract. tough.) Fig.3,4 but for other properties correlations 1 to 1 is not fully satisfied Fig.1,2.

The effects of surface corrosion on the fatigue behaviour of aluminium alloy specimens containing cold expanded holes

N. Glinos,[a] P.G. Wagstaff,[a] R. Cook[b]
[a]*Aeronautical Engineering, Kingston University, Kingston-Upon-Thames, Surrey, UK*
[b]*Defence Research Agency, Farnborough, Hampshire GU14 6TD, UK*

ABSTRACT

The purpose of this work was to determine the influence of prior surface corrosion on the fatigue life improvement produced by cold expansion of fastener holes in an aluminium alloy. Results for corroded and uncorroded material are presented and the effects of cold expansion on fatigue endurance and crack propagation rates are evaluated.

INTRODUCTION

Fatigue damage is one of the most important problems in aircraft structures particularly in structural joints, where cracks frequently initiate at fastener holes. Significant improvements in the fatigue life of joints can be achieved by hole cold expansion which introduces residual compressive hoop stresses around the circumference of the hole (see figure 1 ref. 1) which tend to reduce the growth rates of cracks which subsequently form. The cold expansion technique used in this programme was the split sleeve method marketed by Fatigue Technology Incorporated of Seattle,Washington (ref. 2).

In the UK cold expansion is more commonly used as a repair process than as a manufacturing process. Fastener holes are "cleaned out" and reamed oversize if necessary, prior to cold expansion. There is a possibility, that surface corrosion exists on free or mating surfaces due to service exposure, which is not removed during the rework programme. The plastic deformation which occurs during cold expansion could initiate cracks from corrosion pits in the vicinity of the hole and hence reduce its fatigue life enhancing quality. The balancing tensile residual hoop stress field around the compressive region could also cause acceleration of any cracks so formed and hence reduce the effectiveness of the cold expansion process.

This research programme was proposed in order to investigate the influence of cold expansion on corrosion pits in the vicinity of the open holes and quantify any detrimental effect on fatigue endurance caused by corrosion.

TEST PROGRAMME

Material Specification and Specimen Configuration

The material chosen for this study was the high strength aluminium alloy 7050-T76, used extensively in aircraft structures, the composition and mechanical properties of which are given in Table 1 and 2 respectively (ref 2). A number of sheets of the alloy were protected (uncorroded) whilst others were subjected to normal atmospheric corrosion for a period of approximately two years (corroded) which caused pitting corrosion at the surface of the material. Microscopic investigation revealed corrosion pits varying in size between about 20 μm and 50 μm.

Rectangular specimens, 300 mm x 40 mm, were manufactured from the two material types and each contained a centrally drilled hole. Specimens denoted Plain were drilled and reamed and specimens denoted CX were subjected to about 4% cold expansion prior to the final reaming operation. 4% expansion results in optimum fatigue endurance (ref.4).

The "inlet" and "outlet" sides of the specimen refer to the positions where the mandrel enters and exits the specimen during expansion. For both "inlet" and "outlet" side , a left and a right side have been defined from the hole centre.

	Al	Cr	Cu	Fe	Mg	Mn	Si	Ti	Zn	Zr
Min	90.4	-	1.9	-	1.9	-	-	-	5.7	0.08
Max	87.4	0.04	2.6	0.15	2.6	0.1	0.12	0.06	6.7	0.15

Table 1. Chemical Composition of the 7050-T76

0.2% yield stress	UTS	Elongation	Modulus of Elasticity
555 MPa	582 MPa	11%	69.5 GPa

Table 2. Mechanical Properties of the 7050-T76

Fatigue Testing Programme

Sinusoidal loading at a frequency of 10 Hz was applied to all specimens with a maximum gross stress of 174 MPa at a stress ratio of R = 0.1.

A surface replication technique using cellulose acetate sheet softened with acetone was used to monitor crack initiation and growth both in the bore of the hole and on the specimen surfaces. Crack lengths were measured using a shadow graph projector . Crack growth rates were calculated from these measurements using the ASTM seven point polynomial methods (ref.5).

RESULTS

The distribution of fatigue cracks in corroded cold expanded specimens was quite different to that observed in all of the other test types. Maps of crack distributions for a corroded cold expanded specimen are shown in figures 2a (inlet side) and 2b (outlet side) 50 cycles before final failure. It can be seen that cracks not only initiated at the edge of the hole but also in regions remote from the hole. The largest crack was on the right outlet side (figure 2b) which propagated through the plate causing the large breakthrough crack (shown in fig. 2a) to be initiated at a corrosion pit on the left inlet side. Figures 2c (front surface) and 2d (rear surface) are for a corroded plain specimen and show that cracks initiated only from the edge of the hole. Similar distributions to those shown in figs 2c and 2d were observed in uncorroded plain and cold expanded specimens.

The log mean endurance (cycles) for both plain and cold expanded specimens in the corroded and uncorroded conditions are presented in Table 3. It can be seen that the presence of corrosion pits reduced the fatigue life endurance of cold expanded specimens to about one quarter of that for uncorroded specimens, but had little affect on the fatigue life of plain specimens.

The cold expansion does however have a beneficial effect on the specimens tested. The life improvement ratio (log mean endurance of cold expanded specimens/log mean endurance of plain specimens) shows a net increase for both the uncorroded and the corroded specimens, the ratios being 21.65 and 3.20 respectively.

Hole and surface condition	Log mean endurance (cycles)
Cold expanded corroded	119874
Cold expanded uncorroded	507818
Plain corroded	28418
Plain uncorroded	22420

Table 3. Fatigue endurances in 7050-T76

The growth rates of cracks emanating from the central hole are presented in figure 3 for the plain specimens and figure 4 for the cold expanded specimens. The growth rate of cracks emanating from corrosion pits in cold expanded specimens are presented in figure 5. The solid curve shown in this figure represents long crack data for alloy 7050-T76 at an R ratio of 0.1. Figure 6 shows the location of crack origins which caused failure in the corroded cold expanded specimens and also shows the residual hoop stress distributions on the mandrel inlet and outlet faces.

DISCUSSION

It can be seen from Table that the fatigue endurance of plain hole specimens was little affected by the presence of corrosion pits. This is supported by an examination of Figure 3 which shows that the fatigue crack growth rates in plain specimens were similar for both corroded and uncorroded material. This is to be expected as cracks initiated at the edge of the central hole and grew to failure before any significant cracks were observed from pits in the corroded material.

In contrast to this behaviour, significant reductions in the endurance of cold expanded specimens occurred due to the presence of corrosion (see Table 1). As observed earlier, this was due to the growth of cracks from corrosion pits which caused premature fatigue failures from locations remote from the central hole. For cold expanded corroded material, numerous cracks were observed, about 17% of which initiated at the edge of the hole, less than 2% in the compressive residual stress region and about 81% in the tensile residual stress region. About 90% of the cracks which caused failure initiated from corrosion pits in the tensile residual stress region, the majority of which (about 84%) initiated on the mandrel outlet face. This is to be expected as the tensile residual hoop stresses are greater on the mandrel outlet face than on the mandrel inlet face.

The growth rates of cracks from cold expanded holes are presented in figure 4. For the uncorroded material, fatigue cracks on both inlet and outlet faces rapidly decreased in growth rate as the crack length increased. On the inlet face, cracks reached a minimum in growth rate at a length of about 0.75mm and on the outlet face a minimum was reached at a length of about 0.25mm at which point the cracks on the outlet face arrested. The reduction in growth rate at short crack lengths is because the cracks are growing (initially) into a progressively more compressive residual stress field. The different crack growth behaviour between the inlet and outlet faces is due to the different residual stress distributions on the two faces (see figure 1). For the corroded material, the observed crack growth rates initially were similar to those of the uncorroded material (see figure 4) with cracks on the inlet face reaching a minimum in growth rate at about the same crack length. On the mandrel outlet face, however, cracks in the corroded material did not arrest after reaching a minimum in growth rate, but increased rapidly in growth rate as they interacted with long cracks growing from pits.

The growth rates of cracks initiating from corrosion pits, which caused specimen failure, are presented in figure 5. The growth rates are remarkably consistent despite the different locations of crack origin (see figure 6) and hence different residual and applied stress distributions. The stress intensity factor solution for pit crack growth was for cracks growing from semicircular defects in a uniform stress field and hence did not correct for the different stresses present. The solid curve in figure 5 represents the mean of measured long crack data on this material at a stress ratio of R=0.1. The much faster crack growth rates observed in the corroded cold expanded specimens results partly from the increased applied stress due to the proximity of the hole and partly from the tensile residual stresses in the vicinity of the pit. Further work will be undertaken to model the crack growth from pits in residual stress fields which will aim to predict the observed crack growth.

CONCLUSIONS

Corrosion had no significant effect on the fatigue life of open hole specimens containing plain holes.
Cold expansion increased the fatigue life of 7050-T76 specimens containing open holes however the presence of surface corrosion significantly reduced its beneficial effect.
Cracks which caused failure in uncorroded cold expanded specimens started from the edge of the hole on the mandrel inlet side, whilst in the corroded condition cracks which caused failure initiated remote from the hole and on the mandrel outlet face.
This is a significant consideration for the inspection of aircraft structures where such cracks could be overlooked by conventional inspection programmes.

REFERENCES

1. A.T Ozdemir, R Cook, L Edwards. Durability and Structural Reliability of Airframes. *Residual stress distributions around cold expanded holes.* Proceedings of the 17th ICAF Symposium, Stockholm, Sweden, EMAS, 1993
2. M.A Landy and R.L Champoux *FTI Engineering process specification FTI 8108B*, 1984, Seattle, USA, Fatigue Technology, Inc.
3. A.T Ozdemir and L Edwards *Cold expansion of fastener holes*, Technical Report FRG/93/10/1.
4. R.L Champoux *An overview of cold expansion methodes* in 'Fatigue Prevention and Design', EMAS, Warley, UK (1986).
5. *Standard test method for measurement of fatigue crack growth rates.* ASTM standard, Vol 03.01, E647, 1988.

ACKNOWLEDGEMENTS

The authors would like to thank the Structural Materials Centre, DRA and Kingston University for their support for this work.

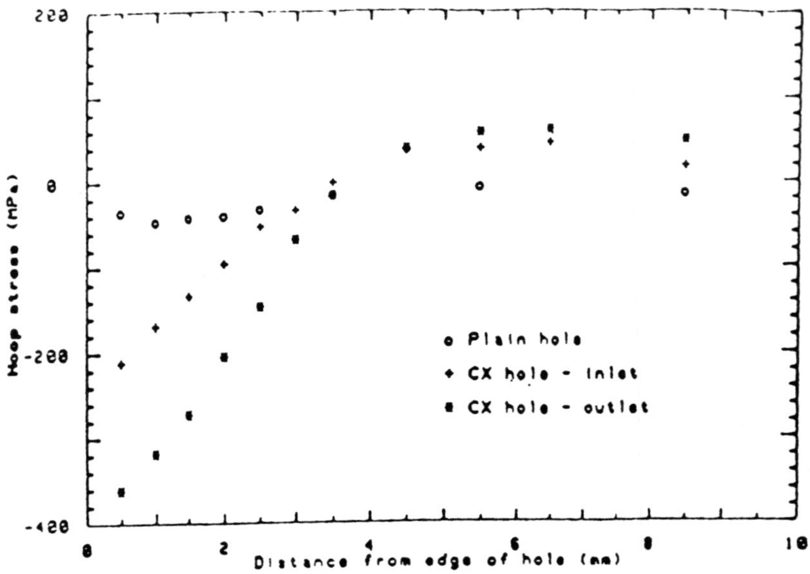

Fig.1 Residual stresses at plain and cold expanded (CX) holes

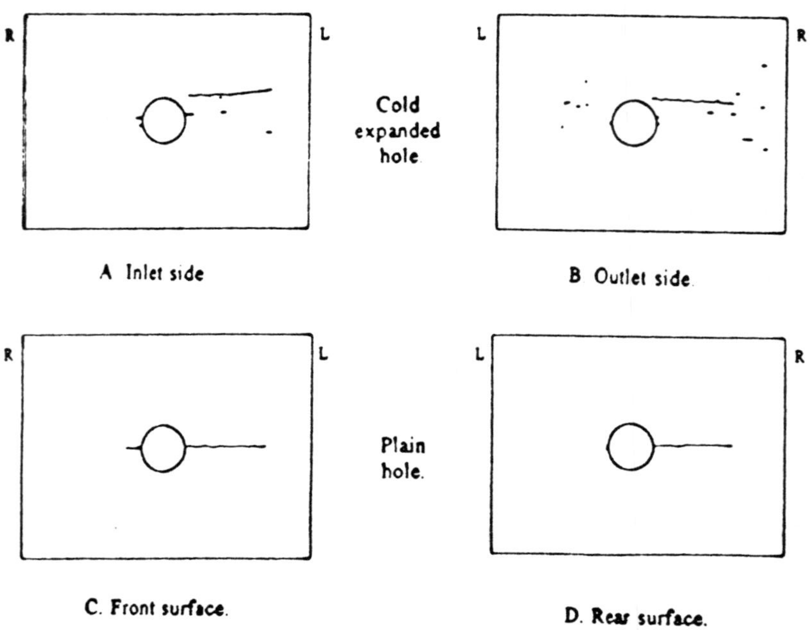

Fig.2 Cracks distribution

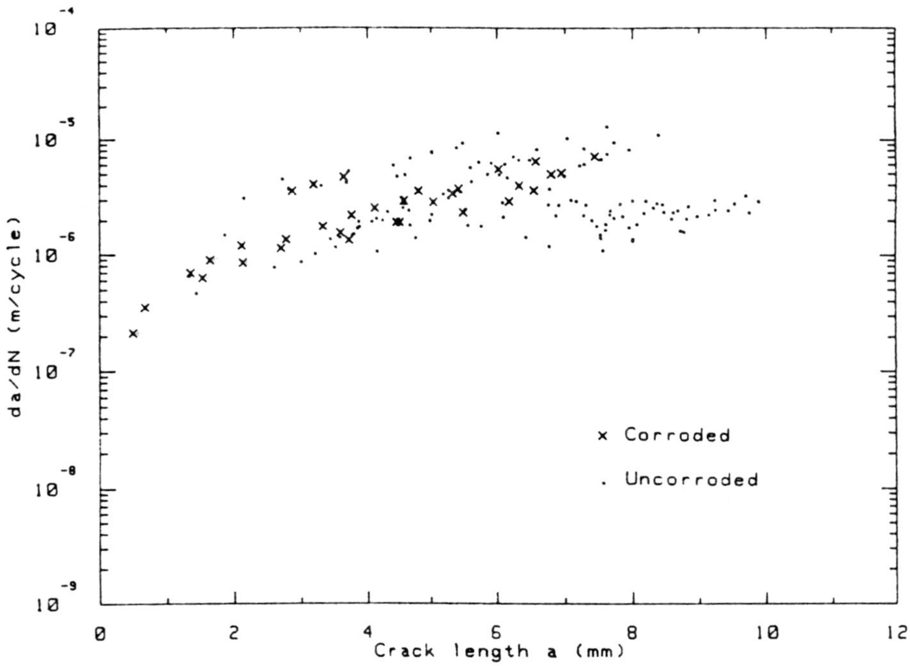

Fig.3 Crack growth rates in plain specimens of corroded and uncorroded material

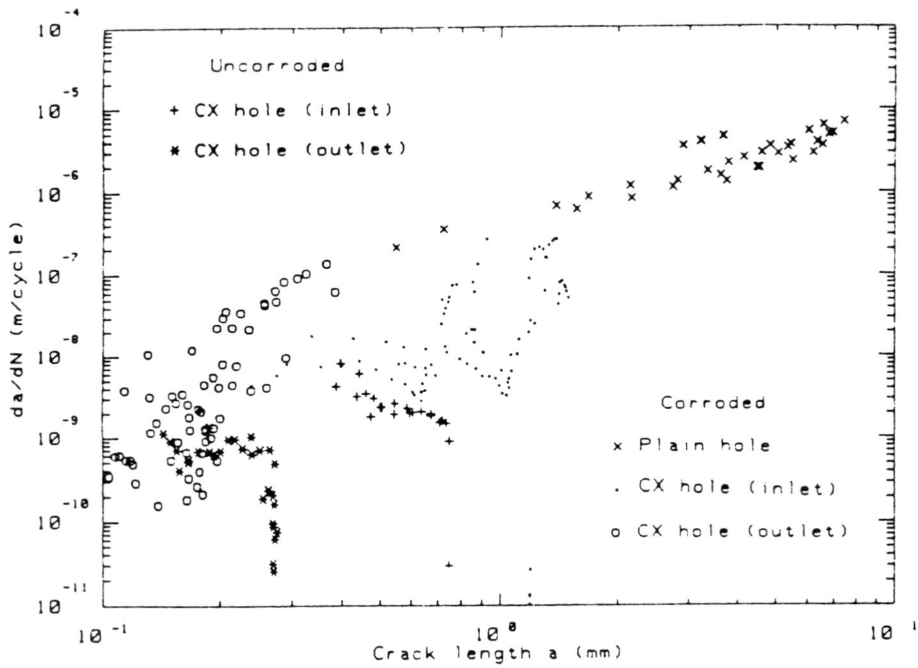

Fig.4 Crack growth rates from plain and cold expanded holes

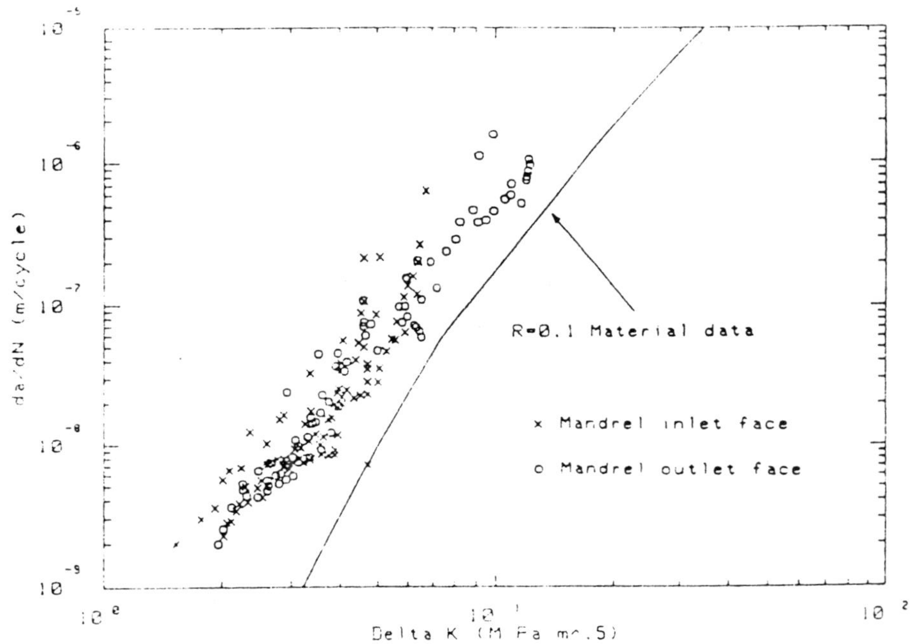

Fig.5 Crack growth rates from pits in cold expanded corroded specimens

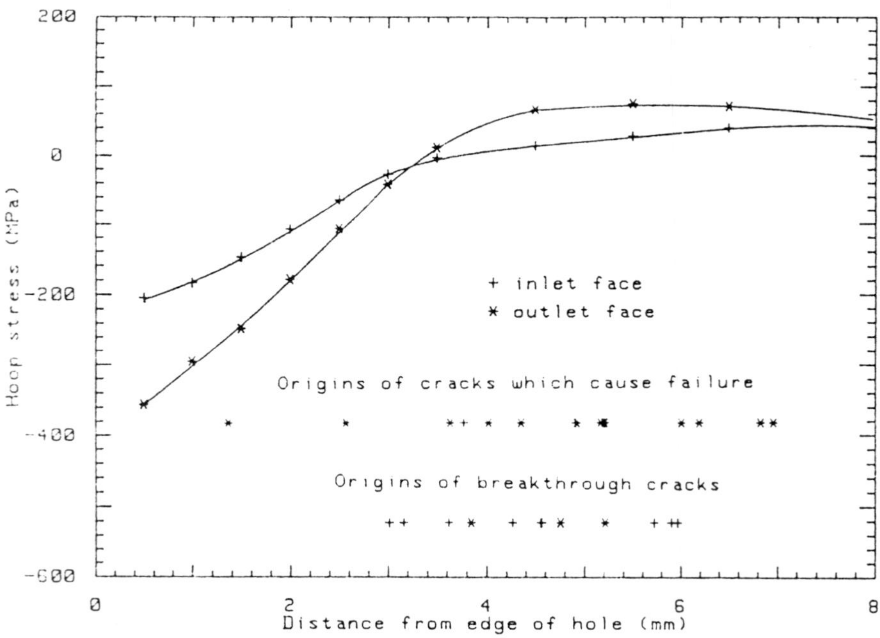

Fig.6 Residual distributions and failing crack origins

A unified treatment of ferrodistortive phenomena stimulated by impurities

J. Kapelewski
Institute of Electronics Materials Technology, 133 Wólczyńska Street, 01-919 Warsaw, Poland

Abstract

The aim of the paper is to outline a consistent method for treating changes in overall macroscopic material parameters of produced by some point centres of a microdistortion.
The paper treats the problem in microscopic terms. Main emphasis is focused at the formulation of a general methodology in which a semiphenomenological treatment, leading to relatively simple closed-form expressions, can be formulated. The method involves the use of localized representation of lattice vibrations with a separate treatment of the central near-defect regions responsible for the phenomena in consideration.

1 Introduction

There is now considerable interest in investigations directed toward inducing a controllable change in material properties of crystals brought about by impurity defects. Especially interesting in this aspect are such sorts of defects as tunnelling and Jahn-Teller centres, which due to their coupling with a host lattice, have potential to cause even a dramatic change in both mechanical and electrical material parameters as well as in the effective symmetry of the medium.
A common peculiar feature of the phenomena in consideration is an inherently microscopic character of their conditioning. It resides in an essential role of configurational symmetry of ligands in a near vicinity of the defect centre with frequent predominance of one of the irreducible representations (irr. repr.) of the crystal point group involved (as it is the case for vibronic-like phenomena), as well as of anharmonic effects associated with relatively large value of

microdisplacements in that region. These circumstances provide a framework for constructing any tractable and realistic model system of the properties in question.

The discreteness of the material may, basically, be accounted for through introducing a nonlocality to the description of microscopic material characteristics. This can be formally performed with making use of a pseudocontinuum-type approach constructed appropriately to the existing size parameter as determining the basic volume element [1]. The latter is usually taken in relation to the unit cell of the crystal or to its extended variety consistent of a group of the cells of a determined macroscopic symmetry. When treating point defects or more complex point irregularities, the natural choice for the volume element is the region surrounding the defect centre with a high enough magnitude of the dominant microdisplacements (with respect to that at asymptotic distance). The mentioned pseudocontinuum approach, in its conventional form, constitutes in fact a method of interpolation based on the Shannon - Kotielnikov theory as applied to a periodic crystal lattice, the eventual microstructure being then treated in terms of Cosserat micromoments used to simulate relative intracell modes of microdisplacements.

In the paper a unified general formulation is proposed. Its main idea lies in employing the lattice discretization method, previously developed by us for treating phenomena related with the defect-induced preasymptotic and asymptotic distortion in complex crystals [2]. The attractiveness of the method in this application stems for its capability to greatly simplify the treatment by transforming the initial microscopic description expressed by means of atomic displacements and forces to the description in terms of a single lattice with a simultaneous projecting on a chosen local distortion mode. Of importance is also a considerable degree of flexibility in adapting the method to various physical situations, involving the possibility to adjust the size of an element volume of the above mentioned lattice to that of the effective support for dominant single - defect perturbations.

2 Localized and symmetrized basis of lattice vibration modes

To construct a localized basis appropriate for treating lattice distortion in a symmetric and uniform way, let us consider general conditions for doing it by way of constructing the appropriate superpositions of eigenfunctions of vibrational modes from a single frequency band. It is assumed to have the following properties:

a) a localization at the respective points of a Bravais lattice,

b) they are transformed according to a definite irreducible representation of a point group of the crystal,

c) they are orthogonal with each other.

For simplicity, the vibration bands of the crystal are assumed to be mutually isolated. Adapting the approach developed in [3, 4] for electron states, main

points of method is below shortly analyzed and their possible symmetry defined for the point group of the crystal in consideration.
Denoting by $t_{l-l'}$ - the translation vector of the crystal, the translation group of the elastic Hamiltonian may be distributed to left-cosets according to the scheme:

$$G_H = \sum_l t_{l-l_0} G_{l_0} \tag{2.1}$$

where $G_{l_0} \equiv G_0$ denotes the point group of symmetry that includes the operations which do not change the lattice site $\mathbf{l}_0$.
The arbitrariness of choosing the "zero" point $\mathbf{l}_0$ (which further on will be adopted as the origin of the coordinate system, i.e. $\mathbf{l}_0=0$), results from the isomorphism of all groups G_l, resulting from the relationship:

$$\alpha_l = (t_l)^{-1} \alpha_0 t_l, \tag{2.2}$$

where t_l designates the translation of the point 0 to the lattice point $\mathbf{l}$, while $\alpha_l \in G_l$.
The distribution (2.1) makes it possible to construct the representation induced in the translation group G_H from any irreducible representation Γ of the group G_0, on the strength of the relationship:

$$\{\Gamma\}_H(\alpha_l) = \Gamma\left[t_l \alpha_l (t_l)^{-1}\right] \tag{2.3}$$

The vibrational modes of the host crystal provide convenient basis functions of the irreducible representations of the translation group H, while their superpositions under analysis should be the basis functions of the representation (2.3) (in view of the properties a) and b)).
The wanted superpositions of vibrational modes for the Bravais lattice is given by the expansion of the representation $\{\Gamma\}_H$ according to the irreducible representations of the group G_H, i.e. :

$$\{\Gamma\}_H = \sum_{a,k \in \Omega_0} n_0(\mathbf{ka}|\Gamma)\Gamma_H^{\mathbf{ka}}, \; (n = fn_0) \tag{2.6}$$

where Ω_0-basic region of the vector star $\mathbf{k}$, while $\Gamma_H^{\mathbf{ka}} = \{\Gamma_{\mathbf{ka}}\}_H$- representation induced in the star $\mathbf{k}$ from the irreducible representation $\Gamma_{\mathbf{ka}}$ of the wave vector group G_k. Inducing is effected here on the strength of the scheme:

$$\Gamma_H^{\mathbf{ka}}(\alpha) = \Gamma_{\mathbf{ka}}(\alpha_i^{-1}\alpha\alpha_j) \equiv \Gamma_{ij}(\alpha) \tag{2.7}$$

where $\alpha_i, \alpha, \alpha_j$ are such operations of the star $\mathbf{k}$, that

$$\alpha_i^{-1}\alpha\alpha_j \in G_k \; (\mathbf{k}_i = \alpha\mathbf{k}_j, \mathbf{k}_i = \alpha_i\mathbf{k}).$$

Since inside the Brillouin zone we have $d_{\mathbf{k}a} = \dim\Gamma_{\mathbf{k}a} = 1$ (save same high symmetry points), then the blocks Γ_{ij} become here the matrix elements, while $\dim\Gamma_H^{\mathbf{k}a} = g_0$ (i.e. the size of the point group G_0).
The application of the correlation (Frobenius) theorem to the relationships (2,4) and (2.6) yields:

$$\left[\Gamma_H^{\mathbf{k}a}\right]_{G_0} = \sum_{\Gamma} n_0(\mathbf{k}a|\Gamma)\Gamma \qquad (2.8)$$

i.e.

$$\dim\left[\Gamma_H^{\mathbf{k}a}\right]_{G_0} = f\sum_{\Gamma} n_0(\mathbf{k}a|\Gamma)d_0 = \sum_{\Gamma} n(\mathbf{k}a|\Gamma)d_0 \quad (d_0 = \dim\Gamma).$$

Since $\left[\Gamma_H^{\mathbf{k}a}\right]_{G_0} = \{\Gamma_{\mathbf{k}a}\}_{G_0}$, then using the Frobenius theorem again one obtains:

$$[\Gamma]_{G_{\mathbf{k}}} = \sum_{a} n_0(\mathbf{k}a|\Gamma)\Gamma_{\mathbf{k}a}. \qquad (2.9)$$

Consequently, on the basis of (2.4) and (2.6):

$$\{\Gamma\}_H = \sum_{\mathbf{k}\in\Omega}[\Gamma]_{G_{\mathbf{k}}} = \sum_{\mathbf{k}\in\Omega}\left\{[\Gamma]_{G_{\mathbf{k}}}\right\}_H. \qquad (2.10)$$

The above relationships permit, as will be seen below, the conditions to be defined for constructing the above presented localized functions from one (or several) bands, and the basis of this construction will be the relationship (2.10). Further on, the base functions of the representation $[\Gamma]_{G_{\mathbf{k}}}$ (reducible in $G_{\mathbf{k}}$) will be denoted by $\varphi_{\Gamma\gamma}^{\mathbf{k}}(\mathbf{r},z)$. As follows from (2.9) they are the superpositions of the eigenfunctions of the elastic Hamiltonian, that correspond to the single value **k** and transform themselves according to the representation $[\Gamma]_{G_{\mathbf{k}}}$. Their number is equal to $\dim\Gamma = d_0$.
Since $d_{\mathbf{k}a} = 1$, then:

$$d_0 = \sum_{a} n_0(\mathbf{k}a|\Gamma) = \frac{1}{f}\sum_{a} n(\mathbf{k}a|\Gamma) = \frac{n_\Gamma}{f} \qquad (2.11)$$

where n_0 are obtained as coefficients of expansion (2.8).
Consequently, the set of functions $\varphi_{\Gamma\gamma}$ can be constructed from the eigenfunctions of a definite number of vibration bands, while their number in the cross-section, corresponding to a single **k**, is:

$$d = \sum_{a} n(\mathbf{k}a|\Gamma). \qquad (2.12)$$

The dependence (2.10) implies the relation between the system $\varphi_{\Gamma\gamma}(\mathbf{k},\mathbf{r};z)$ and the wanted functions $a_{\Gamma\gamma}(\mathbf{r}-\mathbf{l})$.
In order to express it, in an explicit form, let us note that for the confinement on the subgroup G_0 one obtains:

$$\{\Gamma\}_{G_0} = \sum_{k\in\Omega_p} \left\{[\Gamma]_{G_k}\right\}_{G_0}, \quad (2.13)$$

where Ω_p basic region in the star G_0. Consequently, for the basis functions defined in G_0, one obtains the relationship:

$$a^j_{\Gamma\gamma}(\mathbf{r}) = A\sum_{\mathbf{k}\in\Omega_p} F^{kj}_{\Gamma\gamma}(\mathbf{r}) = A\sum_{\mathbf{k}\in\Omega} \varphi^j_{\Gamma\gamma}(\mathbf{k},\mathbf{r}), \quad (2.14)$$

and from the condition of normalizing: $A = \sqrt{\frac{\Omega}{V}}$, Ω - the volume of the unit cell, V-the volume of the crystal, F and φ are the basis functions of reducible representations $\left\{[\Gamma]_{G_k}\right\}_{G_0}$ and $[\Gamma]_{G_k}$, respectively.

The expression (2.14) defined simultaneously the basis functions of all the blocks $\{\Gamma\}_H(\alpha_l)$. In fact, as against (2.1) - (2.2), we have in particular:

$$\{\Gamma\}_H(\alpha_l) = \Gamma\ (T_l\alpha_l T_l^{-1}), \quad (2.15)$$

Consequently,

$$a^{jl}_{\Gamma\gamma}(\mathbf{r}) = T_l a^j_{\Gamma\gamma}(\mathbf{r}) = a^j_{\Gamma\gamma}(\mathbf{r}-\mathbf{l}) \quad (2.16)$$

In view of the definition $\varphi^j_{\Gamma\gamma}(\mathbf{k},\mathbf{r}) \equiv \left[\varphi^j_\alpha\right]_{\Gamma\gamma}$ as the basis function of the representation of group G_k and the essence of T_l the latter equality may be written down in the form:

$$a^{jl}_{\Gamma\gamma}(\mathbf{r}) = a^j_{\Gamma\gamma}(\mathbf{r}-\mathbf{l}) = \sqrt{\frac{\Omega}{\mathbf{V}}}\sum_{\mathbf{k}\in\Omega} e^{i\mathbf{kl}}\varphi^j_{\Gamma\gamma}(\mathbf{k},\mathbf{r}) \quad (2.17)$$

From (2.17), the reverse relationship follows:

$$\varphi^j_{\Gamma\gamma}(\mathbf{k},\mathbf{r}) = \sqrt{\frac{\Omega}{V}}\sum_{\mathbf{l}} e^{i\mathbf{kl}} a^{jl}_{\Gamma\gamma}(\mathbf{r}), \quad (2.18)$$

With employing a suitable interpolation procedure the orthogonality relation for polarisation vectors can be represented in the integral form:

$$\sum_\alpha \int e^{(p)^*}_\alpha(\mathbf{q},r)e^{(p')}_\alpha(\mathbf{q}',r)dr = \delta_{pp'}\delta_{\mathbf{qq}'}\frac{n}{\Omega} \quad (2.19)$$

where n denotes the number of atoms in the unit cell, Ω- its volume, p the branch of vibration, α, β, - the Cartesian components, and $\mathbf{q}$ - the wave vector. The localised representation of the vibration bands can then be introduced by means of the formula [2]:

$$a^{(p)}_\alpha(\mathbf{r}-\mathbf{l}) = \Omega^{1/2}\int e^{-i\mathbf{kl}}e^{(p)}_\alpha(\mathbf{k},\mathbf{r})d\mathbf{k} \quad (2.20)$$

For local symmetry - adapted combinations of (2.20), defined us

$$a_{(\alpha)}\left(\Gamma^{j}\gamma^{j};\mathbf{r}-\mathbf{l}\right)=\Omega^{1/2}\int e^{-ikl}\left[e_{\alpha}(\mathbf{k},\mathbf{r})\right]_{\Gamma^{j}\gamma^{j}}d\mathbf{k} \tag{2.21}$$

the following orthogonality relation holds

$$\sum_{\alpha}\int a_{(\alpha)}\left(\Gamma^{p},\gamma^{p};\mathbf{r}-\mathbf{l}\right)a_{(\alpha)}\left(\Gamma^{p'},\gamma^{p'};\mathbf{r}-\mathbf{l}\right)d\mathbf{r}=\delta_{pp'}\delta_{ll'}\delta_{\Gamma^{p}\Gamma^{p'}}\delta_{\gamma^{p}\gamma^{p'}} \tag{2.22}$$

with $\Gamma^{(p)}$ being the irr. repr. of the polarisation vectors at Γ - point (k=0) for the p-th band and subscript (α) is to label the initial coordinate of the symmetrisation.

The corresponding relationship between the locally symmetrised displacement (Q) and force (F) fields can be written

$$Q_{(\alpha)}\left(\Gamma^{i}\gamma^{i};\mathbf{r}\right)=\sum_{k\beta}\int G_{(\alpha,\beta)}\left(\Gamma^{i}\gamma^{i},\Gamma^{k}\gamma^{k};\mathbf{r},\mathbf{r}'\right)F_{(\beta)}\left(\Gamma^{k}\gamma^{k};\mathbf{r}'\right)d\mathbf{r}' \tag{2.23}$$

with the Green's function

$$\begin{aligned}&G_{(\alpha,\beta)}\left(\Gamma^{i}\gamma^{i},\Gamma^{k}\gamma^{k};\mathbf{r},\mathbf{r}'\right)=\\&=\left[m(\mathbf{r})m(\mathbf{r}')\right]^{-1/2}\sum_{j,l,l'}g^{\Gamma^{j}\gamma^{j},\Gamma^{j}\gamma'^{j}}_{\Gamma^{i}\gamma^{i},\Gamma^{k}\gamma^{k}}(l,l')a_{(\alpha)}\left(\Gamma^{j}\gamma^{j};\mathbf{r}-\mathbf{l}\right)a^{*}_{(\alpha)}\left(\Gamma^{j}\gamma^{j};\mathbf{r}'-\mathbf{l}'\right)\end{aligned} \tag{2.24}$$

where

$$g^{\Gamma^{j}\gamma^{j},\Gamma^{j}\gamma'^{j}}_{\Gamma^{i}\gamma^{i},\Gamma^{k}\gamma^{k}}(\mathbf{l}-\mathbf{l}')=\frac{d_{\Gamma^{i}}d_{\Gamma^{k}}}{g^{2}}\sum_{R,R'}g(R\mathbf{l},R\mathbf{l}')\Gamma^{i}_{\gamma_{i}\gamma_{i}^{0}}(R)\Gamma^{k}_{\gamma_{0}^{k}\gamma_{k}}(R'^{-1})\Gamma^{j}_{\gamma'_{j}\gamma'^{j}_{0}}(R'^{-1}) \tag{2.25}$$

with R denoting the group operation and

$$g(l,l')=N^{-1}\sum_{\mathbf{q}}\frac{e^{i\mathbf{q}(\mathbf{l}-\mathbf{l}')}}{\omega^{2}-\omega_{j}^{2}(\mathbf{q})}\quad\text{(for bulk bands)} \tag{2.26}$$

where N stands for the total number of unit cells. It is worth to notice the simple scalar form of g independent of the actual crystal symmetry.

The above relations allows one to introduce the lattice representation Q(l), F(l) by means of the formulas

$$Q_{(\alpha)}\left(\Gamma^{i}\gamma^{i};r\right)=\left[m(r)\right]^{-1\ 2}\sum_{j,l}m_{j}^{1/2}Q^{\Gamma^{j}\gamma^{j}}_{\Gamma^{i}\gamma^{i}}(l)a_{(\alpha)}\left(\Gamma^{j}\gamma^{j};r-l\right) \tag{2.27a}$$

$$F_{(\alpha)}\left(\Gamma^{i}\gamma^{i};r\right)=\left[m(r)\right]^{-1/2}\sum_{j,l}m_{j}^{1/2}F^{\Gamma^{j}\gamma^{j}}_{\Gamma^{i}\gamma^{i}}(l)a_{(\alpha)}\left(\Gamma^{j}\gamma^{j};r-l\right) \tag{2.27b}$$

where m_j is a dimensional mass parameter.

Inverting (2.27), we obtain

$$m_j^{1/2} Q_{\Gamma^i\gamma^i}^{\Gamma^j\gamma^j}(l) = \sum_{\alpha} \int Q_{(\alpha)}(\Gamma^i\gamma^i; r)[m(r)]^{1/2} a_{(\alpha)}^{*}(\Gamma^j\gamma^j; r - l)dr \tag{2.28a}$$

$$m_j^{1/2} F_{\Gamma^i\gamma^i}^{\Gamma^j\gamma^j}(l) = \sum_{\alpha} \int F_{(\alpha)}(\Gamma^i\gamma^i; r)[m(r)]^{1/2} a_{(\alpha)}^{*}(\Gamma^j\gamma^j; r - l)dr \tag{2.28b}$$

The counterpart of (2.23) in the lattice representation takes then the form

$$Q_{\Gamma^i\gamma^i}^{\Gamma^j\gamma^j}(l) = \sum G_{\Gamma^i\gamma^i, \Gamma^k\gamma^k}^{\Gamma^j\gamma^j, \Gamma^j\gamma'^j}(l - l') F_{\Gamma^k\gamma^k}^{\Gamma^j\gamma'^j}(l') \tag{2.29}$$

As it was mentioned above, in the foregoing formulation, the case of mutually isolated dispersion curves is assumed.
In general, the above procedure of constructing the function $a_{\Gamma\gamma}^{l}(\mathbf{r})$ cannot be replaced by using in (2.17) the system $\varphi_{\mathbf{k}}(\mathbf{r})$ instead of $\varphi_{\Gamma\gamma}(\mathbf{k},\mathbf{r})$ with the symmetrization of the superposition thus obtained. This follows from the fact that the dependence $\varphi_{\mathbf{k}}(\mathbf{r})$ upon $\mathbf{k}$ is not analytical throughout the Brillouin zone, since the points of degeneracy become, for this zone, the branch points (such a situation takes place e.g. for $d^m > 1$, as well as in the presence of incidental degeneracy). This is associated with a jump change in the size of the representation $\Gamma_{\mathbf{k}a}$ at these points and with the change resulting there from both transformation properties of $\varphi_{\mathbf{k}}(\mathbf{r})$ and their basis functions.
This means, that the above mentioned superposition does not fulfil the conditions of a uniform convergence. Attention has been drawn to this fact in Ref. [3] for the case of the electron complex bands. This case does not apply to the system $\varphi_{\Gamma\gamma}(\mathbf{k},\mathbf{r})$, where to each vector $\mathbf{k}$ (in the degeneracy points as well) corresponds only one of these functions, while the transformation properties change in a continuous manner.

3 An outline of the modelling changes in elastic behaviour of high symmetry SAW substrates due to substitutional defect centres.

The spatial homogeneity of the lattice framework allows us to directly employ the pseudocontinuum procedure [1], which formally leads here to replacing the discrete variable **l** by the continuous one (**R**).
In view of inherently local character of $a_{\Gamma\gamma}(\mathbf{r})$ for the assumed isolated bands in the lattice framework having symmetry centre (as it is the case for a single Bravais lattice) one can contract all the vector and two point tensor quantities to the components indexes by a one and two irr. repr. Having this in mind, we will denote them, respectively as follows:

$$u_{\Gamma\gamma}^{\Gamma\gamma}(\mathbf{R}) \equiv u_{\Gamma\gamma}(\mathbf{R}) \text{ and } T_{\Gamma\gamma,\Gamma'\gamma'}^{\Gamma\gamma,\Gamma'\gamma'}(\mathbf{R},\mathbf{R}') = T_{\Gamma\gamma,\Gamma'\gamma'}(\mathbf{R},\mathbf{R}')$$

Following the method developed in [5, 6], the system potential energy, in the harmonic approximation, is then modelled by

$$U_0 = \int \Big\{ -\sum P_{\Gamma\gamma}\varepsilon_{\Gamma\gamma}(\mathbf{R}) - \sum w_{\Gamma\gamma}(\mathbf{R})Q_{\Gamma\gamma}(\mathbf{R})c(\mathbf{R}) + \frac{1}{2}\sum C_\Gamma \varepsilon^2_{\Gamma\gamma}(\mathbf{R}) + -\sum \mathbf{h}_\Gamma Q_{\Gamma\gamma}(\mathbf{R})\varepsilon_{\Gamma\gamma}(\mathbf{R})c(\mathbf{R}) + \frac{1}{2}\sum Q_{\Gamma\gamma}(\mathbf{R})c(\mathbf{R})\int K_{\Gamma\Gamma'}(\mathbf{R},\mathbf{R}')Q_{\Gamma'\gamma'}(\mathbf{R}')c(\mathbf{R}')d\mathbf{R}' \Big\} d\mathbf{R} \tag{3.1}$$

In (3.4), Q denotes symmetry adapted displacements averaged over unit cell or extended cell centered at **R**, represented here by volume element (**R**), and the meaning of particular terms is directly given by their analytical form [6].

From the equilibrium condition

$$\frac{\partial U}{\partial Q} = \frac{\partial U}{\partial \varepsilon} = 0 \tag{3.2}$$

one then obtains

$$\mathbf{C}\varepsilon(\mathbf{R}) - \mathbf{hQ}(\mathbf{R})c(\mathbf{R}) = \mathbf{P}(\mathbf{R}) \tag{3.3}$$

$$\int \mathbf{K}(\mathbf{R},\mathbf{R}')\mathbf{Q}(\mathbf{R}')c(\mathbf{R}')d\mathbf{R}' - \mathbf{h}\varepsilon(\mathbf{R}) = \mathbf{W}(\mathbf{R}), \tag{3.4}$$

where symmetry indices have been dropped and **P** and **W** designate, respectively, the elastic dipole moment and global field of forces in the region(**R**).

By eliminating Q and taking the statistical average over spatial distribution of the defect centres in consideration it leads to the formula

$$\mathbf{C}_{\text{eff}}(\mathbf{R},\mathbf{R}') = = \mathbf{C}\delta_B(\mathbf{R}-\mathbf{R}') - \mathbf{K}^{-1}(\mathbf{R},\mathbf{R}')\mathbf{h}^2 < c(\mathbf{R})c(\mathbf{R}') > \equiv \mathbf{C}\delta_B(\mathbf{R}-\mathbf{R}') - \delta\mathbf{C}(\mathbf{R},\mathbf{R}') , \tag{3.5}$$

with $\delta_B(\mathbf{r})$ being the pseudocontinuum delta function [1].

In the case of impurity-modified near surface layer it was shown to be reasonable to assume an expotentially decreasing model distribution of the form [7]

$$c(\mathbf{R}) = c(\mathbf{R}_\parallel)\exp(-\alpha Z), \ \langle c(\mathbf{R}_\parallel)\rangle = c \tag{3.6}$$

where $\mathbf{R} = (\mathbf{R}_\parallel, \mathbf{Z})$,

with the correlation function:

$$\langle c(\mathbf{R},\mathbf{R}')\rangle = \sigma W(\mathbf{R}_{\parallel}-\mathbf{R}_{\parallel}';Z,Z') = \frac{\Omega\sigma}{\pi\rho^{2}}\exp\left[\frac{-(\mathbf{R}_{\parallel}-\mathbf{R}_{\parallel}')^{2}}{\rho^{2}}-\alpha(Z+Z')\right],\ (\sigma = c^{2}) \tag{3.7}$$

exhibiting a Gaussian type of the distribution along the surface plane.

The foregoing treatment does not involve the case of multilevel centres introducing temperature dependence arised from thermal jumps between equivalent positions.

To allow for this effect one should include into the adiabatic potential energy in (3.1) the Jahn - Teller contribution given, generally, by $U_{JT} = \left[V_{\Gamma\gamma}Q_{\Gamma\gamma} + P_{\Gamma\gamma}\varepsilon_{\Gamma\gamma}\right]A_{\Gamma\gamma}$, where $\mathbf{A}$ is the matrix of the Clebsh-Gordon coefficients [8].

The system free energy can then be calculated, according to the formula

$$\mathcal{F} = U - \beta^{-1}\int c(\mathbf{R})\ln Z(\mathbf{R})d\mathbf{R} \tag{3.8}$$

where the last term describes the entropy contribution corresponding to thermal equilibrium conditions. In (3.8), $\beta = (k_B T)^{-1}$ with k_B being the Boltzmann constant, and $Z(\mathbf{R}) = \mathrm{Tr}\, e^{-\beta U(\mathbf{R})}$ - the statistical sum over the equivalent equilibrium positions.

For two-level centres, where $\mathbf{A} = \sigma_i$, with σ_i being the respective Pauli matrix, it reduces to the form

$$Z(\mathbf{R}) = 2\cosh[\beta\Delta(\mathbf{R})] \tag{3.9}$$

with $\Delta(\mathbf{R})$ given by the difference of potential energies of the centre as situated in the two equilibrium positions.

In view of statistical uniformity of the distribution of centres along the surface the set of equilibrium equations, after spatial averaging, can be reduced to a simplified form with all system field coefficients being invariant with respect to the point group of the crystal surface. Such a form is composed of blocks which transform under the group operation according to particular irreducible representations. For small values of Q and , the subset corresponding to $=_1$, with $_1$ (denoted, hereafter, by 1)being the active irreducible representation of the double-well centres in consideration, takes the form

$$\int \widetilde{K}_{11}(Z,Z')Q_1(Z')dZ' + \widetilde{h}_1(Z)\varepsilon_1(Z) - \widetilde{w}_1(Z)\beta\Delta_v(Z) = 0 \tag{3.10}$$

$$\sigma_1(Z) = C_{11}\varepsilon_1(Z) + \widetilde{h}_1(Z)Q_1(Z) - \widetilde{P}_1(Z)\beta\Delta_p(Z) \tag{3.11}$$

where, we have denoted

$$\Delta_v(Z) = \left[K_{11}(0)\right]^{-1} \int \widetilde{K}'_{11}(Z,Z')Q_1(Z')dZ' \tag{3.12}$$

$$\Delta_p(Z) = P_l(Z)\varepsilon_1(Z) \tag{3.13}$$

$$\widetilde{K}(Z,Z') = \int K(\mathbf{R}_{\parallel};Z,Z')\langle c(\mathbf{R}_{\parallel},Z)c(0,Z')\rangle d\mathbf{R}_{\parallel} \tag{3.14}$$

$$\widetilde{K}'(Z,Z') = K(Z,Z') - K(0)c^2(Z)\delta_B(Z-Z') \tag{3.15}$$

$$\left[\widetilde{h}_1(Z), \widetilde{w}_1(Z), \widetilde{P}_l(Z)\right] = c(Z)\left[h_1, w_1, P_1\right] \tag{3.16}$$

and the subscript 1 is used as a shorthand notation for the composite index (Γ_1,γ_1).

Combining (3.10) with (3.14) one readily obtains the change of the elastic tensor given by:

$$\delta C(Z,Z') = \left[h^2\widetilde{K}_1^{-1}(Z,Z') + \beta\left[P_1^2\right]c(Z)\delta_B(Z-Z')\right] \tag{3.17}$$

while the defect induced equilibrium strain is found to be

$$\varepsilon_1(Z) = \left[C_1 - \beta P_1^2(Z)\right]^{-1} h_1^2(Z)Q_1(Z) \tag{3.18}$$

The foregoing closed-form formulas allows one to evaluate (from the complete set (3.1-11)) microscopic local near surface ferrodistortion given here by means of internal strain (represented by Q_1) associated with the double-well defect system in consideration. It also provides a simple prescription for determining its temperature dependence from that of Q_1.

References

[1] Kröner E., Datta B.K., *Zs. f. Phys.*, 196, 203, 1966.
Rogula D., *Arch. Mech.* 31, 1, 15, 1979.

[2] Kapelewski J., Nonlocal Theory of Surface Defects, WAT, Warsaw 1981.

[3] Cloiseaux J., *Phys. Rev.*, 129, 2, 1963.

[4] Kapelewski J., *Journ. Techn. Phys.*, 20, 3, 341, 1979.

[5] Kapelewski J., *Ferroelectrics*, 81, 157, 1988.

[6] Kapelewski J., *Computer Aided Innovations of New Materials* 571, ed. M. Doyama et al., North Holland, 1993.

[7] Nakayama T., Sakuma T. in *Phonons Scattering in Solids*, 52, ed. L. J. Hallis et al., Plenum Press 1995.

[8] Pytte E., *Phys. Rev.*, B3, 10, 3503, 1971.

* The paper is supported by the Commitee of Scientific Research.

SECTION 2:
EXPERIMENTAL AND COMPUTATIONAL SIMULATION

Fatigue crack growth life predictions for single and dual amplitude loadings

S.P. Claridge, B.E. Powell
Mechanical Behaviour of Materials Laboratory, Department of Mechanical and Manufacturing Engineering, University of Portsmouth, Portsmouth, Hampshire PO1 3DJ, UK

Abstract

Fatigue crack growth lives at room temperature have been measured for a forged titanium aeroengine alloy. The experimental loadings used simulate two features of the engine flight pattern: the major stress cycles representing the start-stop operation, and the minor stress cycles representing the in-flight vibrations. The fatigue crack growth lives for the same conditions have also been predicted using the nCode software package "Kraken" in conjunction with materials data established by the experimental programme. The comparison of experimental and predicted lives shows good correlation, with the majority of predictions being within ±50% of the actual observed life. It is concluded that for the loading configurations considered in this study, particularly those with high mean-low amplitude cycles within the loading block, the accuracy of the computer prediction is very sensitive to the material threshold data for the range of stress ratios under consideration.

1 Introduction

In the design and operation of gas turbine engines safety is paramount, since the failure of a critical component can lead to human and financial losses which are unacceptable. For this reason, the majority of aeroengine discs are retired from service well before fatigue cracks develop; the service life being sufficient to develop a small crack in only 1 in a batch of 750 discs. With this approach, the fatigue crack growth (FCG) life represents a safety margin.

The discs are expensive components, using valuable and limited resources; thus it is desirable to obtain a safe but economic life from them. With this in mind, it has been suggested that each disc might be examined frequently and retired from service only if found to be cracked. In this case, the inspection period is related to the maximum FCG life, which is the number of flights required to grow a crack from a size which is just below the limit of non-destructive testing to a larger critical size. Within this life span there must be fitted an appropriate number of economic inspection intervals.

Thus it is that irrespective of whether the lifing strategy is the former "life to first crack" or the latter "retirement for cause" there is a need for the measurement, modelling and prediction of the FCG lives of disc materials. Previous studies [1,2] have shown that FCG rates can be increased, and lives decreased, by the presence of superimposed vibrational stresses. The aim of the present study is therefore to establish the effectiveness of FCG life predictions for a disc material, particularly when subjected to fatigue loadings involving superimposed vibrational stresses.

2 Experimental details

Compact tension testpieces having a width of 26 mm, thickness of 13 mm, height of 31.2 mm and conforming to the ASTM specified design [3] have been cut from a fan disc of Ti-6A-4V which has an alloy composition conforming to the British Standard for Ti-6Al-4V forgings [4]. The specimen orientation, following ASTM notation [5],was tangential radial (TR).

Prior to the fatigue testing, the material was solution treated at 900°C, water quenched, and aged at 700°C. This heat treatment produced a microstructure containing 47% primary α by volume, the particles having a wide range of aspect ratio, with their average size being approximately 15 μm wide and 45 μm long. The matrix was transformed β, consisting of colonies of aligned α separated by β laminations. The width of these laminations was typically 15 μm with a maximum α plate thickness of 4μm. In this condition, the typical room temperature properties correspond to a tensile strength of 990 MPa, a 0.2 proof stress of 865 MPa, an elongation to fracture of 12%, and a reduction in area of 38%.

For civil engines, a major cycle corresponds to a flight pattern of take-off, cruise and landing, and may be approximated in the fatigue testing by a trapezoidal stress wave having the asymmetric form shown in Figure 1a and a stress ratio of 0.1. This loading is applied by a servohydraulic machine at a frequency of 0.1 Hz. The minor cycles are high-frequency vibrations induced by gas pressure. In the fatigue experiments, they were simulated by 150 Hz sinusoidal stress waves of constant amplitude, generated by an electromagnetic vibrator which was positioned between the servohydraulic actuator and the test specimen. This high mean-low amplitude stress cycle is shown in Figure 1b. With this arrangement, major and minor stress cycles could be applied separately or in combination. In the latter case, the minor cycles were generally superimposed on that part of the major cycle corresponding to the cruise condition only, that is the dwell at maximum load (Figure 1c,d,e). Additionally, a number of tests were also performed with minor cycles superimposed on the rise and fall of the major cycle (Figure 1f)

The parameter used to describe the relative magnitude of the major to minor stress cycles is the amplitude ratio (Q), that is to say, the ratio of the minor to major stress amplitudes. For a cracked component this corresponds to the ratio of the minor to major stress intensity ranges. Four levels of combined major and minor cycle loading, typical of some service conditions were adopted; namely, minor/major amplitude ratios of 0.06, 0.12, 0.22 and 0.32. The term cycle ratio (n) is simply the number of minor cycles per major cycle. In the present study cycle ratios of 1000, 10 000 and 100 000 were generated by extending the major cycle dwell period at maximum load on which the high-frequency minor cycles were superimposed. A summary of the complete test programme is given in Table 1.

Loading	Specimen Numbers	R Major	R Minor	Amplitude Ratio Q	Cycle Ratio
Majors only	A1,A2,A3	0.1			
Minors only	B1		0.74		
Minors only	B2,B3,B4,B5		0.84		
Minors only	B6		0.90		
Major/Minor	C1,C2,C3,C4,C5	0.1	0.74	0.32	1,000
Major/Minor	C6	0.1	0.74	0.32	10,000
Major/Minor	D1,D2,D3,D4,D5	0.1	0.82	0.22	1,000
Major/Minor	D6	0.1	0.82*	0.22	1,300
Major/Minor	D7,D8	0.1	0.82	0.22	10,000
Major/Minor	E1,E2,E3	0.1	0.90	0.12	1,000
Major/Minor	E4	0.1	0.90*	0.12	1,300
Major/Minor	E5	0.1	0.90	0.12	10,000
Major/Minor	E6	0.1	0.90	0.12	100,000
Major/Minor	F1	0.1	0.95	0.06	1,000
Major/Minor	F2	0.1	0.95	0.06	10,000

* For cycles superimposed on dwell at maximum load.

Table 1: .Summary of 31 Test Conditions

The test procedures for precracking, and the subsequent determination of fatigue crack growth rates and lives, met the requirements of the appropriate ASTM specification [3], and have been fully described elsewhere[1,6].

3 Life Prediction Method

The life predictions were made using the nCode software suite, KRAKEN, which allows rapid prediction of fatigue crack growth and propogation life based on modelling crack extension using a cycle by cycle approach. Uniquely, the analysis procedure is coupled with the modelling of crack closure, stress history effects, static fracture modes, notches and environmental influences, although in this study the latter two were not neccessary. The methodology is based on standard linear elastic fracture mechanics (LEFM) concepts and uses the following materials behaviour model (7,8,9):

$$da/dN = C\ (\Delta K^{eff})^{n} \quad (1)$$

$$\Delta K^{eff} = (K_{max})^{eff} - (K_{min})^{eff} \quad (2)$$

$$(K_{max})^{eff} = (K_{max})^{app} + K_{fs} \quad (3)$$

$$K_{f} = (K_{max})^{app} / (K_{c} - (K_{max})^{app}) \quad (4)$$

$$(K_{min})^{eff} = (K_{min})^{app} \text{ or } K_{cl} \quad (5)$$

$$K_{cl} = \Delta K_{th} / (1 - R) \quad (6)$$

where: da/dN = fatigue crack growth rate (m/cycle)

ΔK^{eff} = effective stress intensity range (MPa $m^{½}$)

$(K_{max})^{eff}$ = maximum effective stress intensity (MPa $m^{½}$)

$(K_{min})^{eff}$ = minimum effective stress intensity (MPa $m^{½}$)

$(K_{max})^{app}$ = maximum apparent stress intensity (MPa $m^{½}$)

$(K_{min})^{app}$ = minimum apparent stress intensity (MPa $m^{½}$)

K_{cl} = stress intensity at closure (MPa $m^{½}$)

ΔK_{th} = threshold stress intensity range (MPa $m^{½}$)

R = stress ratio

K_{fs} = additional stress intensity from static fracture modes (MPa $m^{½}$)

K_{c} = critical stress intensity (MPa $m^{½}$)

The software is constructed to create working files in three principal areas namely:

a. Material Data

Basic material data such as modulus, yield stress and fracture toughness are required with effective crack growth data in the form of equation 1 above, with infomation relating ΔK_{th} to stress ratio (Knee)

The following data was used for the Ti-6A-4V material under consideration:

Property	Value
C	7.13E-12 m/cycle
n	3.49
$\Delta K_{th\ (@R=1)}$	2.1 MPa $m^{1/2}$
$\Delta K_{th\ (@R=0)}$	5.01 MPa $m^{1/2}$
R at the Knee	0.52
K_{IC}	75 MPa $m^{1/2}$
σ_y	865 MPa
σ_{UTS}	990 MPa
E	120 MPa

Table 2: Materials Data for Forged Ti-6Al-4V Disc Material.

b. Geometry

A stress intensity calibration is required for the geometry under consideration. Calibrations for a number of standard specimen and geometrical configurations, such as welds and contained cracks in plates, are available in the software database.

A standard specimen - CT Geometry (B = 13 mm, W = 26 mm) was sourced in NSoft software database.

c. Loading (Time History)

Load or stress time histories can be generated and classified with standard reduction techniques, such as rainflow analysis. The stress time histories used in this study are shown schematically in Figures 1a to 1f.

4 Results

The results are presented in the form of a Predicted Life -v- Experimental Life graphs. The axes in Graph 1, are presented in terms of blocks i.e. a typical flight cycle comprising of one major cycle and a number (1,000, 10,000, 100,000) of minor cycles. In Graph 2, the axis are shown in terms of actual number of cycles counted and analysed.

5 Discussion

The accuracy of the results is encouraging with good correlation between the experimental and predicted lives. The majority of predictions are within ±50% of the experimental life. The results which differ greatly can be explained by the sensitivity of the software model to threshold data for the range of stress ratios under consideration. The model clearly struggles when attempting to accurately determine very slow growth rates, in the order if 1×10^{-11} m/cycle, at low R values and near threshold condition. Minute errors in the modelling at this stage are going to accumulate over the cyclic lives of the test (some in excess of 10,000,000 cycles). This threshold data is also vital in predicting the onset of minor cycle damage, which when initiated is going to greatly accelerate the crack growth.

This work will continue and move onto correlating experimental and predicted lives for similar loading configuration for square section CN specimens. It is now recognised that cracks in CN specimens are more representative of the conditions that prevail in service.

6 Conclusions

The nCode software package "Kraken", in conjunction with materials data established by the experimental programme, has been successfully used to predict fatigue lives of standard specimens under complex loading comnditions

The comparison of experimental and predicted lives shows good correlation, with the majority of predictions being within ±50% of the actual observed life. It is concluded that for the loading configurations considered in this study, particularly those with high mean-low amplitude cycles within the loading block, the accuracy of the computer prediction is very sensitive to the material threshold data for the range of stress ratios under consideration.

Acknowledgements

The authors would like to thank the University of Portsmouth and Rolls Royce Plc for their generous support during the data generation and analysis phases of this work. Also to Dr Ian Austen (of nCode International Ltd.) for his invaluable assistance.

References

1 Powell, B.E., Duggan, T.V. and Jeal, R.H. "The influence of minor cycles on low cycle fatigue crack propagation",*Int. J Fatigue* 4 (1982) pp 4-14

2. Powell, B.E., Hall, R.F. and Hawkyard, M. "The effect of superimposed vibrational stress on fatigue crack propagation life". *1st Int Conf on Structural Integrity Assessment,* Manchester, (1992) pp336-345

3. "Standard Test Method for Measurement of Fatigue Crack Growth Rates", ASTM Standard E-647-83 (1983) Part 3, pp 710-730

4. "Specification for forgings of titanium-aluminium-vanadium alloy (tensile strength 900-1160 MPa)(Limiting ruling section 150mm)", BSI Standard 2TA 13 (1974)

5. "Standard Test Method for Plane-Strain Fracture Toughness of Metallic Materials", ASTM Standard E-399-83 (1983) Part 3, pp 518-551

6. Powell, B.E, and Henderson, I. "The conjont action of high and low cycle fatigue", *AFWAL-TR-83-4119* Air Force Wright Aeronautical Laboratories,USA. (November 1983)

7. Austen I.M., "Quantitative Understanding of Corrosion Fatigue Crack Growth Behaviour", ECSC Contract 7210.KE/806, Final Report, July 1982, (HMSO)

8. Austen I.M., "Quantitative Understanding of Corrosion Fatigue Crack Growth Under Variable Amplitute Loading", ECSC Contract 7210.KE/810, Final Report, July 1987, (HMSO)

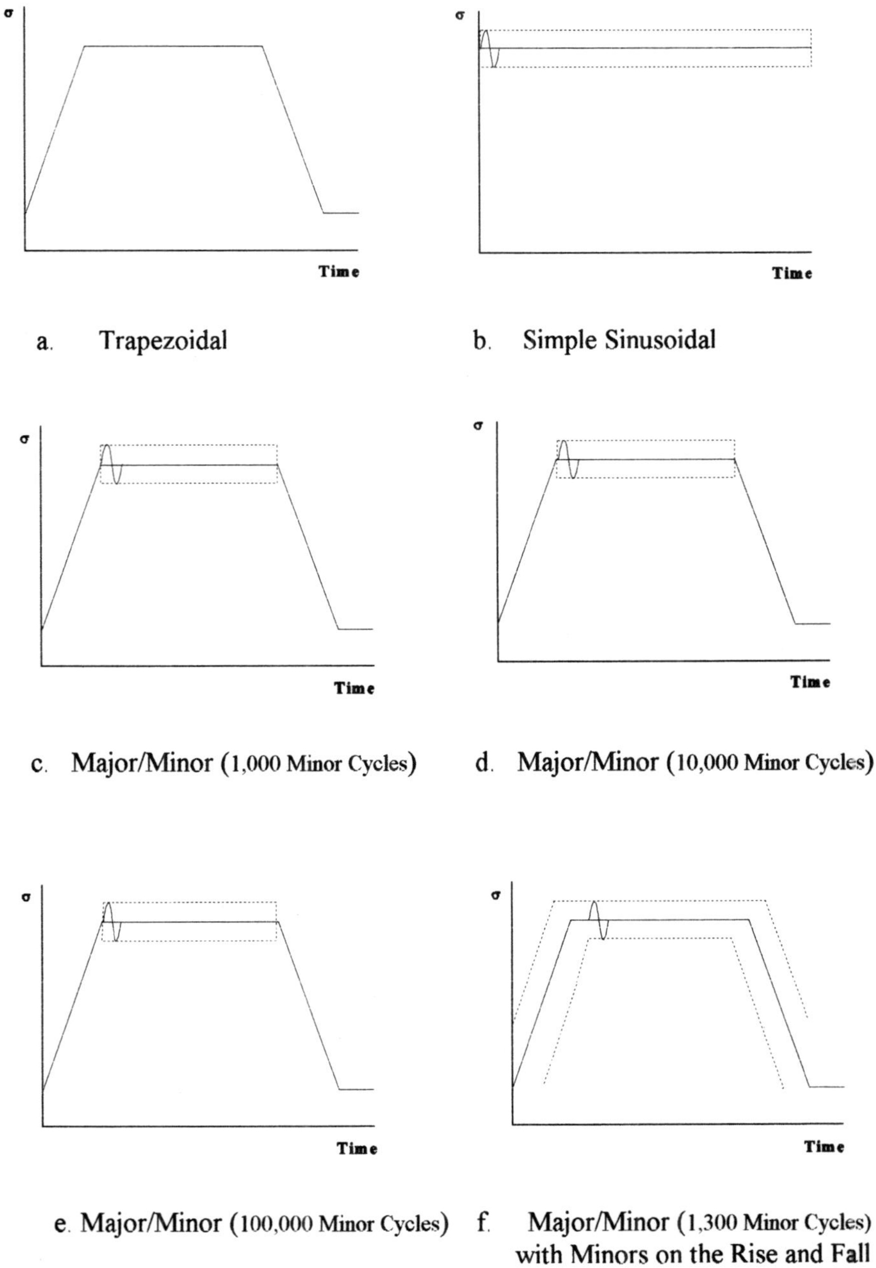

a. Trapezoidal
b. Simple Sinusoidal
c. Major/Minor (1,000 Minor Cycles)
d. Major/Minor (10,000 Minor Cycles)
e. Major/Minor (100,000 Minor Cycles)
f. Major/Minor (1,300 Minor Cycles) with Minors on the Rise and Fall

Figure 1. Schematic Representation of the Stress - Time Histories Analysed

GRAPH 1. COMPUTER PREDICTIONS - RESULT
(C=7.14E-12,n=3.5,Knee=.52,DELKth=2.1)

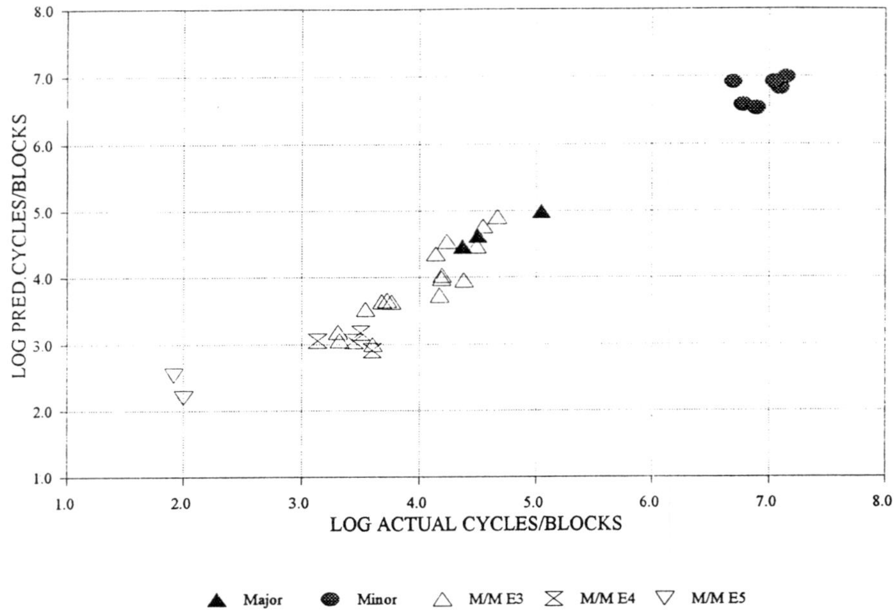

GRAPH 2. COMPUTER PREDICTIONS - RESULT
(C=7.14E-12,n=3.5,Knee=.52,DELKth=2.1)

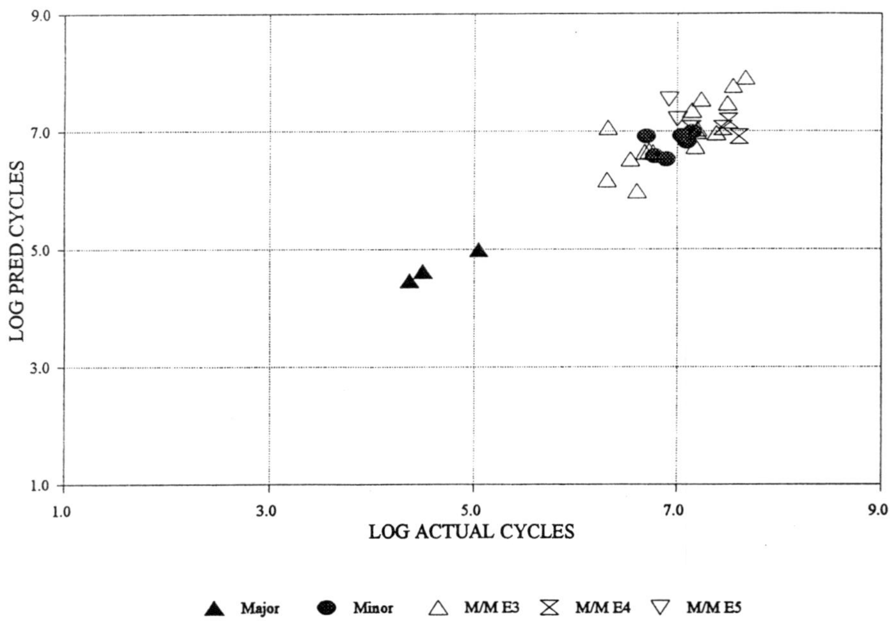

Stability of tunnel heading using coupled modeling

P. Procházka,[a] J. Skořepová[b]
[a]*Department of Structural Mechanics, Czech Technical University, CZ-166 29 Prague, Czech Republic*
[b]*Institute of Structural and Rock Mechanics, Czech Academy of Science, CZ-182 00 Prague, Czech Republic*

Abstract

In the past, the coupled modeling (mutual influence of mathematical and physical models from equivalent materials) proved several advantages when dealing with geotechnical structures. In the paper a new procedure will be proposed to evaluate internal parameters of the mathematical physically nonlinear model using partial results of the physical model. The procedure may be considered a special case of inverse analysis. An example shows an application of the procedure in detail.

1 Introduction

A modeling in geotechnics investigates the mechanism of geotechnical effects, a prediction of the stress and deformation changes during processes of underground constructions on some model samples (mathematical and physical models, in site measurement etc.), and the application of preventive measures, [4].

A number of numerical methods are now available. They are very often used for solving of geotechnical problems under some simplifying assumptions. This is done because of a less accurate description of physical behavior of a trial model, incorrect input data, reduction of dimension from 3D to 2D due to a high consumption of computer time, etc. Consequently, experimental physical methods are now being revived worldwide. The physical experiments can support the conceptional assumptions and results of numerical methods and/or, on the contrary, the results of the mathematical treatment can affect the decision how to prepare the physical model. A combination of both methods and mutual application of their results enables one to restrict the disadvantages of both methods and avoid results that principally do not agree with the reality. The coupling also significantly widen our current knowledge of rock mechanics.

The advancement of the coupling of mathematical and physical modeling is very useful for an explanation of behavior of rock mass under consideration. This is very desirable for the basic research in geotechnics and geomechanics and also for practical use.

In the paper, a special case of method of inverse analysis is proposed. 2D plasticity is solved mathematically and both the plastic law and the third dimension is involved in eigenstrain tensor $\boldsymbol{\mu}$. The distribution of these internal parameters are assumed subdomain-wise constant and are to be stated from comparison of stresses obtained from both the physical and the mathematical models at a finite number of trial points (point of comparison) from the domain describing the body of the simulated rock.

2 Physical modeling

Principles of the physical modeling are based on the relationship between the geometrical and the physical similarity derived from dimensional analysis. To determine the changes of the stress state in the rock mass, the following relevant factors are taken into consideration:

- volume weight γ
- elasticity modulus E
- Poisson ratio ν
- cohesion C_p
- angle of internal friction ϕ_p
- time t

Dimensional equation which is an implicit function of the relevant factors describes the behavior of the rock mass in a simplified form given by a choice of these factors. According to Buckingham's theorem, [5], the dimensional equation expressing a relation between the reality and the model can be reduced to the problem of finding the critical dimensionless parameters. The corresponding parameters are numerically identical for the model and the reality. This condition and the condition of the geometrical similarity between the model and the reality have to be fulfilled, [2].

Rock medium in the model is substituted by equivalent materials which are prepared in such a way that the relevant factors computed from the above described procedure are respected.

The models are constructed in stands of various dimensions offering different possibilities of loading of the model. The geometrical scale of the model is determined by a size of the stand, extent of the modeled problem, introduction of external forces and by other factors.

For measuring of the changes of the stress states in the model body, semiconductors sensors are used, [6]. As the basis for analyzing the data of measuring system, the functions of calibration curves determined prior to the model experiment serve. Due to the forces acting on the model during the experiment, the values of stresses in a discrete set (points of comparison) are determined.

3 Mathematical modeling

The triangular finite elements with 2D linear distribution of displacements and with eigenstrain influences has been considered in the comparative computation. From the stress state of the physical model involving an opening (tunnel, collector etc.), the vertical normal stresses at selected points were extracted and by virtue of the mathematical treatment the internal parameters were improved in such a way that the variance of differences of stresses from both methods at the selected points reaches its minimum on the set of admissible internal parameters. Note that this procedure avoids the necessity of introduction of hypothesis expressing the kind of plastic behavior of the material.

Let us consider a bounded domain Ω in R_3 with a boundary Γ. The transformation field analysis, [3], consists in expressing the stress $\boldsymbol{\sigma}$ at an arbitrary point $\boldsymbol{\xi}$ of the domain by virtue of superposition of stress tensor $\boldsymbol{\sigma}^0(\boldsymbol{\xi})$ due to the external loading, which is applied to a homogeneous, isotropic body, and a linear hull of the eigenstrains $\boldsymbol{\mu}$ at another point $\mathbf{x}$. Here, a special relation between the above mentioned tensors, being usable for our next considerations, is used.

Assume that there are subdomains $\Omega_1, ..., \Omega_n$ from Ω and in each such subdomain a constant distribution of eigenstrain field is considered. There also is a set of points of comparison $N_k, k = 1, .., m \in \Omega$ & Γ at which the stress is measured. Then it can be proved, [3], that the relation between stress field $\boldsymbol{\sigma}^k$ at points $N_k, k = 1, ..., m$, and eigenstrain field $\boldsymbol{\mu}^l, l = 1, ..., n$ reads

$$(\sigma_i)^k = (\sigma_i^0)^k + \sum_{j=1}^{6} \sum_{l=1}^{n} (G_{ij})^{kl} (\mu_j)^l, i = 1, ..., 6, k = 1, ..., m \tag{3.1}$$

where σ_i is i-th component of stress tensor, μ_j is j-th component of eigenstrain tensor and $\mathbf{G}$ is the influence matrix of constants stated apriori, relating the stresses at ξ_k and the eigenstrains in subdomain Ω_l. Stresses $(\sigma_i^0)^k$ are also known quantities from a prior computation.

A natural requirement is to assure that both measured stresses $(\sigma_i^{meas})^k$ and computed stresses $(\sigma_i)^k$ be as close as possible. This means that the variance

$$I[(\mu_j)^l] = \left[\sum_{i=1}^{6} \sum_{k=1}^{m} (\sigma_i^{meas})^k - (\sigma_i)^k \right]^2 \rightarrow \text{minimum}. \tag{3.2}$$

Differentiation of I with respect to $(\mu_\alpha)^\beta$ leads to the following system of linear algebraic equations for unknown $(\mu_j)^l$

$$A_{\alpha j}^{\beta l} (\mu_j)^l = b_\alpha^\beta, \tag{3.3}$$

$$A_{\alpha j}^{\beta l} = \sum_{i=1}^{6} \sum_{k=1}^{m} (G_{i\alpha})^{k\beta} (G_{ij})^{kl}, \quad b_\alpha^\beta = \sum_{i=1}^{6} \sum_{k=1}^{m} (G_{i\alpha})^{k\beta} \left[(\sigma_i^0)^k - (\sigma_i^{meas})^k \right]$$

It can be proved that for $n \leq m-1$, the system (3.3) is uniquely solvable. For $n \geq m$, the optimization solution attains more minimum points.

4 Comparison computations

The tunnel was excavated horizontally at the side of the model at the height of 500 mm above the stand bottom. For this purpose, the special equipment for excavation was developed (Fig. 2).

At the beginning of the model experiment, the stress in the axis of tunnel due to volume weight of the model material and to the external load (metal weight) reached 18.76 kPa.

To determine stress states with respect to the position of an advancing tunnel heading and, mainly, to obtain the stress distribution in the plane perpendicular to the axis of the driven tunnel (being identical with the plane in which the computation is carried out), the physical model was constructed. The model was designed as simple as possible to enable one both testing and comparison with the mathematically modeled form of the solution. The tunnel section was taken as circular (diameter of 100 mm), the rock medium was considered to be both isotropic and homogeneous.

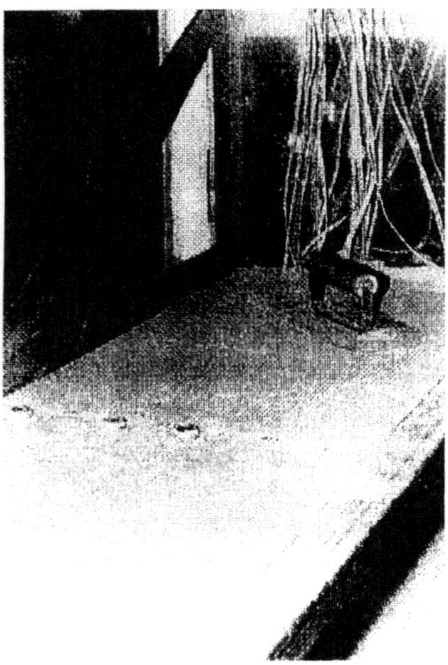

Figure 1: Construction of model deposition of sensors

Figure 2: Equipment for tunnel driving

For construction of the model, the modeling stand with bottom dimension of 480 x 1460 mm^2 and with the wall height of 1200 mm was used. The rock medium in the model was substituted by equivalent material consisting from bentonite (34%), sand (63%) and fat (3%). The material properties were examined by laboratory methods usual in soil mechanics, [1], with the following results:

$\gamma = 1.53$ g/cm^3
$E = 4.5$ MPa

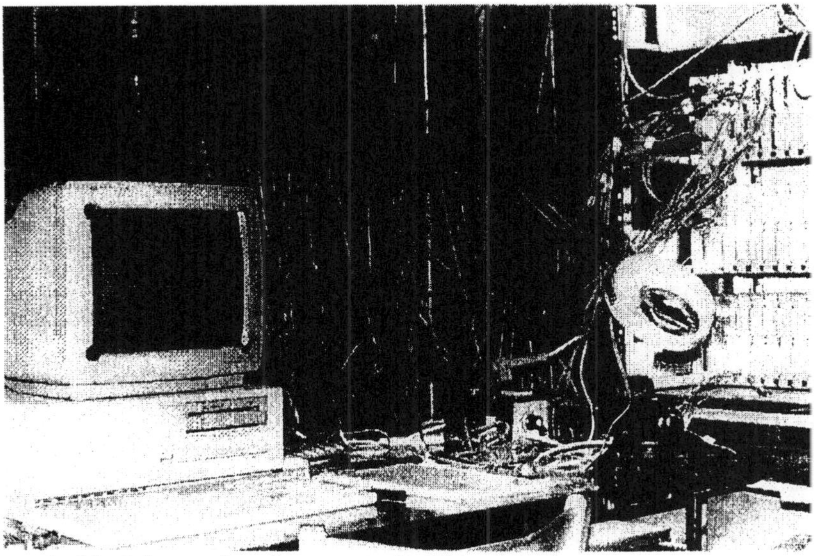

Figure 3: Model stand and support instrumentation

$\nu = 0.284$
$C_p = 0.0039$ MPa
$\phi_p = 43^0 30'$

The model material was deposited in the modeling stand and was shaped and compacted.

Monitoring of the changes of stress state in the model body was performed on the basis of tensometric measurements. Semiconduct sensors TM 440/2 of the nominal range of 200 kPa (Fig. 1) were used for these measurements. During the construction of the model they were put in the vertical plane passing thru the center of the modeling stand, perpendicularly to the longitudinal axis of the tunnel. The data-acquisition and automated measurement equipment Unilog 2500 was used for registration of the stress state changes.

The model stand and the support instrumentation prepared just for testing is shown in Fig. 3.

Simulation of driving of the tunnel was disposed into six phases of model experiment in dependance of stress changes: three positions of the advancing tunnel heading ahead of sensor plane and, additionally, three states of tunnel heading positioned behind the sensor plane (tunnel intersects this plane) were measured and studied. The position of sensors (points of comparison) is illustrated in Fig. 4a.

As from the physical model the values of stresses in a discrete set are given, the mathematical model has to adjust the chosen eigenstresses in such a way that the values from both the models in the points of comparison should be as close as possible. The procedure described in Chapter 3 were employed. Points N_i coincide with the points where vertical stresses were measured (denumeration see Fig. 4a). The positioning and shape of the subdomains is depicted in Fig. 4b for the tunnel heading ahead the sensor plane while in Fig. 4c the subdomains are depicted for the tunnel heading

N	x	y	1	2	3	4	5	6
1	.000	−400.000	24.761	24.751	24.328	23.921	23.681	23.049
2	75.000	−300.000	23.465	23.478	23.693	23.253	22.603	22.592
3	150.000	−300.000	23.260	23.267	23.267	23.052	22.835	22.829
5	100.000	−200.000	21.972	22.181	22.198	21.543	20.479	20.469
6	200.000	−200.000	21.757	21.759	21.961	21.959	21.759	21.759
8	75.000	−100.000	21.117	21.342	21.996	19.172	13.505	13.444
9	150.000	−100.000	20.266	20.468	20.684	20.899	21.105	21.317
10	250.000	−100.000	20.062	20.062	20.258	20.265	20.271	20.466
11	.000	−50.000	23.676	24.553	26.095	10.505	10.505	10.505
12	35.360	−35.360	20.575	20.801	21.659	10.662	10.068	10.068
13	50.000	.000	20.041	20.473	21.542	24.333	19.182	19.188
14	125.000	.000	18.966	18.976	19.392	20.231	20.874	21.301
15	200.000	.000	18.756	18.757	18.962	18.977	19.183	19.388
16	300.000	.000	18.749	18.560	18.563	18.749	18.749	18.751
17	35.360	35.360	20.211	20.653	21.986	9.377	9.377	9.377
18	.000	50.000	17.783	18.008	18.684	9.282	9.282	9.282
19	75.000	100.000	17.364	17.466	17.687	15.372	10.954	10.747
20	150.000	100.000	17.460	17.466	17.679	17.896	18.107	18.316
21	250.000	100.000	17.247	17.243	17.247	17.247	17.247	17.247
22	.000	150.000	15.857	16.074	16.479	11.346	7.513	7.513
23	100.000	200.000	15.545	15.749	15.758	14.900	13.413	13.183
24	200.000	200.000	15.756	15.758	15.765	15.768	15.769	15.959
25	.000	250.000	14.812	14.817	15.000	13.818	12.032	11.217
26	75.000	300.000	14.250	14.252	14.254	14.051	13.843	13.837
27	150.000	300.000	14.051	14.052	14.051	13.416	12.566	12.129
28	.000	400.000	12.972	12.749	12.525	12.300	12.067	11.825
29	100.000	450.000	12.003	12.002	12.001	11.994	11.793	11.786
30	.000	550.000	10.499	10.288	10.286	10.282	10.278	10.069

Table: Coordinates of sensors and actual measured vertical stresses for six stages of excavation.

after crossing the sensor plane.

In the Table, the values of vertical stresses are given for the same distances of tunnel heading from the sensor plane, as introduced in Fig. 5. In the first column of the Table, the numbers of sensors (see. Fig. 4a) are given, the second and third column contain the coordinates of these sensors (the origin is positioned in the center of the tunnel), the rest columns contain the values of vertical stresses according to Fig. 5 (column #1 contains values from the state were tunnel heading is 20 cm ahead of sensor plane, etc.).

The stress distribution and the occurance of loading and unloading zones in the plane perpendicular to the axis of the driven tunnel were determined after the tunnel heading passed thru the plane where the sensors were positioned, Fig. 5. In Fig. 6 the computed eigentrains are shown: a)

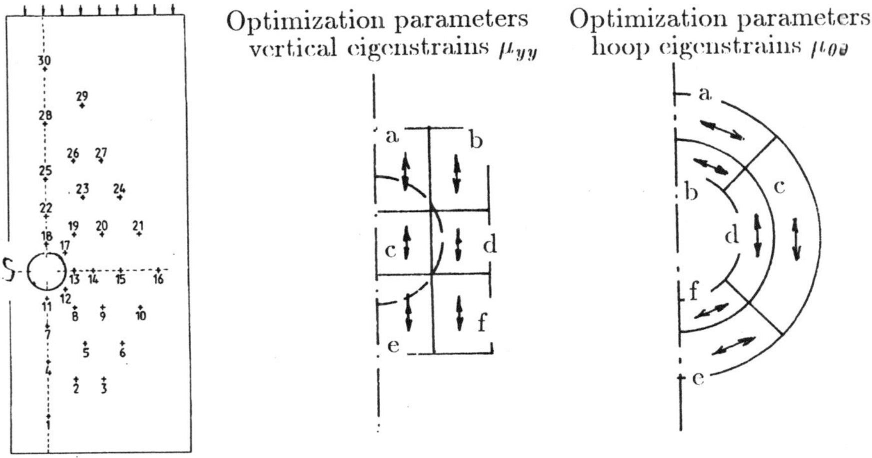

Figure 4: Vertical section of the model. Position of the semiconduct sensors

ahead of sensor plane and b) after tunnel heading crossed the sensor plane.

Conclusion

In the paper, a new procedure for testing the behavior of a rock mass in front of tunnel heading has been proposed. It consists in adjusting of eigenstrains in selected subdomains in such a way that both measured stresses and computed stresses are in "the best" agreement. As the eigenstrains cover both plastic behavior and the third dimenstion, they, among others, may be taken as a measure for assesment of stability of tunnel heading.

References:

[1] Head, K. H. *Manual of soil laboratory testing*. John Wiley & sons. Inc. New York, Toronto, 1992.

[2] Kožešník, J. *Theory of similarity and modelling*. Academia, Prague, 1983.

[3] Procházka, P., Šejnoha, J. Behavior of composites on bounded domain. *OPTI'95*, Key Biscane, Florida, USA, 1995.

[4] Skořepová J. Use of physical models in geomechnics. In: Proc. *32th U.S. Symp. Rock Mech. as a multidisciplinary Sc.*, Balkema-Rotterdam, 1991, 521-530.

[5] Stilborg, B., Stephensson, O., Swan, G. Three-dimensional physical model technology applicable to the scaling of underground structures. Proc. *40th Int. Conf. on Rock Mech.* **2**, Montreux, 1979, 655-662.

[6] Vencovský M., Málek J. Assesment of the stress state of a physical rock mass by means of tensometric sensors. *Acta Montana* **Ser.A**, No.5(92), *Geodynamics*, 1994, 19-32.

ACKNOWLEDGEMENT: This paper was supported by GAČR, grant projects # 103/95/1265 and # 205/94/1788

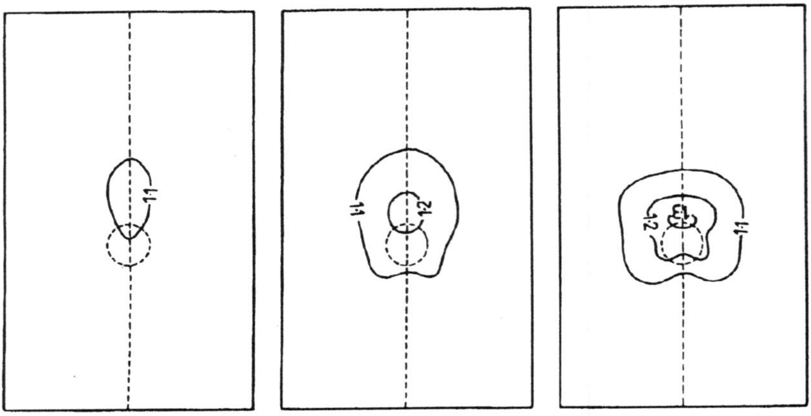

Tunnel heading ahead of sensor plane (by 20 cm, 10 cm, 5 cm)

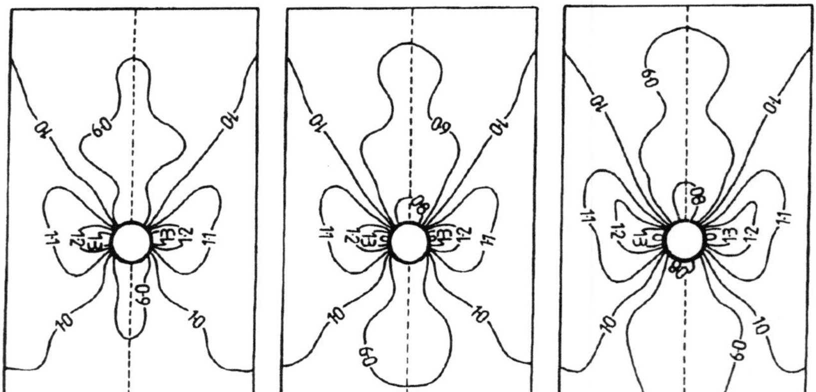

After sensor plane crossed by tunnel heading (close, 5 cm, 15 cm)

Figure 5: Relative stress changes (Rates of current stresses and original stresses)

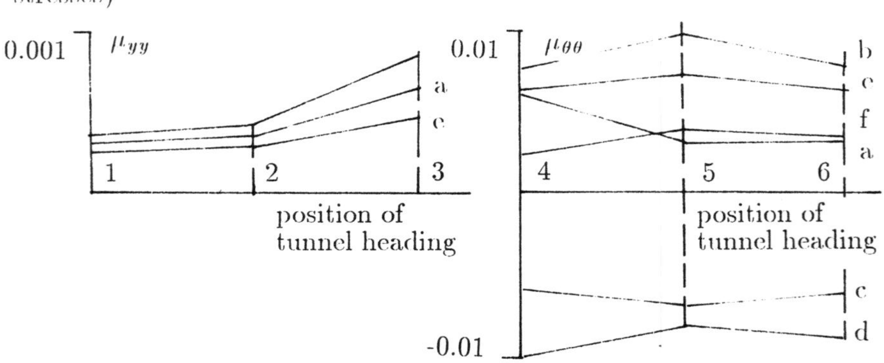

Figure 6: Computed eigenstrains for tunnel heading ahead of and after crossing the sensor plane

Buckling of circumferentially corrugated cylinders under uniform external pressure

C.T.F. Ross,[a] G. Apor,[b] S.P. Claridge[a]
[a]*Department of Mechanical and Manufacturing Engineering, University of Portsmouth, Portsmouth, Hampshire PO1 3DJ, UK*
[b]*Xavier University, Cagayan de Oro City, Philippines*

Abstract

The paper describes a theoretical and an experimental investigation into the buckling of three circumferentially corrugated circular cylinders under uniform external pressure. The experimental results showed that the vessels buckled through general instability.

The theoretical analysis was of a semi-empirical nature, which involved the finite element method and an elastic knockdown factor to cater for inelastic instability.

1 Introduction

The pressure hulls of submarines and the legs of off-shore drilling rigs are often constructed from thin-walled circular cylinders. Now under external water pressure, a thin-walled circular cylinder can buckle at a small fraction of the pressure to cause axisymmetric yield. This mode of failure is known as shell instability, and it is very structurally inefficient. One method of improving the structural efficiency of these vessels is to stiffen them with circumferential ring-stiffeners, as shown in Figure 1.

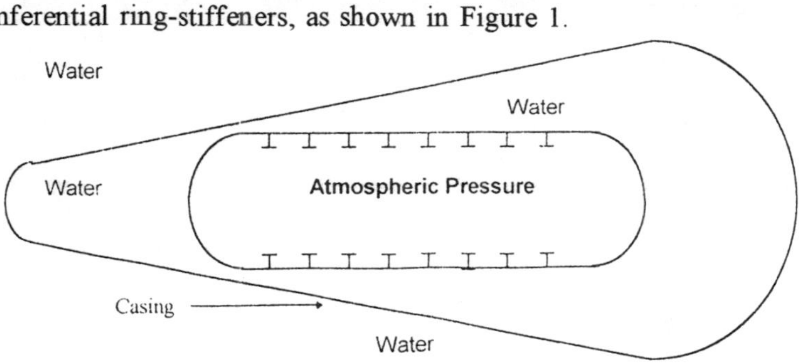

Figure 1 - Ring-stiffened circular cylinder

In Figure 1, it can be seen that the ring-stiffened pressure hull is surrounded by a hydrodynamic hull, where the latter has water pressure on both its internal and external surfaces. Thus, as the hydro-dynamic hull is in a state of hydrostatic stress, it is unlikely to fail due to the effects of hydrostatic pressure alone.

If the ring stiffeners are not strong enough, the entire ring-shell combination can buckle bodily, through general instability. In 1987, Ross [1] presented an alternative design to the ring-stiffened vessel, as shown in Figure 2.

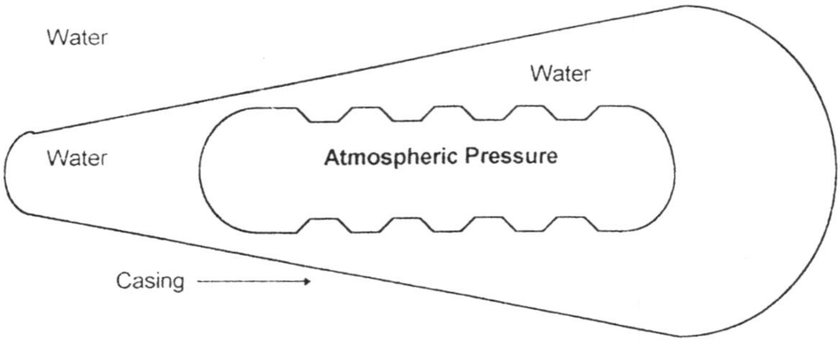

Figure 2 - Circumferentially Corrugated Pressure Hull

From Figure 2, it can be seen that the pressure hull was a circumferentially corrugated vessel. Using two mathematical models, Ross showed that the corrugated pressure hull was structurally more efficient than a ring-stiffened vessel of the same weight and volume. In one case, the corrugated vessel had nearly twice the strength of its ring-stiffened equivalent.

In fact, Kuan-Ya et al [2] showed that the corrugated cylinders could be made even more structurally efficient by increasing the cone angles to certain optimum values. The work reported so far is of a theoretical nature, and experimental work to verify the theoretical mode of failure was carried out by Ross et al [3 to 5], see Figure 3.

Figure 3 - General instability of corrugated circular cylinders

These experimental investigations showed that like unstiffened and stiffened circular cylinders, the vessels were prone to suffer elastic knockdown due to initial out-of-circularity. To cater for this effect, Ross and Palmer [3] presented a thinness ratio λ', which was based on the thinness ratio of Windenburg and Trilling [6]. The thinness ratio λ' is defined as follows:-

$$\lambda' = [(L_b/D')^2/(t'/D')^3]^{0.25} \times \sqrt{(\sigma_{yp}/E)} \qquad (1)$$

where $D' = R_i + R_o$
t' = equivalent wall thickness
σ_{yp} = yield stress
E = Young's modulus
R_i, R_o and L_b are defined in Figures 4 and 5

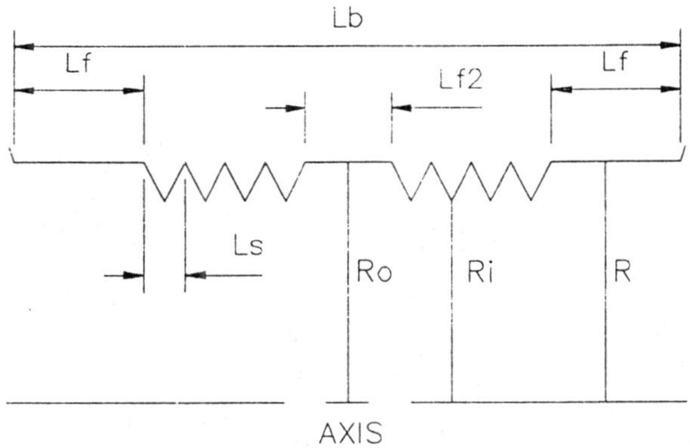

Figure 4 - Geometry of DML1

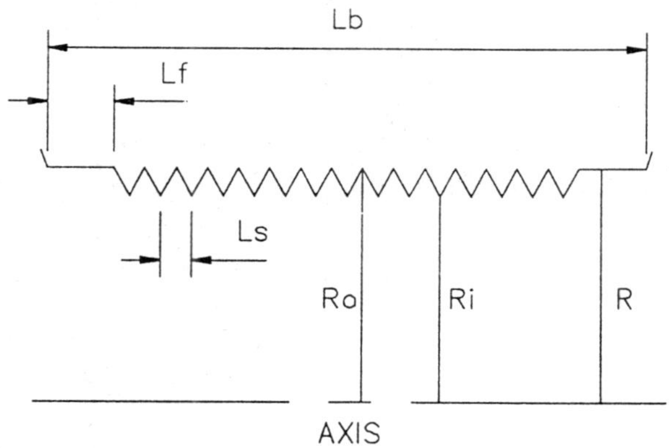

Figure 5 - Geometry of DMM3 & DMS3

In this paper a theoretical and an experimental investigation will be made on three corrugated cylinders of different geometries.

2 Experimental Apparatus

Three steel models were tested until failure. These models were named DML1, DMM3 and DMS3. The models are shown in Figures 6 to 8, and their geometrical details are given in Table 1, where

N = number of corrugations
t = wall thickness

Table 1 - Geometrical details (mm)

	N	L_b	L_f	L_{f2}	L_s	R	R_o	R_i	t
DML1	8	173.0	25.10	17.84	12.48	76.53	76.33	75.23	0.310
DMM3	15	111.6	23.30	-	4.333	49.81	49.97	49.48	0.222
DMS3	15	101.3	18.23	-	4.373	41.99	42.42	41.85	0.208

Figure 6 - Model DML1

Figure 7 - Model DMM3

Figure 8 - Model DMS3

All models were manufactured by Del Monte Phillipines Incorporated; their material properties were as follows:-

Modulus (E)	= 190 GPa	σ_{yp} (DML1) = 344MPa
Density (ρ)	= 7800 kg/m^3	σ_{yp} (DMM3) = 453 MPa
Poisson's ratio (ν)	= 0.3 (assumed)	σ_{yp} (DMS3) = 475 MPa

The longitudinal profiles of the vessels were measured by a Mitutoyo co-ordinate measuring machine. The profiles were of a sinusoidal nature, but for the theoretical analysis, it was convenient to represent each corrugation with two truncated finite element conical shell elements; previous work [7] has found this to be satisfactory. The ends of the vessels were blocked off by end disc caps, sealed by a household silicone sealant, as shown in Figures 6 to 8.

The out-of-circularity of the vessels is given in Table 2. Out-of-circularity is defined as maximum difference between the maximum inward and outward deviations of the external surface from a mean circle (using a least squares fit)

Table 2 - Out-of-circularity measurements (mm)

DML1	DMM3		DMS3	
Middle	Top	Bottom	Top	Bottom
1.7752	1.6511	1.6458	0.6550	1.2767

3 Experimental Procedure

The models were tested in a test tank shown in Figure 9. Water was used as the pressure raising fluid, with the aid of a manually-operated hydraulic pump.

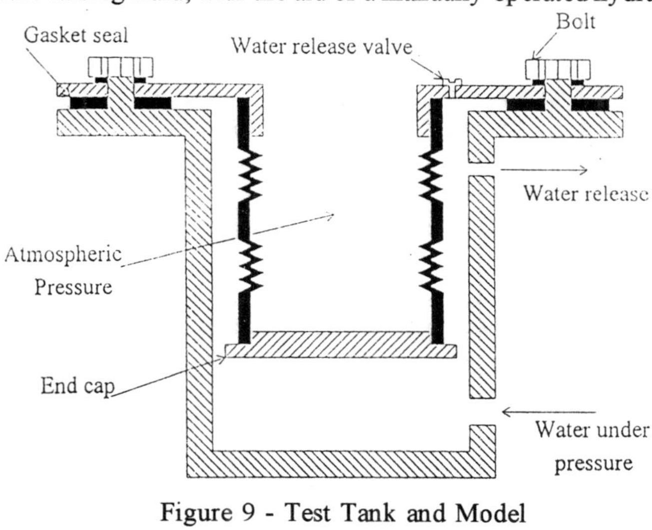

Figure 9 - Test Tank and Model

Line losses were negligible, as the hose was only 2m in length, and the pump was manually operated. Prior to raising the pressure, a bleed screw, was left open, so that any trapped air could be pumped out. When the trapped air was removed, the bleed screw was sealed and the pressure tests commenced. The experimental boundary conditions for each vessel was probably in-between clamped and simply-supported.

3.1 Testing the Models

Each one of the three models had attached to it, several linear strain gauges placed in the circumferential direction. The purpose of these strain gauges was to observe the circumferential buckling patterns, throughout each test. The strain gauges were spaced at equidistances apart. Ten strain gauges were attached to DML1, the largest of the three models, eight strain gauges to DMM3 and six strain gauges to DMS3, the smallest of the three models.

Each model was tested to destruction. Model DML1 collapsed at a pressure of 2.55 bar with the formation of four lobes. Model DMM3 collapsed to a pressure of 1.931 bar and model DMS3 collapsed at a pressure of 3.172 bar, the number of lobes corresponding to each of these two buckling pressure was not as easily determined as for model DML1.

Plots of strain recording for various values of pressure are shown in Figures 10 to 12, and pictures of the collapsed models are shown in Figures 13 to 15.

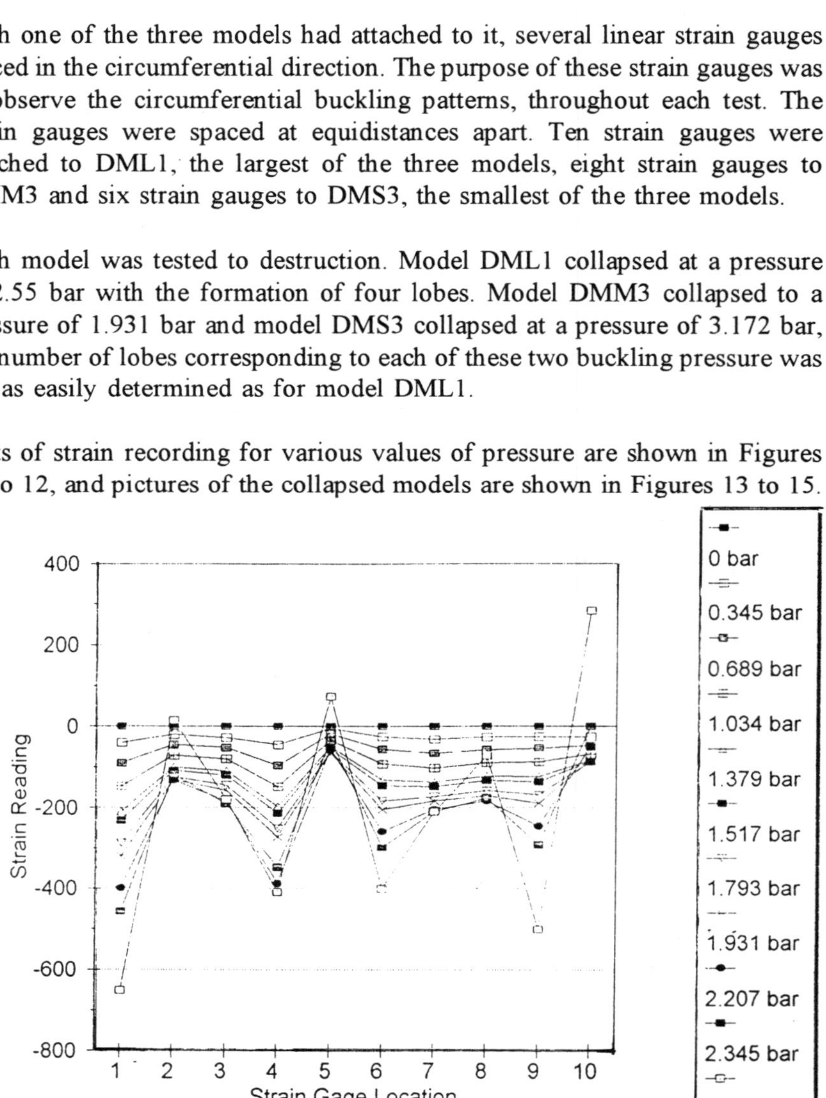

Figure 10 - Strain recordings for DML1 Buckling Test (micro-strain)

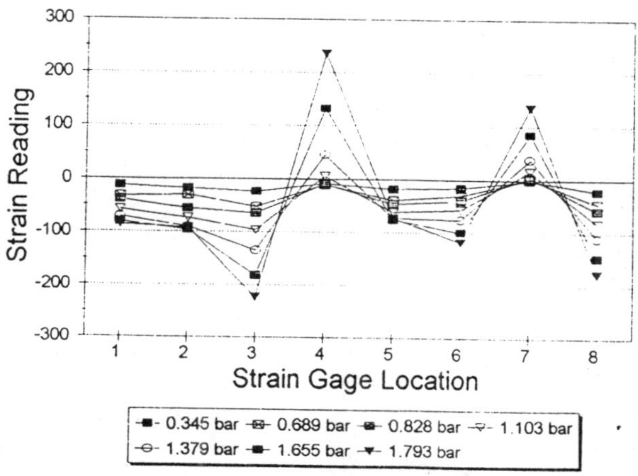

Figures 11 - Strain recordings for DMM3 Buckling Test (micro-strain)

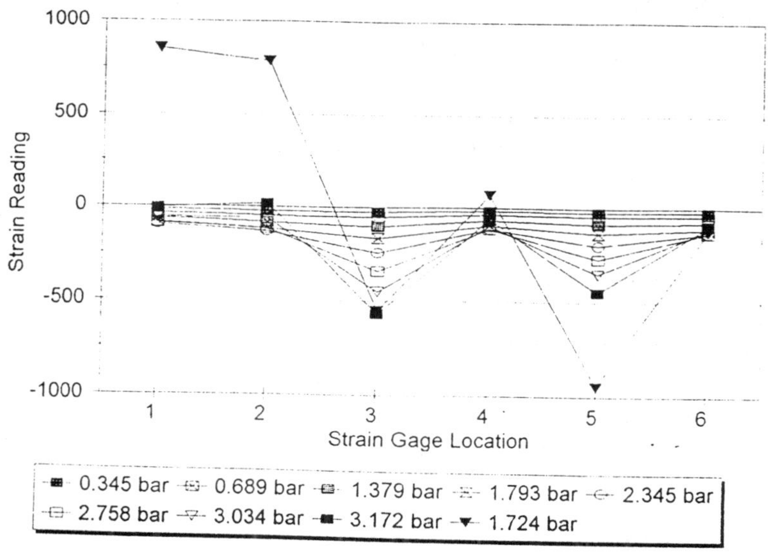

Figure 12 - Strain recordings for DMS3 Buckling Test (micro-strain)

Figure 13 - Collapsed model DMM3

Figure 14 - Collapsed model DML1

Figure 15 - Collapsed model DMS3

As the strain recordings for DML1 were well below the yield strain, it is likely that these vessels buckled elastically. In the case of DMM3 and DMS3, the strain recordings were well past the yield strain, hence, these vessels buckled inelastically, and were more likely to suffer elastic knockdown.

4 Theoretical Analysis

The theoretical buckling pressures were calculated by a computer program called CONEBUCK [8], which is based on the finite element method. The thin-walled truncated conical element was used and is reported elsewhere (Ross [9]). The theoretical buckling pressures (P_{cr}) were found to be 3.267 bar, 3.113 bar and 4.0 bar for DML1, DMM3 and DMS3 respectively.

4.1 Design procedure

The design procedure, therefore, is to calculate the theoretical buckling pressure P_{cr} and the thinness ratio λ', and to divide P_{cr} by the elastic knockdown factor, namely (P_{cr}/P_{exp}), to obtain the actual collapse pressure. The elastic knockdown factor will be obtained from a design curve, when sufficient experimental results are available, where λ' will be plotted against P_{cr}/P_{exp} for each model, where, P_{exp} = experimental buckling pressure

To obtain the design pressure, it will be necessary to divide the collapse pressure by a large safety factor.

4.2 Calculation of thinness ratio λ'

The thinness ratio λ' is based on representing the corrugated cylinder by an equivalent unstiffened circular cylinder, with equivalent thickness, namely t', such that the flexural stiffness of the equivalent cylinder is the same as that of the corrugated vessel; that is:

$$(L_b \times t'^3)/12 = (I_{NA} \times N) + \sum [(L_f \times t^3)/12 \times 2] \qquad (2)$$

where,

$I_{NA} = ((t_w \times D_s^3)/12) \times 2$

N = number of corrugations, t = wall thickness

t_w = width of a leg of the corrugation (see Figure 24)

D_s = depth of a leg of the corrugation

$\alpha = \tan^{-1} [(2 \times D_s)/ L_s]$

L_s = length of an individual corrugation

L_b = length of model

L_f = length of flat or non-corrugated section

t' = thickness of equivalent unstiffened circular cylinder

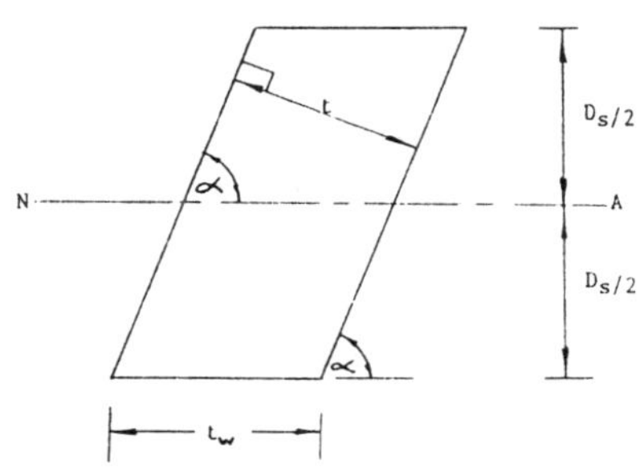

Figure 16 - leg of the corrugation (assumed to be a parallelogram)

Table 3 gives the values for λ' and P_{cr}/P_{exp}, together with e/t', where, e = out-of-circularity measurements (mm)

Table 3 - Theoretical and experimental buckling pressures (bar) and λ'

Model	λ'	e/t	e/t'	P_{exp}	P_{cr}	P_{cr}/P_{exp}
DML1	2.82	5.726	2.877	2.55	3.267	1.281
DMM3	3.71	7.425	4.988	1.931	3.113	1.612
DMS3	3.28	4.644	2.683	3.172	4.0	1.333

From Table 3, it can be seen that the model DMM3 had the largest value of elastic knockdown, and this was due to the fact that it also had the largest value of e/t'. Similarly, as both λ' and e/t' were similar for DML1 and DMS3, their elastic knockdown factors were similar. Table 4 compares these results with results obtained from other recent tests.

Table 4 - Comparisons of λ', e/t' and P_{cr}/P_{exp}

Model	e/t'	λ'	P_{cr}/P_{exp}
DML1	2.877	2.82	1.28
DMM3	4.988	3.74	1.61
DMSB	2.683	3.28	1.33
DF	0.185	3.57	1.15
MBS	0.185	2.67	1.61
MBL	0.365	3.53	1.69
CA	0.46	2.73	1.45

5 Conclusions

The results have shown that it is possible to replace a ring stiffened pressure hull of a submarine with a similar corrugated vessel. This should be particularly attractive if the hull is to be made from carbon or glass fibre composite material, as it should be easier to construct such a vessel in corrugated form, rather than as a ring stiffened vessel. Further research is required to determine the loss of buckling resistance due to initial out-of-cicularity of the cylinders.

References

1. Ross,CTF, A novel submarine press. hull design, J.Ship Res.,31,186-188, 1987.
2. Kuan-Ya Yuan, Cho-Chung Liang and Yung-Chung Me, Investigation of the cone angle of a novel swedged-stiffened pressure hull, J Ship Research, 35, 83-86, 1991.
3. Ross,CTF and Palmer,A, General instability of swedge-stiffened circular cylinders under uniform external pressure, J Ship Research, 37, 77-85, 1993.
4. Ross,CTF and Humphries,M, The buckling of corrugated circular cylinders under uniform external pressures, J. Thin-Walled Structures, 17, 259-271, 1993.

5. Ross,CTF and Heigl,T, Buckling of corrugated axisymmetric shells under uniform external pressure, Proc ASME Conf on Structural Dynamics and vibrations, PD-Vol 70, 199-205, 1995.
6. Widenburg,DF and Trilling, C, Collapse by instability of thin cylindrical shells under ext. press., trans ASME, 11, 819-825, 1934.
7. Ross,CTF and Johns, T, Vibration of hemi-ellipsoidal axisymmetric domes submerged in water, Proc IMechE, 200, 389-298, 1986.
8. Ross,CTF, Finite element programs for axisymmetric problems in engineering, Ellis Horwood, Chichester, 1984.
9. Ross,CTF, Pressure vessels under ext. pressure, Chapman and Hall, London, 1990.

Analytical and experimental verification of a flight article for a Mach-8 boundary-layer experiment

W. Lance Richards, R.C. Monaghan
National Aeronautics and Space Administration, Dryden Flight Research Center, P.O. Box 273, Edwards, CA 93523, USA

Abstract

Preparations for a boundary-layer transition experiment to be conducted on a future flight mission of the air-launched Pegasus® rocket are underway. The experiment requires a flight-test article called a glove to be attached to the wing of the Mach-8 first-stage booster. A three-dimensional, nonlinear finite-element analysis has been performed and significant small-scale laboratory testing has been accomplished to ensure the glove design integrity and quality of the experiment. Reliance on both the analysis and experiment activities has been instrumental in the success of the flight-article design. Results obtained from the structural analysis and laboratory testing show that all glove components are well within the allowable thermal stress and deformation requirements to satisfy the experiment objectives.

Introduction

The prediction of boundary-layer transition for three-dimensional flow is essential to the design and development of hypersonic vehicles. Accurate transition assumptions must be made in the vehicle design because of extreme differences in the aerodynamic heating between fully-turbulent and fully-laminar flow at hypersonic speeds. If the conservative (but unrealistic) assumption of fully-turbulent flow is assumed throughout the mission, the resulting design will be extremely inefficient. Conversely, fully-laminar flow cannot be assumed for the safe design of the vehicle. Therefore, accurate assumptions concerning boundary-layer transition must be made for the vehicle design to be both efficient and robust.

Preparations for a future flight experiment are currently underway to improve crossflow-transition prediction methods for actual hypersonic flight conditions because wind-tunnel testing has not been historically feasible, e.g. Bertilrud, et al.[1] Crucial to the success of the experiment is the ability to design, analyze, and fabricate a fully-instrumented flight-test fixture that can provide a smooth, three-dimensional, structurally stable aerodynamic surface; and to determine when and where boundary-layer transition occurs during the flight.

To achieve these objectives, a flight-test article called a glove has been designed, analyzed, manufactured, and installed on the wing of the air-launched Pegasus® (Orbital Sciences Corporation, Fairfax, Virginia) space booster. A comprehensive structural test and analysis program has been successfully completed to validate the integrity of the complex glove design. Complexity of the design stemmed from the use of nontraditional engineering materials, geometrically complex components, transiently varying boundary conditions, and nontrivial manufacturing processes. This paper presents the glove design and describes the test and analysis details that were essential in the glove verification process.

Design Requirements

The glove was designed to provide valuable boundary-layer transition data during a typical Mach-8 flight of the Pegasus® space booster. This section describes the key requirements for the glove design.

Vehicle and Mission Requirements

The Pegasus® is a multistaged, air-launched rocket designed to economically place small payloads into low Earth orbit. Figure 1 shows the physical dimensions of the rocket and the location of the glove. Figure 2 shows a photograph of the Pegasus® space booster mounted under the wing of a B-52 carrier aircraft.

The booster follows a fixed trajectory that is predetermined for a particular payload requirement. For a typical mission, the booster separates from the carrier aircraft at Mach 0.8 and an altitude of approximately 13,000 m and descends for 5 sec. Then, the first stage ignites. After approximately 70 sec, the vehicle has accelerated to Mach 8 and an altitude of approximately 61,000 m. After first-stage burnout occurs, the glove experiment concludes as the entire first stage is jettisoned and second-stage ignition occurs.

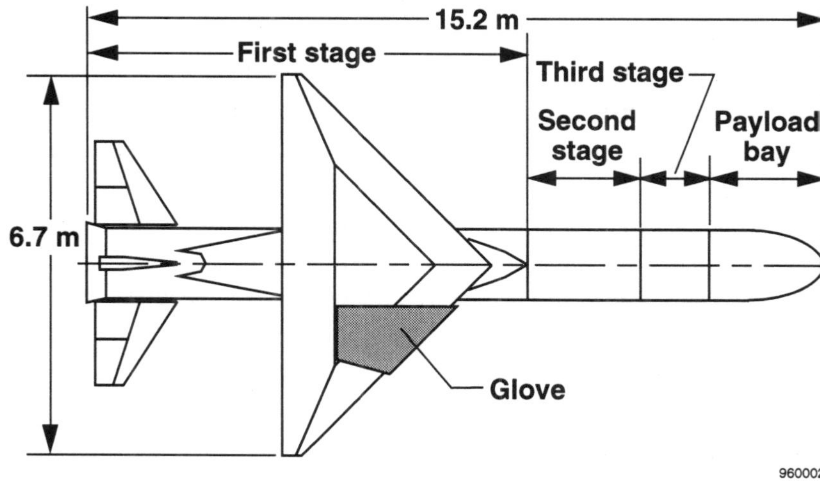

Figure 1: Plan view of the Pegasus® XL space Booster with glove.

EC89 309-3

Figure 2: Pegasus® space booster mounted under the wing of the B-52 aircraft.

Aerothermal Loads Analysis

A vehicle trajectory, similar to those previously described, e.g. Noffz, et al.[2] and Noffz, et al.,[3] was used as input to an aero/thermal analysis to determine aerodynamic heating loads applied to the glove during the flight. An engineering aerodynamic heating code was used to calculate heat-transfer coefficients, heating rates, skin friction, and surface-static pressures at discrete locations during the transient-heating profile. The program permitted the use of different theories for calculating heat transfer at various glove locations and flow conditions. Additional details of the aerothermal analyses performed in the glove design have previously been provided, e.g. Gong, et al.[4] Several combinations of glove skin thickness and materials were used in the analysis to begin defining an appropriate glove configuration.

Structural Design Requirements

Structural design requirements included a surface waviness criterion not to exceed 0.008 cm over a 5-cm length throughout the flight envelope; a requirement to minimize any step discontinuities at the inboard edge of the test surface, especially near the leading edge; and a requirement to have a thermally conductive thin skin. Such a skin was needed so that thermocouples installed on the inside skin surface would sense heating-rate changes caused by boundary-layer transition. Weight was a secondary design consideration, and the structure was not optimized in this regard.

Preliminary studies showed that if the glove was rigidly constrained during heating, the test skin would buckle, or the rigid attachments would yield, or both. Thermal expansion and contraction capability was therefore regarded as an important feature in the glove design.

Aerodynamic Design Requirements

Aerodynamic requirements such as those previously described, e.g. Godil & Bertilrud,[5] were also important to the design but are beyond the scope of this paper.

Preliminary Design and Thermal Analysis

Figure 3 shows the design and verification process for the design of the hypersonic glove test article. After the design requirements were defined, a preliminary design of the glove was initiated. Such a design required the selection of materials, definition of boundary constraints, sizing of skin thicknesses and fasteners, and laying out of instrumentation and sensor wiring. The details of the

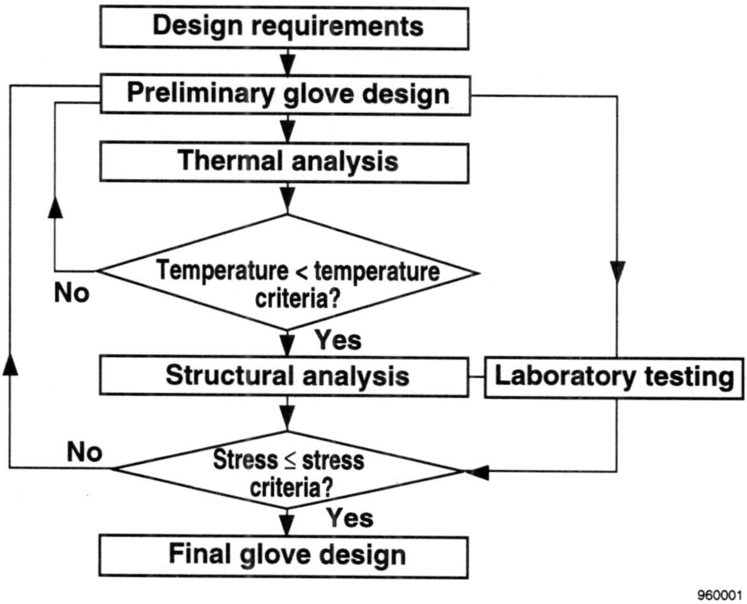

Figure 3: Glove design process.

initial glove design and preliminary results from the one- and two-dimensional thermal analyses that supported these tasks have previously been documented, e.g. Gong, et al.[4]

Thermal analyses of the preliminary glove design were performed to predict the glove transient-temperature response during the mission. The applied heating rates from the aerodynamic heating analysis described previously were coupled with the thermal response of the preliminary glove design. The transient-temperature distributions were then evaluated to determine if any area of the glove exceeded maximum temperature limits. Those areas in the glove design that exceeded these criteria were modified and reanalyzed until all the glove materials and structures were below maximum operating temperature criteria.

Glove Design

The iterative process between the preliminary design and thermal analyses resulted in a glove with a relatively thick, metallic outer skin and a large leading-edge heat sink to accommodate the extreme leading-edge heating during flight. The aerodynamic fairing that blends the metallic portion of the glove to the existing Pegasus® composite wing has previously been described, e.g. Gong, et al.[4]

As previously mentioned, the ability to accommodate thermal growth and contraction during flight was an important requirement for the glove. To achieve this capability, the leading edge was rigidly fixed at a single point and the metallic skin was allowed to freely expand over a balsa wood support foundation during heating.

Figure 4(a) shows a top view of the glove. Figure 4(b) shows the inboard leading edge highlighted. The single rigid attachment at the inboard leading edge and the expansion joints in the outboard direction are also shown. The preliminary design constrained the movement of the glove leading edge in the direction normal to the Pegasus® wing leading edge. However, a series of slotted attachments allowed movement in the direction parallel to the Pegasus® wing leading edge. The rigid attachment of the glove leading edge forced all thermal growth of the glove aft and parallel to the leading edge. This approach ensured glove dimensional stability at the leading edge where boundary-layer transition is the most sensitive.

The glove outer skin consists of a nickel-plated low-carbon steel sheet that has a plan view area of approximately 1 m^2 (fig. 4(a)). The upper and lower skins are 0.23- and 0.35-cm thick, respectively. These skins are attached to the wing at 270 independent locations by spring-loaded swivel studs designed specifically for this flight program. Figure 4(c) shows a cross-sectional view of the glove and

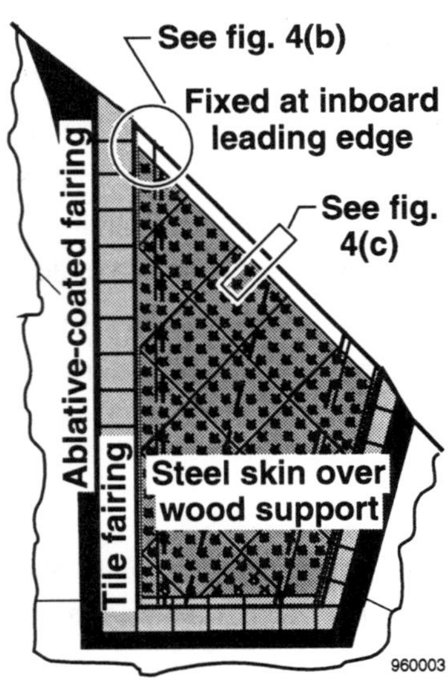

(a) Plan view of glove assembly.

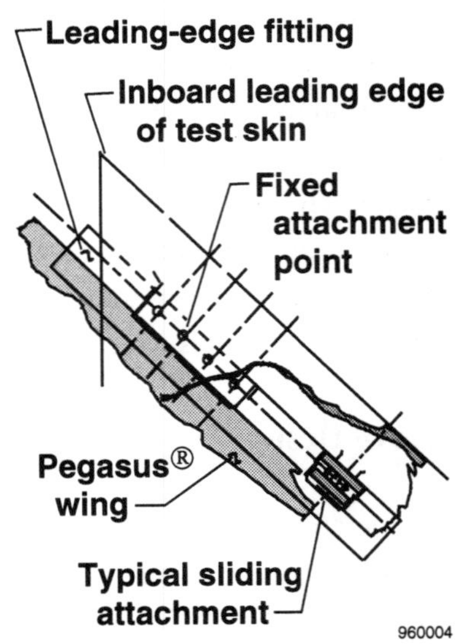

(b) Glove leading-edge fixed and sliding attachment.

Figure 4: Glove assembly.

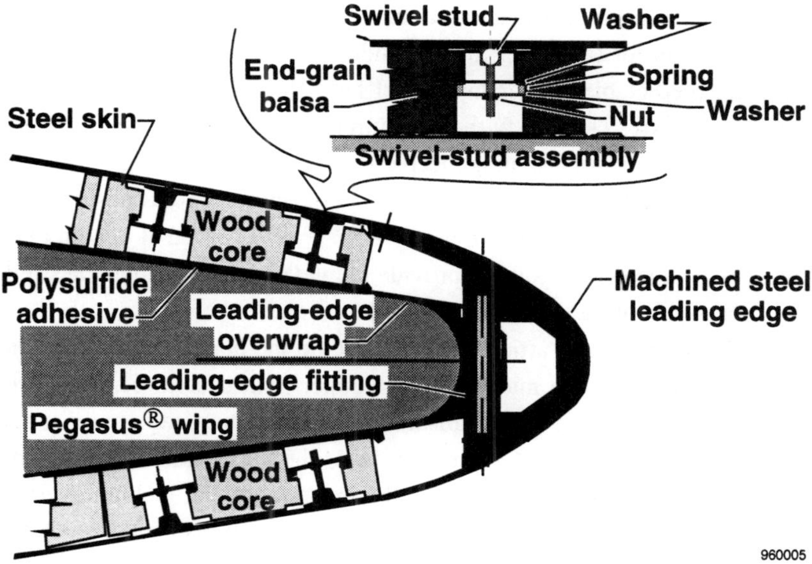

(c) Glove internal assembly.

Figure 4: Concluded.

an enlargement of a typical swivel-stud mechanism. The swivel studs, spaced 6.4 cm apart, are bonded to the steel skins with high-temperature epoxy. To react against aerodynamic forces, the studs are preloaded to 111 N through the spring attachments to the balsa substructure. The balsa glove structure is preshaped and bonded to the outer skin of the basic Pegasus® wing. This attachment method holds the skin securely to the surface but also allows the skin to expand thermally with only a small resistance caused by sliding friction.

Small-Scale Laboratory Testing

The structural design of the glove incorporated unconventional construction methods and materials. Consequently, extensive small-scale laboratory testing was performed to verify those areas in the glove that possessed the greatest design uncertainty or areas that were impractical to model analytically. This section describes the two main areas on the glove that were evaluated: the joint between the upper skin and leading-edge mass (the upper joint), and the sliding attachment of the test skin to the wing. In all, more than 20 small-scale laboratory tests were performed.

Solder-Joint Strength and Performance Tests

The requirements for the upper joint included smoothness of the aerodynamic test surface, predictable heat transfer, and fail-safety design under potentially high thermal stresses. To satisfy these requirements, the upper joint was redundantly designed with two fastening mechanisms. Solder was used to provide predictable heat transfer and a means of smoothing the joint without causing thermal conduction discontinuities. Two rows of 3-mm screws spaced at 1.3 cm in each row were also added to the joint to provide redundancy and predictable strength values for analysis of the primary structural load path.

Figure 5 shows the test article used to develop the manufacturing technique and gain confidence in the integrity of the upper-joint design. This test article consisted of a 25 cm–long section of the full-scale leading-edge mass attached by a representative joint to a 65-cm^2 section of the upper skin. This test article was used to develop the solder process, the joint smoothing technique, and the nickel-plating process. A pass-fail thermal stress test was also conducted on this test article.

Preliminary joint analysis had indicated that the thermal stress in the joint would be only slightly less severe on the 25-cm specimen than on the full-scale article. A representative thermal gradient was applied to this test article to demonstrate survivability of the joint. This test produced temperatures in the soldered and bolted joint similar to the predicted flight condition and produced a similar thermal gradient across the joint. The leading-edge mass was heated to 102 °C,

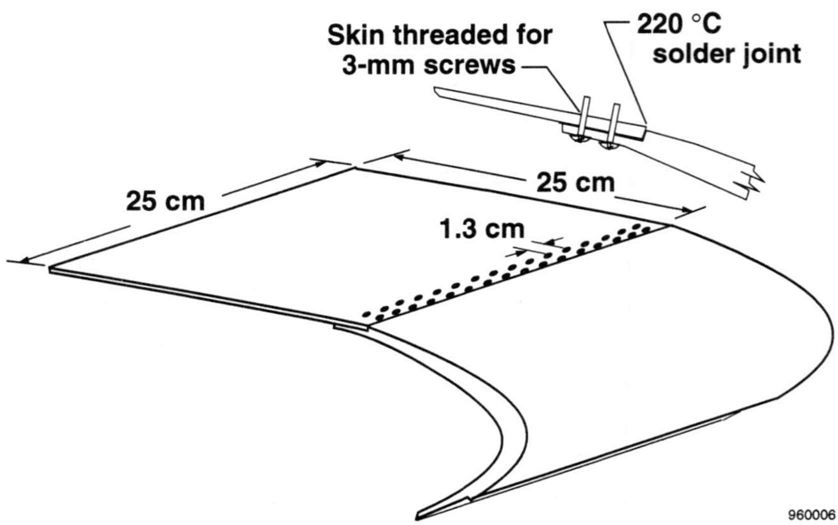

Figure 5: Upper joint test article.

and the thin skin was heated to 193 °C. Post-test inspection using visual and dye-penetrant techniques indicated no damage to the joint.

Bolt-Shear Tests in the Upper-Skin Joint

The upper-joint design called for bolts to be shear-loaded in the threaded section, which is not standard engineering practice, and therefore, no strength data were available from existing sources. Bolt-shear tests using the actual joint materials (low-carbon steel and 793-MPa tensile-strength bolts) were conducted to determine shear strength as well as to observe the yielding characteristics. The test results indicated large margins of safety (40 percent) and significant yielding in the skin material prior to bolt failure. Bolt failure occurred from a combination of shear and bending stresses. The bolt strength was adequate to satisfy thermal stress requirements. Actual bolt loads will be significantly lower than those analytically predicted because of local yielding in the area of the bolt that is not included in the analysis.

Swivel-Stud Strength Tests

Figure 6 summarizes four component tests that were done to demonstrate the strength of the various elements of an individual swivel-stud attachment point.

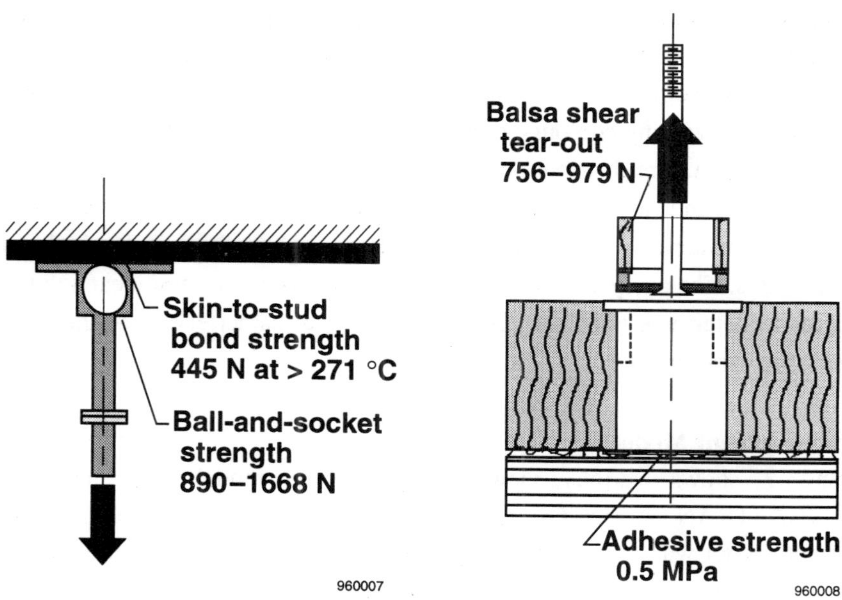

(a) Swivel-stud bonding integrity test.

(b) Balsa flange shear test.

Figure 6: Small-scale laboratory tests.

The testing of the swivel-stud-to-skin bond (fig. 6(a)) was used to select and qualify an adhesive that would maintain strength at high temperatures. A 445-N load was selected as the static-load criterion. The test results for the selected adhesive indicated a 95-percent probability of holding a 445-N tensile load to a temperature greater than 271 °C using a normal statistical distribution.

The weak link in the balsa attachment was the shearing of the balsa flange that supports the swivel-stud preload springs. Figure 6(b) shows the configuration of the coupons used to test this area. The test results indicated good strength for the typical balsa flange. In some local areas, the flange depth was reduced significantly and basswood inserts were installed to increase the shear strength.

The balsa substructure of the glove is bonded to the Pegasus® wing with a polysulfide adhesive. Testing done for this joint, which includes a real wing upper-surface skin, indicated strengths well in excess of the balsa and swivel-stud strength. The demonstrated high strength ratio of the Pegasus® wing skin to foam core provides a fail-safe feature. In the unlikely event of a local overload from the glove outer skin, the studs and balsa would fail before the loads could be high enough to damage the basic wing structure.

Structural Analyses

In addition to the laboratory tests, analyses were required in the design verification process to understand those features that were nearly impossible to represent in the laboratory. The analyses were used to represent such design features as geometrically complex glove shapes, such as Bezier curves and three-dimensionally varying contours; nontrivial load conditions, such as nonuniform transient-temperature distributions; nonlinear boundary conditions, such as expansion joints and surfaces with contact and sliding friction; and nontraditional engineering materials, such as balsa wood, mechanical springs, and swivel-stud mechanisms. The analyses used a computationally intensive nonlinear solution process to determine transient stress and buckling characteristics of the glove during the experiment.

Finite-Element Model

Figure 7 shows various views of the glove model generated with MSC/NASTRAN (version 68), Lahey, et al. (ed).[6] A three-dimensional finite-element model was constructed that consisted of 3777 hexahedron elements, 187 pentahedron elements, and 1194 nonlinear contact-friction elements. The model was meshed in such a way as to match the location of each swivel-stud assembly in the balsa support system. The leading-edge heat sink is modeled

(a) Top view.

(b) Isometric view.

(c) Forward view.

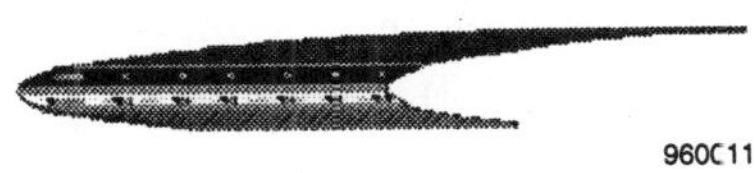

(d) Outboard view.

Figure 7: Finite-element model

with solid elements to incorporate the large temperature drop through the thickness of the leading edge. The fixed restraint is located inboard (fig. 7(c)).

Linear stress analysis assumes that displacements are linearly related to the forces applied to the structure. However, this relationship does not apply at the sliding attachment boundary between the thin skin and the balsa wood support. Nonlinear, contact-friction elements, formulated to represent the balsa wood stiffness; swivel-stud assemblies; wavy springs; and the wood/skin interface were used to model this complex behavior. These elements are inherently nonlinear because at each load increment in the solution sequence, the code determines if the contact surface is stationary and resisted by static friction, if the contact surface has slipped and is resisted by kinetic friction, or if the internal element forces have exceeded the preload and the surface has opened.

Temperature Distributions

Initially, the temperature distribution at a single time in the thermal analysis was input to a structures model to predict the stress field and buckling characteristics of the glove. The temperature distribution at 69 sec (first-stage burnout time) was initially chosen because it corresponded to the end of the heating profile when thermal stresses are usually assumed to be the most extreme. A more rigorous analysis showed that multiple times in the temperature-time history were necessary because the temperature gradients varied considerably throughout the trajectory. Therefore, analyses were also performed for times at 45, 75, and 90 sec (second-stage ignition) into the mission profile. Times of 45 and 75 sec correspond to the points at which the thermal gradient between the thick leading edge and thin skin is the most extreme.

Boundary Conditions

The fixed boss located at the inboard area of the leading edge is the only place where the glove is rigidly attached to the wing. All degrees of freedom at this location were therefore constrained in the analysis. Displacements of the leading-edge slotted attachments were initially prevented in the direction normal to the leading edge, but were allowed in the parallel direction to accommodate the thermal expansion in this direction. However, after the stress analyses were performed at the three additional profile times, it was discovered that the original attachment scheme produced stresses greater than the acceptable yield stress criteria.

Various attachment schemes were analytically studied to help identify an acceptable attachment technique. Based on these studies, glove design modifications were made at the leading edge at the attachments and slotted joints. The displacement in the direction normal to the leading edge was allowed everywhere at the leading edge except the most inboard and most outboard leading-edge connections. Consequently, the constraints at the leading edge were similar to a simply supported beam that is able to rotate at both ends. These changes allowed greater thermal distortion at the leading edge, but were necessary to ensure the thermal stresses were less than the allowable yield stress criteria.

Results and Discussion

Stresses and displacements were predicted at multiple points in the transient heating profile. This section highlights some of the significant results from the structural analysis.

Stress Results

Table 1 summarizes the transient stress results and factors of safety of the most critical components to the glove design. The maximum stresses induced during the trajectory are highlighted. These results were used to determine the component safety factor. As table 1 shows, the maximum stress in each of the glove components occurs at different times during the mission. This result underscores the need for transient analyses; conventional analyses techniques that use steady-state, linear assumptions to simplify the problem could produce erroneous predictions and nonconservative design results.

Figure 8 shows one of the component stress results shown in table 1. The shear stress in the upper joint is mapped on the glove model. The area of highest shear occurs at the inboard end of the soldered and bolted joint. The analysis did not attempt to account for additional strength provided by the solder. Despite this, all critical glove components are well below the allowable stress with a factor of safety of at least 1.5.

Deformation Results

Figure 9 shows top views of the leading-edge distortion at 45, 69, 75, and 90 sec in the trajectory. This figure also illustrates why a transient structural analysis was required. The 69-sec deflection, assumed as the worst-case condition for the

Table 1. Stress results.

Glove component	Time, sec: 45	69	75	90	Allowable stress	Factor of safety
	Stress results, MPa					
Upper skin						
Max. shear stress	82	80	84	55	241	2.9
Lower skin						
Max. shear stress	113	121	117	89	241	2.0
Upper joint						
Max. skin-shear stress	90	96	82	62	241	2.5
Lower joint						
Max. skin-shear stress	136	134	129	97	241	1.8
Leading edge						
Max. shear stress	120	104	112	118	241	2.0
Boss bearing stress	48	49	63	66	241	3.7

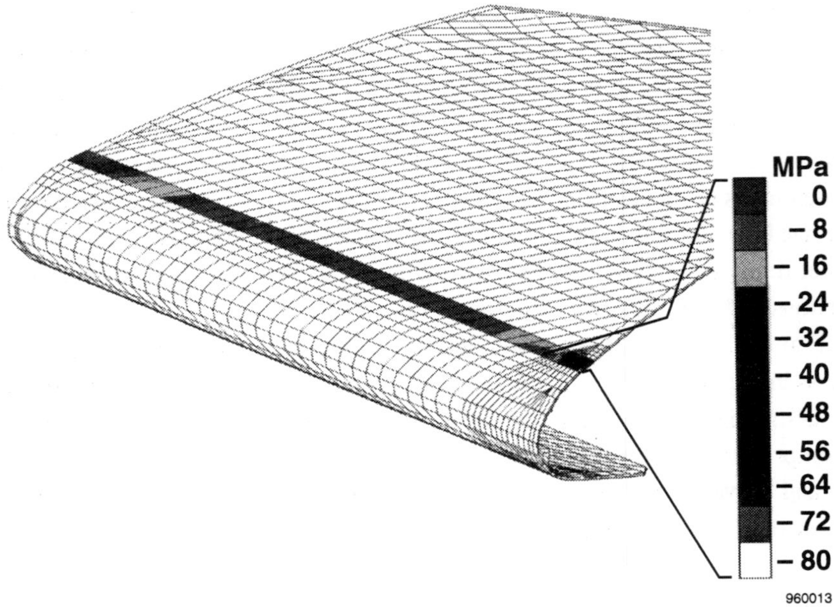

Figure 8: Shear stress in upper joint.

preliminary design, appears to be the most benign of the four times evaluated. The center leading edge thermally distorts aft at 45 sec in the profile. At 69 sec, the stagnation heating at the leading edge forces the leading-edge deflection through its initial starting point. At 75 sec, the gradient between the leading edge and thin skin has reached a displacement nearly equal in magnitude and opposite in sign to the displacement at 45 sec. At 90 sec, the displacement has bowed out to a maximum displacement of 0.03 cm. Although the maximum center deflections vary from 0.02 cm in the aft direction to 0.03 cm in the forward direction, these are well within the experiment deflection limits of 0.005 cm over a 5-cm in-plane length.

Concluding Remarks

A comprehensive structural test and analysis program has been successfully completed to validate the design of a flight-test article for a Mach-8 boundary-layer experiment. The test article consisted of nontraditional engineering materials and geometrically complex components, had transient boundary conditions, and was constructed using unique manufacturing processes. Extensive small-scale laboratory testing was used to verify those areas in the glove that possessed the greatest design uncertainty or were impractical to model analytically. A

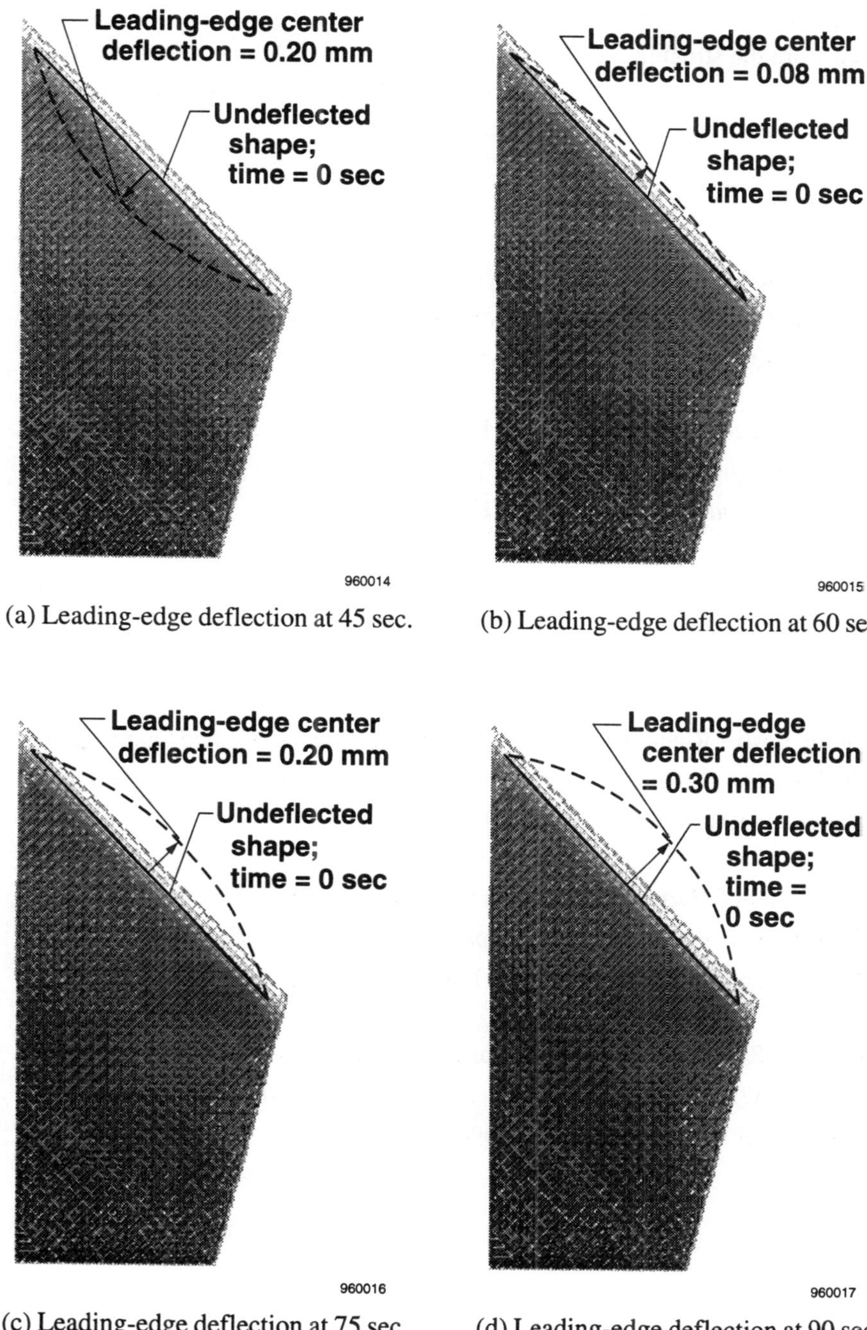

(a) Leading-edge deflection at 45 sec.

(b) Leading-edge deflection at 60 sec.

(c) Leading-edge deflection at 75 sec.

(d) Leading-edge deflection at 90 sec.

Figure 9: Leading-edge deflections at several profile times.

three-dimensional nonlinear structural analysis was instrumental in accurately representing the complex glove characteristics during the transient flight profile reaching Mach 8. Test and analysis activities were worked in concert to help design and validate the glove test article required for the experiment. Results obtained from the structural analysis and laboratory testing were presented. These results show that all glove components are well within the allowable stress and deformation limits to satisfy experiment objectives.

References

1. Bertilrud, Arild, Kolodziej, Paul, Noffz, Greg K., and Godil, Afzal. *Plans for In-Flight Measurement of Hypersonic Crossflow Transition on the Pegasus® Launch Vehicle*, AIAA-92-4104, Aug. 1992.

2. Noffz, Gregory K., Curry, Robert E., Haering, Edward A., Jr., and Kolodziej, Paul. *Aerothermal Test Results From the First Flight of the Pegasus Air-Launched Space Booster*, NASA TM-4330, 1991.

3. Noffz, Gregory K., Moes, Timothy R., Haering, Edward A., Jr., and Kolodziej, Paul. *Aerothermal Test Results From the Second Flight of the Pegasus® Booster*, NASA TM-4391, 1992.

4. Gong, Leslie, Richards, W. Lance, Monaghan, Richard C., and Quinn, Robert D. *Preliminary Analysis for a Mach 8 Crossflow Transition Experiment on the Pegasus® Space Booster*, NASA TM-104272, 1993.

5. Godil, A. and Bertilrud, A. Design of a Wing Shape for Study of Hypersonic Crossflow Transition in Flight, *Computing Systems in Engineering*, Vol. 3, Nos. 1–4, pp. 115–130, Pergamon Press Ltd, Great Britain, 1992.

6. Lahey, Robert S., Miller, Mark P., and Reymond, Michael (ed), *MSC/NASTRAN Reference Manual: Version 68, Volume 1*, The MacNeal-Schwendler Corporation, USA, 1994.

Experimental investigation on plastic rotational capacity of prestressed hollow core panels

C. Bosco, P.G. Debernardi
Department of Structural Engineering, Politecnico di Torino, Corso Duca degli Abruzzi 24, 10129 Torino, Italy

Abstract

In this paper the authors present the results of an experimental investigation on plastic rotational capacity of prestressed hollow core concrete panels. In particular, this parameter is evaluated in prestressed precast elements at the ultimate loading capacity and the interaction between plastic behaviour, shear and bending resistance is taken into account.
The tests were carried out on fifteen simply supported panels 220 mm deep, having three reinforcement percentages and five different slendernesses.

1 Introduction

The investigation concerned normal weight precast prestressed hollow core concrete panels without any transverse reinforcement and shear reinforcement.
These elements are usually employed to make simply supported floors or floors weakly restrained at the ends.
On the other hand, it is also possible to produce a structural continuity between the panels so as to obtain continuous floors with the supports consisting of panels either outside or inside the slab thickness. In the latter case, continuity is achieved by means of ordinary steel reinforcement inserted in special slots and bonded with a supplementary casting (FIP Recommendations,[1]; Bosco and Debernardi,[2]).
As a rule, these structures must possess a certain degree of ductility if they are to withstand seismic actions or intrinsic imposed deformations (e.g., temperature variations); at all events they must be able to provide some warning before failure.
In this connection, the tests were aimed at assessing the ductile behaviour of the structures with reference to three basic aspects: plastic rotational capacity, shear resistance and bond of the prestressing steel.

2 Testing programme

The testing programme was carried out at the Laboratory of the Department of Structural Engineering of the Politecnico of Turin on specimens taken from standard factory production and tested according to the three (3P) or four (4P) point bending method. The specimens had the same cross–section, all of them being 220 mm deep and 1105 mm wide, but were different in length and had three different reinforcement percentages. The cross–section of the panels is shown in Figure 1, while loading conditions, type of steel, actual span length and span to depth ratio (L/H) of the specimens are listed in Table 1.

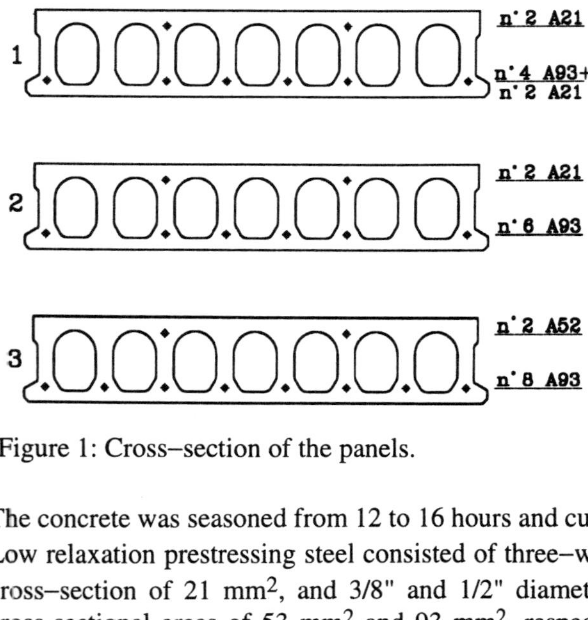

Figure 1: Cross–section of the panels.

The concrete used in these tests was cast using four different types of aggregate with maximum grain size of 12 mm; with 380 kg/m^3 of 425 Portland cement and a W/C ratio of 0.48. Characteristic cube strength at the time of prestressing was 40 N/mm^2 and cube strength after 28 days came to 55 N/mm^2.

The concrete was seasoned from 12 to 16 hours and cured in steam at 55 °C.

Low relaxation prestressing steel consisted of three–wire strands with nominal cross–section of 21 mm^2, and 3/8" and 1/2" diameter strands with nominal cross-sectional areas of 53 mm^2 and 93 mm^2, respectively. The strands used had mean strength f_{ptm} = 1947 N/mm^2 and a minimum elongation at failure = 4.83%. Prestressing steel was pretensioned at 1400 N/mm^2 and the release of prestressing reinforcement was controlled by means of tensioning hydraulic jacks. The steel was cut with a disk saw at the cuts made into the concrete with air after casting it onto the bed of prestressing.

The loads were applied by means of a servocontrolled MTS machine in controlled displacement conditions and the deformations at the lower and upper edges of the panels were measured by means of gauges of length equal to the depth of panel, i.e. 220 mm.

The loads were applied in two or more stages. During the initial stage the midspan displacement was increased until the opening of cracks was reached and then the load was removed. In the second stage, specimen cracking was obtained again by controlling the midspan displacement, and then specimen failure was reached, either directly or, in some cases, after an additional loading and unloading cycle. Failure could come about for different reasons as discussed below. Measurement starting was fixed at the load corresponding to the dead weight of the panel plus the weight of the testing equipment.

In the second loading stage, the state of decompression of the intrados is identified in load–deflection diagrams by a change in inclination allowing the determination of the moment of decompression, when the section shifts from uncracked to cracked conditions. The initial cracking moment depends on the state of prestressing and concrete tensile strength and it is higher than the subsequent cracking moment since at the latter stage tensile strength in the section is nil.

3 Experimental results

The results were strongly affected by bond between strands and concrete, especially in the panels with low L/H ratio. In such cases, an evident slippage of strands was observed and the elements did not reach their ultimate theoretical capacity.

In panel L–180–3 slanted cracks opened resulting in failure by shear. In other cases, failure occurred due to concrete crushing at an initially vertical crack produced by bending which, upon further load increments, spread in a slanted direction and reached into the upper edge of panel and reduced its compression resistant section. This type of failure must be ascribed to a combined shear–bending effect.

In the specimens with higher slenderness, failure took place by concrete crushing in bending. As for the behaviour of the specimens in serviceability conditions, its evolution was seen to be regular up to crack formation; in elements with reduced slenderness, however, prestressing steel slippage occurred and extended to the ends of the panel.

Some aspects of experimental behaviour are illustrated below by a few diagrams regarding deflection, crack formation and plastic rotation. Figures 2 and 3 refer respectively to midspan deflection of specimen L–440–2A and L–440–2B having different degree of prestressing.

In both cases, two transversal cracks were formed at the intrados astride midspan and spaced about 200 mm apart. In the panel with lower reinforcement percentage (type 1), the cracks propagated towards the compression zone – in an initially vertical direction which then became only slightly slanted upon further load increments – until failure. In type 2 reinforcement panels, the cracks became markedly slanted in the plastic stage.

Compared to the two previous elements, panels L–440–2A2 and L–440–2B were loaded with two concentrated loads (four point bending) astride midspan and spaced five times the depth of the panel. These panels displayed widespread cracking over the entire portion of constant bending moment. Failure occurred for bending moment values virtually identical to those recorded in panels loaded at a single point at midspan (3P). Shear, at panel failure, was rightly found to be higher than the value recorded for the other two panels.

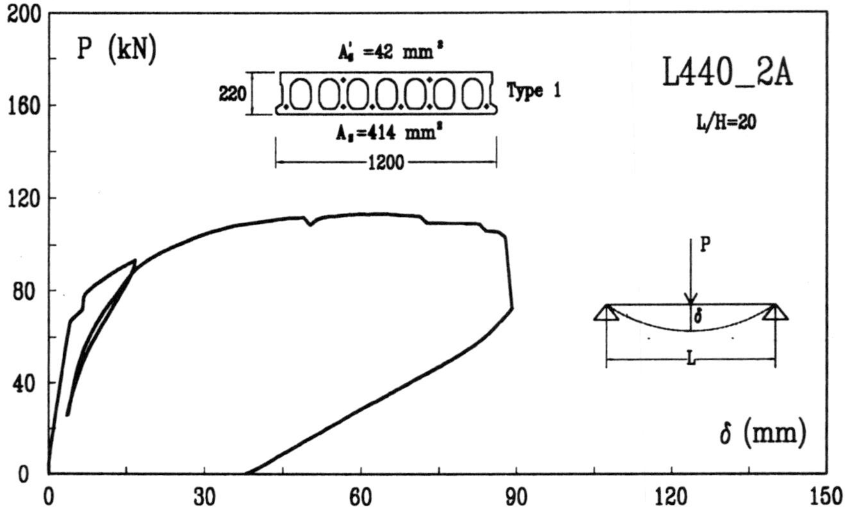

Figure 2: Load versus midspan deflection for panel L–440–2A.

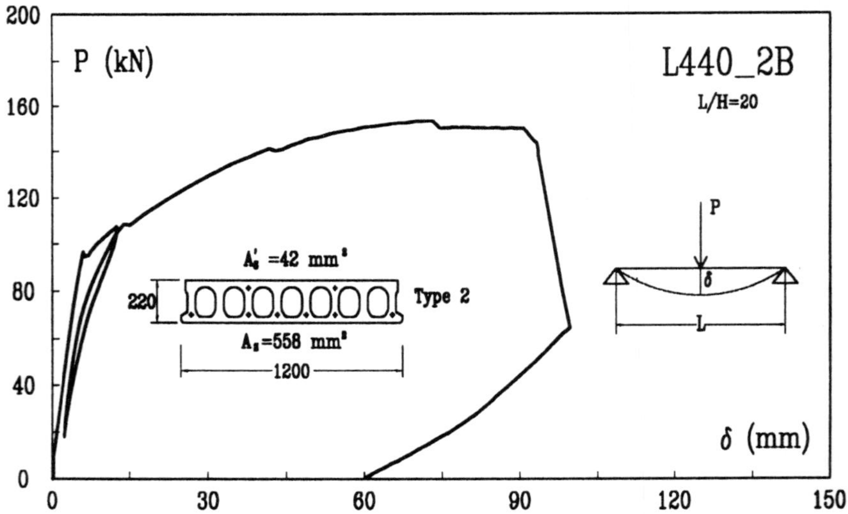

Figure 3: Load versus midspan deflection for panel L–440–2B.

To describe the behaviour of the panels with a low L/H ratio, Figure 4 shows the midspan deflection for panel L–176–2A. Behaviour is seen to be regular up to the formation of a crack at midspan. After that, reinforcement slippage begins and develops progressively up to the ends of the panel. Obviously, crack formation no longer occurs and the bending moment at failure, lower than the theoretical moment, turns out to be just above the cracking moment.

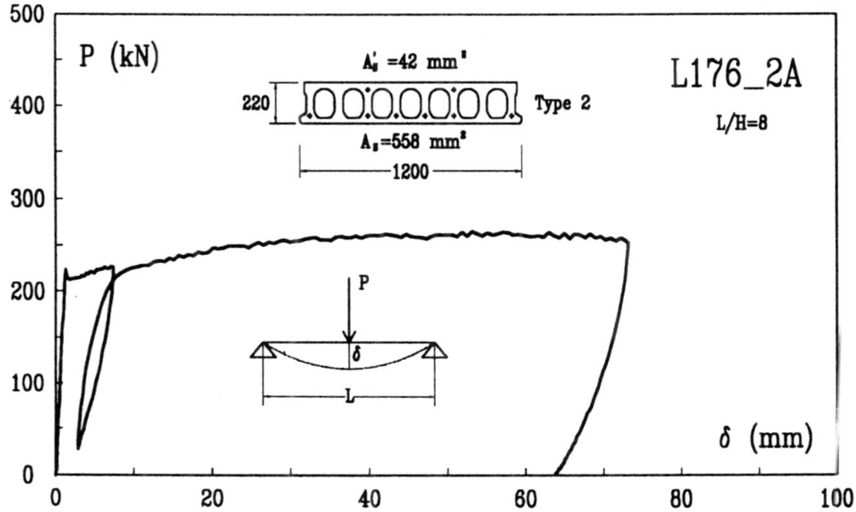

Figure 4: Load versus midspan deflection for panel L–176–2A.

4 Theoretical comparison with the experimental results

The theoretical evaluation of structural behaviour both in serviceability conditions and at the ultimate limit state was performed according to the provisions of Eurocode 2,[3,4,5].

a) At serviceability conditions the determination of the initial state of stress, during the test, takes into account the prestressing losses due to shrinkage and creep, relaxation and the interaction between these factors.

Starting from these values and as a function of concrete tensile strength, both the cracking moment and the moment of decompression were theoretically determined. The comparison with the experimental results, obtained in terms of bending moments at midspan, is shown in Figure 5.

As for the cracking moments, it can be seen that test results are in agreement with theoretical predictions save for panels with spans up to 1.80 in which high shear force plays a decisive role in the formation of cracks. In all cases, instead, the experimental value of the decompression moment was found to be higher than the theoretical one, but it should be noted that the markedly rounded shape of the moment–curvature diagram may lead to inaccurate evaluations.

b) The determination of the failure load in the individual specimens has been performed taking into account the lowest value given by the relationship of shear strength in elements with no shear reinforcement and that of bending resistance, as set forth in the Eurocode 2. The values of material strength introduced in such relationships were assumed with reference to mean values.

Figure 5 shows a comparison between the theoretical results and those obtained by assuming as reference values the maximum moments reached during the tests.

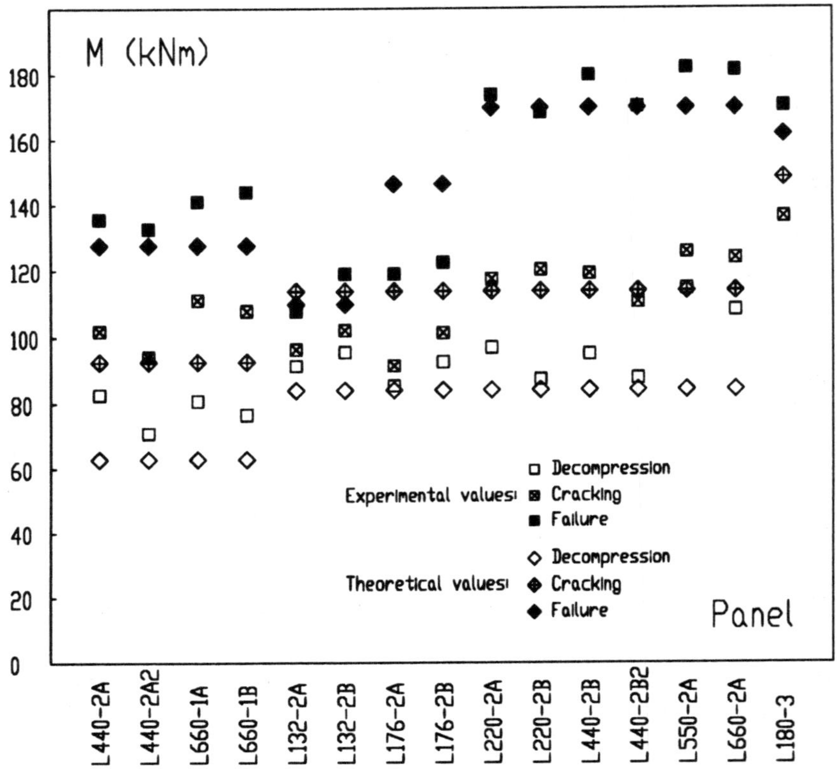

Figure 5: Theoretical and experimental values of bending moment of cracking, decompression and failure.

Here again a good agreement appears between test results and theoretical predictions, save for a few short span panels, in which slippage of prestressing reinforcement resulted in failure for load levels lower than the theoretical ones.

c) Plastic rotation was experimentally determined in two ways:

– by integrating the mean curvatures obtained from strain measurements;

– from the values of experimental deflection by assuming that plastic rotation was concentrated at midspan.

The results obtained by these two methods are very close to one another. Table 1 shows the values of plastic rotation for each specimen, together with shear and bending stresses at failure.

In panels with span up to 1.80 m, the value of plastic rotation is still high, even in the presence of a single central crack, but this is due exclusively to the extensive slippage of the strands relative to the concrete.

Table 1: Values of plastic rotation, shear and bending moments at failure.

Specimen	Type of reinf.	L/H	Loading condi–tions	Shear force (kN)	Bending moment (kNm)	Plastic rotation (rad)	Type of failure
L440–2A	1	20	3P	66.0	13.57	0.095	bending
L440–2A2	1	20	4P	83.3	13.28	0.092	shear–bending
L660–1A	1	30	3P	49.2	14.11	0.112	bending
L660–1B	1	30	3P	50.1	14.40	0.099	shear–bending
L132–2A	2	6	3P	164.5	10.77	0.143	slippage
L132–2B	2	6	3P	181.5	11.90	0.123	slippage
L176–2A	2	8	3P	136.8	11.89	0.107	slippage
L176–2B	2	8	3P	140.8	12.24	0.093	slippage
L220–2A	2	10	3P	160.1	17.37	0.116	shear–bending
L220–2B	2	10	3P	155.1	16.82	0.083	shear–bending
L440–2B	2	20	3P	86.0	17.97	0.103	shear–bending
L440–2B2	2	20	4P	106.0	17.02	0.200	shear–bending
L550–2A	2	25	3P	71.5	18.19	0.126	bending
L660–2A	2	30	3P	61.3	18.12	0.106	shear–bending
L180–3	3	8.2	3P	190.8	17.02	0.087	shear

5 Conclusions

The aim of this investigation was to evaluate behaviour in the plastic field of normal weight precast prestressed hollow core concrete panels with no transverse reinforcement and no shear reinforcement, through the experimental determination of plastic rotation and by assessing the interaction between plastic rotational capacity, bending and shear resistance (see Figure 6).
Test results revealed a minor decrease in bending resistance with increasing shear; a linear evolution is observed for the panels with slenderness between 30 and 10 with either type 1 and type 2 reinforcement.
For lower slendernesses, strand slippage was seen to be considerable and to spread all the way to the ends of the panels. In this connection it should be kept in mind that bond is closely associated with the pitch of the strands which, in this case, came close to the upper permissible safety limit (about 18 diameters). Low slendernesses are not commonly used and should be handled with extreme caution as they can give rise to brittle failure which may be further enhanced by the presence of strong variable loads repeated several times.
On the basis of the test results it proves much harder to find a correlation between shear resistance and plastic rotation. However, the absolute value of plastic rotational capacity was found to be high regardless of the slenderness of the elements and the reinforcement percentage.

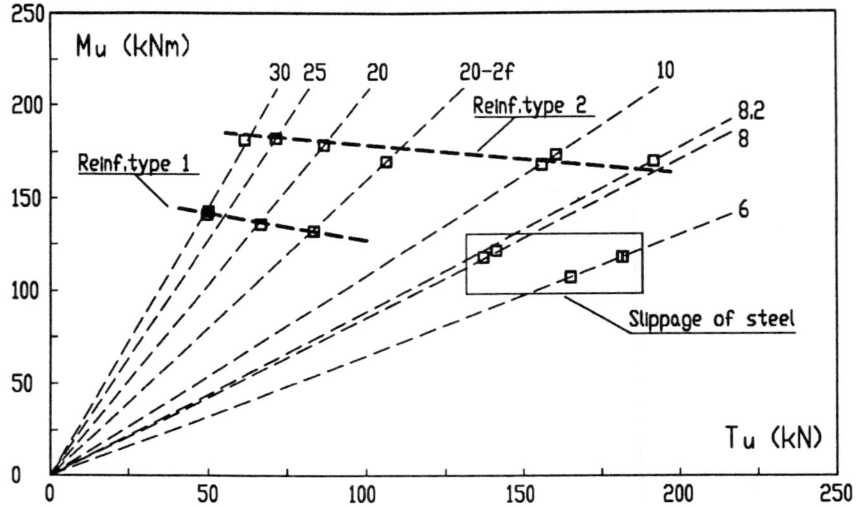

Figure 6: Interaction between bending and shear resistance for slenderness varying from 6 to 30.

The values were invariably higher than 0.09 radians, whilst the values of admissible rotation specified in the Eurocode 2 are lower than 0.02 radians, for reinforced concrete elements with high ductility steel.
Here again, it may assumed that such high values can be ascribed mostly to poor bond of the prestressing reinforcement.
While extended experimental investigations on bearing capacity are available [6], the interaction between bending and shear resistance is not sufficiently considered. The authors wish that this topic should be investigated further.

References

1 FIP Recommendations, *Precast Prestressed Hollow Core Floors,* Ed. Thomas Telford, London, 1988.

2 Bosco C. and Debernardi P. G. Further investigation on structural continuity of hollow core floors produced by slipform technique, *Betonwerk Fertigteil Technik*, 1991, **9**, 99–106.

3 ENV 1992–1–1, Eurocode 2 *Design of Concrete Structures, Part 1. General Rules and rules for buildings*, Dec. 1991.

4 ENV 1992–1–3, Eurocode 2 *Supplementary rules for Precast Concrete Elements and Structures*, 1993.

5 CEN TC 229, European Standard *Floors of precast prestressed hollow core elements*, 1992.

6 Walraven J.C. and Mercx W.P.M. The bearing capacity for prestressed hollow core slabs, *Heron*, Volume 28, 1983, **3**.

On the development of joint element for flexural-torsional analysis of thin-walled axially-discontinuous frame structures

A.H. Hanizah, A.B. Sahol Hamid, A. Ibrahim
Department of Civil Engineering, MARA Institute of Technology, 40450 Shah Alam, Selangor, Malaysia

Abstract

When a thin walled open section is subjected to torsion it has a tendency to warp. If connected to another section at an angled joint, this warping can cause significant twisting and bending of the second section. The direction and amount of this twisting is related to the type of joint used.

This form of structure cannot be analysed using classical analytical technique due to the complexity and nature of the problem. This structural form can however be analysed using the finite element method. However, in a design office, carrying out a full finite element analysis on this type of structure is not economical. To overcome this problem an approximate analysis method has been developed, and documented in the present paper for 90° joints, which allows this structural form to be analysed more economically. The method is based on the stiffness method with a special joint element being used within a frame analysis program which is based upon the one dimensional beam element. A finite element method teaching package CALFEM was used whilst developing the method but any frame analysis program can be used provided that it has the capability of allowing for user defined elements. Design charts are provided which allow the joint element stiffness matrix to be defined for three different joint types. Estimates of maximum errors that may arise when using the approximate method are also included. The results of this approximate analysis method are used to illustrate the design procedure which checks the stress levels developed in the structural members.

1 Introduction

A thin walled beam can be described as being a beam composed of several plates monolitically connected along their edges with the thickness of the plates being small compared with the other dimensions of the beam. Under the action of torsional loads, structures composed of thin walled members with open sections, are susceptable to high warping displacements and high torsional shear stresses (e.g. Hanizah[1], Hanizah & Ibrahim[2], Morrell[3]).

A thin walled structure often has both complex frame geometry and complex section geometry. As a result, classical analytical technique cannot usually be used to analyse these structures since these methods can only handle elastic analyses and very simple structures. These structures can however be analysed using the finite element computer method. For a detailed and relatively accurate structural analysis using the finite element technique, a thin walled structure can be modelled using flat plate elements. In the case where the sections of the member and the type of joint at a member intersection are complex, the total number of elements and degrees of freedom of the problem required to undertake the analysis can be extremely large. It would not therefore be normally economical to undertake this type of analysis for routine design work.

Should it be possible to use a simpler model to represent the structure adequately, considerable economic savings could be made. The simple model would have to be able to represent completely the behaviour of the thin walled structure with respect to bending and torsion. If such a model was formed using beam elements, it would have significantly fewer degrees of freedom than a model formed from shell elements and could be used with typical stiffness method computer programs. This would result in a significant saving in the modelling and computing effort required, compared with a detailed discretization of shell element model.

The proposed approximate method is based on the stiffness method of frame analyses using a modified one-dimensional beam element, which accounts for both St. Venant's and warping torsional effects, combined with a special joint element (Hanizah[1]). The finite element method teaching package *CALFEM*, that allows elements of an elemental stiffness matrix to be defined, was used whilst developing the method.

2 Derivation of the joint element stiffness matrix

In Hanizah[1], an approximate method was developed to analyse a three-dimensional thin-walled frame structure using a discretised finite element model consisting of modified one-dimensional beam elements combined with a special joint element. Detailed descriptions of the elements' formulations are given therein. Only the formulation of the joint element is described in the present paper.

In the proposed simple model (see Figure 1), a joint element 3-4 is introduced. This special and ficticious element that has two nodes represents the structural joint configuration. Its purpose is to represent the influence of the detail of the joint on the transmission of twist in terms of magnitude as well as direction. In order to tie in with the adjoining elements, this joint element requires six degrees of freedom, namely (u_x, u_y, u_z, ϕ_x, ϕ_y, ϕ_z), at each of the two nodes. The degrees of freedom with directions are shown in Figure 2.

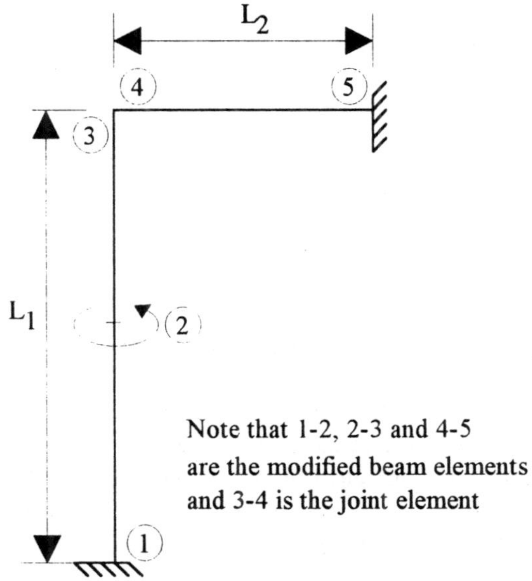

Figure 1: Idealised model of one-dimensional beam elements and a joint element.

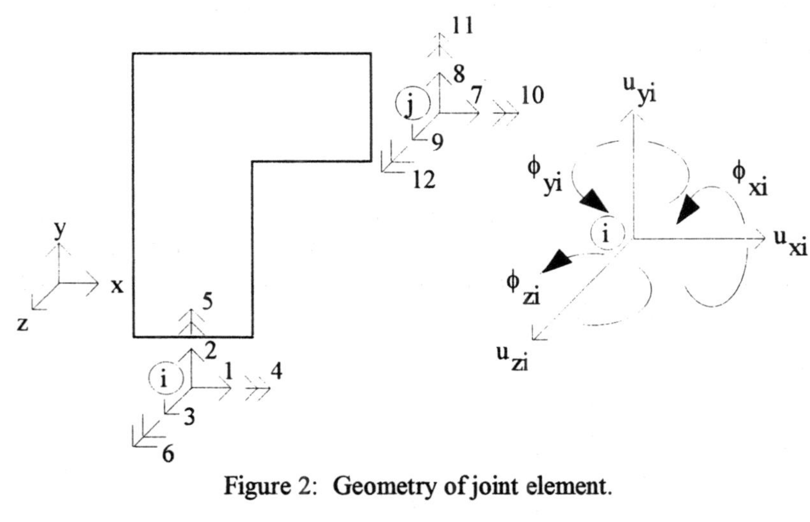

Figure 2: Geometry of joint element.

Consideration of the 12 x 12 stiffness matrix for the joint element now follows. Since the actual joint form is made up of thin walled open sections, the joint rotates and warps when the structure is subjected to torsion. Hence, the structural actions of interest in the joint are associated with rotations u_4, u_5, u_{10} and u_{11}. It can be assumed that the two nodes can be considered to be rigidly connected together with respect to the translations u_1, u_2, u_3, u_7, u_8 and u_9 and the rotations u_6 and u_{12}.

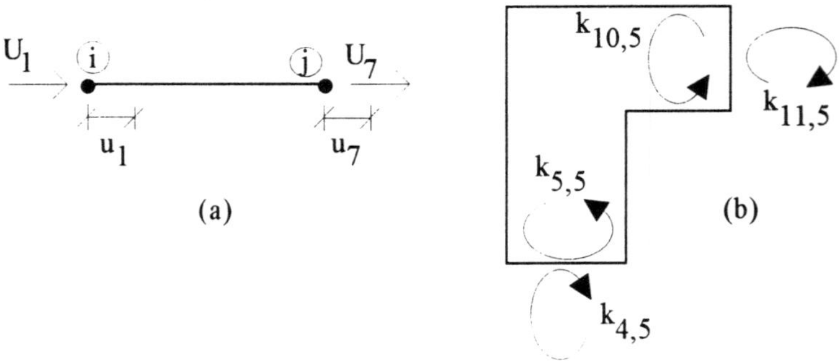

Figure 3: Nodal forces corresponding to a unit value of (a) u_1 and (b) u_5.

To establish the element stiffness matrix, the extension u_1 is first given the value unity and all other displacements have the value zero as shown in Figure 3(a). Since the movement of the joint element in *x*-direction is assumed to be very stiff, the corresponding axial forces U_1 and U_7, which form the terms $k_{1,1}$ and $k_{7,1}$, due to the unit displacement are very high. Let the values be denoted by H. These forces are equal in magnitude but in the opposite direction to each other. Similarly, if u_7 is given the value unity and the other displacements are all zero, then the corresponding axial forces, which form the terms $k_{1,1}$ and $k_{7,1}$ then become as high. The above relationships can be represented in matrix form and are given in equation (1).

$$\begin{Bmatrix} U1 \\ U7 \end{Bmatrix} = \begin{bmatrix} H & -H \\ -H & H \end{bmatrix} \begin{Bmatrix} u1 \\ u7 \end{Bmatrix} \tag{1}$$

The above procedure is repeated for the following pairs, i.e. u_2 and u_8, u_3 and u_9, u_6 and u_{12}. The same sets of equations are obtained because their displacements are as rigid as u_1 and u_7. Their combined stiffness matrix is expressed as in equation (2).

$$\begin{Bmatrix} U2 \\ U3 \\ U6 \\ U8 \\ U9 \\ U12 \end{Bmatrix} = \begin{bmatrix} H & 0 & 0 & -H & 0 & 0 \\ 0 & H & 0 & 0 & -H & 0 \\ 0 & 0 & H & 0 & 0 & -H \\ -H & 0 & 0 & H & 0 & 0 \\ 0 & -H & 0 & 0 & H & 0 \\ 0 & 0 & -H & 0 & 0 & H \end{bmatrix} \begin{Bmatrix} u2 \\ u3 \\ u6 \\ u8 \\ u9 \\ u12 \end{Bmatrix} \tag{2}$$

It should be noted that in the above derivation it has been assumed that development of either node in any of the above directions does not generate forces in any other direction. Considering now the other degrees of freedom u_4, u_5, u_{10} and u_{11}. These degrees of freedom are related to twisting (i.e. u_5 and u_{10}) and bending (i.e. u_4 and u_{11}) actions around the joint. The behaviour of the joint was described in detail in Hanizah[1] and Hanizah & Ibrahim[2].

For a unit twist at node i with all other displacements having the value zero, the forces induced are as follows (see Figure 3(b)):

- twisting moment at node i gives the stiffness term $k_{5,i}$
- bending moment about minor axis at node i gives the stiffness term $k_{4,i}$
- twisting moment at node j gives the stiffness term $k_{10,i}$
- bending moment about minor axis at node j gives the stiffness term $k_{11,i}$

For equilibrium, the terms corresponding to $k_{5,i}$ and $k_{11,i}$ have to be equal in magnitude but in the opposite direction. Let their values be denoted as M. A similar condition applies to $k_{4,i}$ and $k_{10,i}$, whose values are denoted as N. The above relationships are expressed in matrix form in equation (3).

$$\begin{Bmatrix} U4 \\ U5 \\ U10 \\ U11 \end{Bmatrix} = \begin{bmatrix} M & N & -M & -N \\ N & M & -N & -M \\ -M & -N & M & N \\ -N & -M & N & M \end{bmatrix} \begin{Bmatrix} u4 \\ u5 \\ u10 \\ u11 \end{Bmatrix} \tag{3}$$

Consequently the stiffness matrix $\boldsymbol{K}$ of the joint element in its local coordinate system can be assembled directly from the stiffness matrices of equations (1), (2) and (3) with appropriately reordering of rows and columns to suit a standard form for a system of equilibrium equations written in the form of $\boldsymbol{P} = \boldsymbol{Kd}$ where $\boldsymbol{P}$ is the vector of generalised nodal forces, $\boldsymbol{K}$ is the structure stiffness matrix, and $\boldsymbol{d}$ is the generalised nodal (translational and rotational) displacements. The resulting matrix equation is given in equation (4).

$$\begin{Bmatrix} U1 \\ U2 \\ U3 \\ U4 \\ U5 \\ U6 \\ U7 \\ U8 \\ U9 \\ U10 \\ U11 \\ U12 \end{Bmatrix} = \begin{bmatrix} H & 0 & 0 & 0 & 0 & 0 & -H & 0 & 0 & 0 & 0 & 0 \\ 0 & H & 0 & 0 & 0 & 0 & 0 & -H & 0 & 0 & 0 & 0 \\ 0 & 0 & H & 0 & 0 & 0 & 0 & 0 & -H & 0 & 0 & 0 \\ 0 & 0 & 0 & M & N & 0 & 0 & 0 & 0 & -M & -N & 0 \\ 0 & 0 & 0 & N & M & 0 & 0 & 0 & 0 & -N & -M & 0 \\ 0 & 0 & 0 & 0 & 0 & H & 0 & 0 & 0 & 0 & 0 & -H \\ -H & 0 & 0 & 0 & 0 & 0 & H & 0 & 0 & 0 & 0 & 0 \\ 0 & -H & 0 & 0 & 0 & 0 & 0 & H & 0 & 0 & 0 & 0 \\ 0 & 0 & -H & 0 & 0 & 0 & 0 & 0 & H & 0 & 0 & 0 \\ 0 & 0 & 0 & -M & -N & 0 & 0 & 0 & 0 & M & N & 0 \\ 0 & 0 & 0 & -N & -M & 0 & 0 & 0 & 0 & N & M & 0 \\ 0 & 0 & 0 & 0 & 0 & -H & 0 & 0 & 0 & 0 & 0 & H \end{bmatrix} \begin{Bmatrix} u1 \\ u2 \\ u3 \\ u4 \\ u5 \\ u6 \\ u7 \\ u8 \\ u9 \\ u10 \\ u11 \\ u12 \end{Bmatrix} \tag{4}$$

Since the sum of all coefficients in any one column of the element stiffness matrix in equation (4) is zero, these coefficients satisfy the overall equilibrium requirement of the element. It should be noted that in the formation of the above stiffness matrix it has been assumed that only the rotations about the x- and y-axes interact with each other, with all other translations and rotation about the z-axis being independent of each other.

3 Applications and limitations of the method

In order to apply the approximate method, values of H, M, and N for the joint element stiffness matrix (see equation (4)) need to be determined. The value of H has to be very high for the reason mentioned earlier and was therefore given a value of $1E12$. Since M and N govern the transmission of bending and twisting around the joint, their magnitudes are influenced by the section and material properties as well as the joint detail. M and N are however independent of member length.

The nonloaded member of the frame structure shown in Figure 1, having a box joint configuration, has been shown by Hanizah[1], using two-dimensional shell elements in three-dimensional space, to experience a twist in the clockwise direction. Hence, the submatrix expressed in equation (3) is suitable for structures with such a joint form. Whereas for structures having the other joint forms (i.e. stiffened and unstiffened mitre) whose corresponding twist has been shown to be in the reverse direction, the submatrix concerning the structural behaviour around the joint is as in equation (5).

$$\begin{Bmatrix} U4 \\ U5 \\ U10 \\ U11 \end{Bmatrix} = \begin{bmatrix} M & -N & -M & N \\ -N & M & N & -M \\ -M & N & M & -N \\ N & -M & -N & M \end{bmatrix} \begin{Bmatrix} u4 \\ u5 \\ u10 \\ u11 \end{Bmatrix} \tag{5}$$

The values of M and N needed for thin walled structures with right angled joint, having plain channel cross sections, with one of the three joint types, are given in Figures 4 to 6. These values were obtained using a trial and error method. Results obtained using the beam model with values of M and N taken from the plots in these figures, and with the modified torsional constant developed by Hanizah[1], when compared to those obtained from the shell element model, would have a maximum error of not more than 10 per cent. Note that the term λ in Figures 4, 5 and 6(a) is the beam parameter which is defined as the ratio of St. Venant's torsional stiffness to the warping stiffness.

For box and stiffened-mitre joints, their (M-N) values are function of the term as shown in Figures 4 and 5, respectively. Whereas for unstiffened-mitre joint, its (M-N) values vary according to the aspect ratio of the channel section as shown in Figure 6(b). This figure shows that the (M-N) values vary linearly with the given aspect ratio. Only aspect ratios between $^1/3$ and $^1/2$ are considered because this approximate analysis method has been tested only for aspect ratios within that range.

The proposed method was applied to a thin-walled structural frame of the form shown in Figure 1 with a stiffened-mitre joint and 90° axial-discontinuity. Table 1 compares results for torsion and bending moment at the section where the joint is connected to the nonloaded member obtained from the shell, and the beam element models. It can be seen that the error in torsion, and bending moment, value calculated is only 3 per cent.

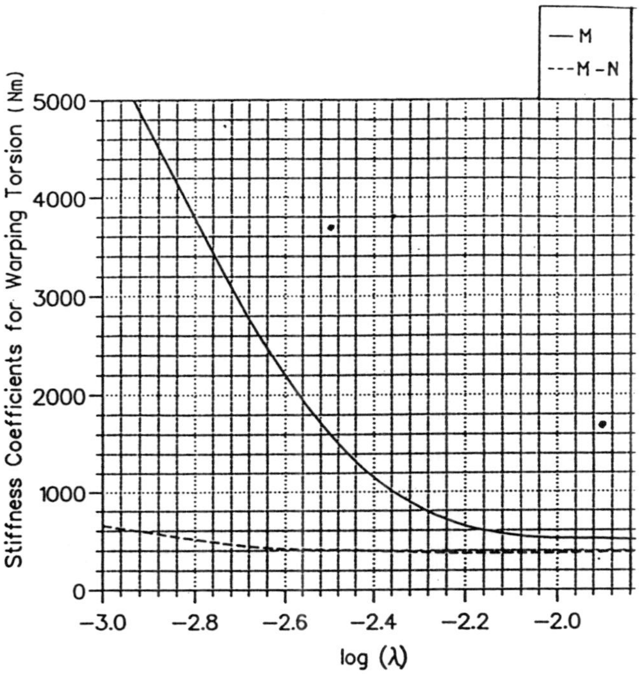

Figure 4: Values of M and $(M\text{-}N)$ for box joint.

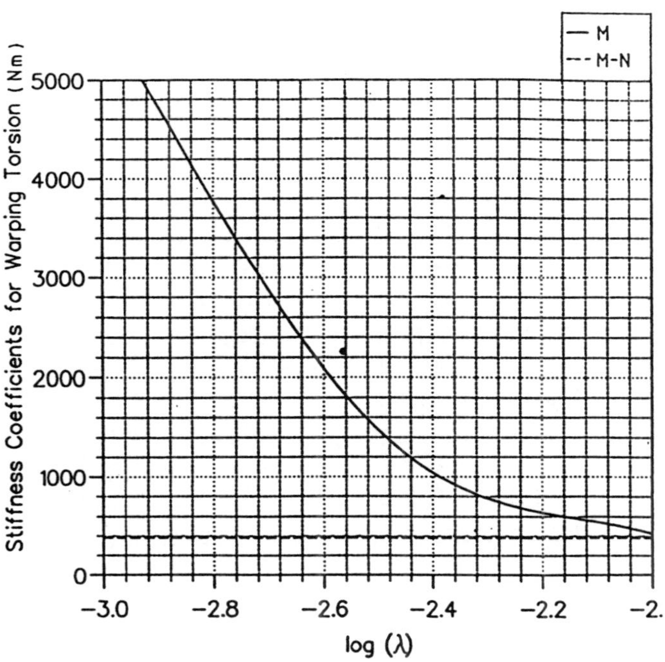

Figure 5: Values of M and $(M\text{-}N)$ for stiffened-mitre joint.

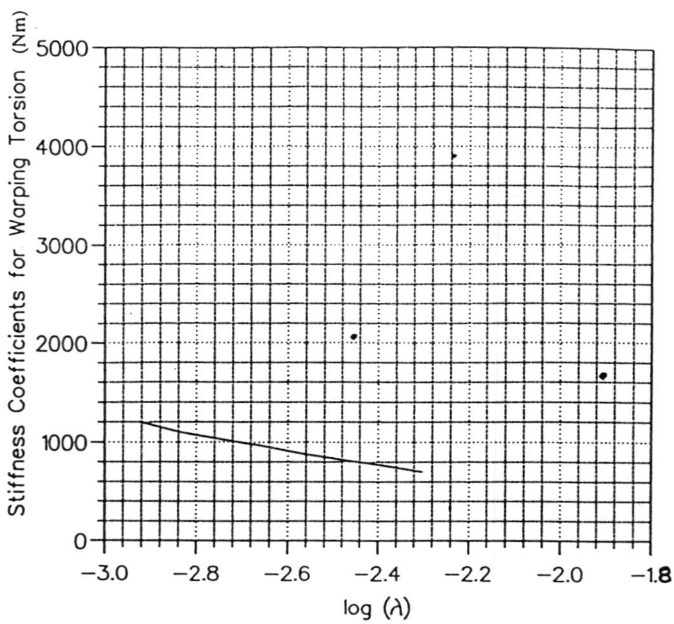

Figure 6(a): Values of M for unstiffened-mitre joint.

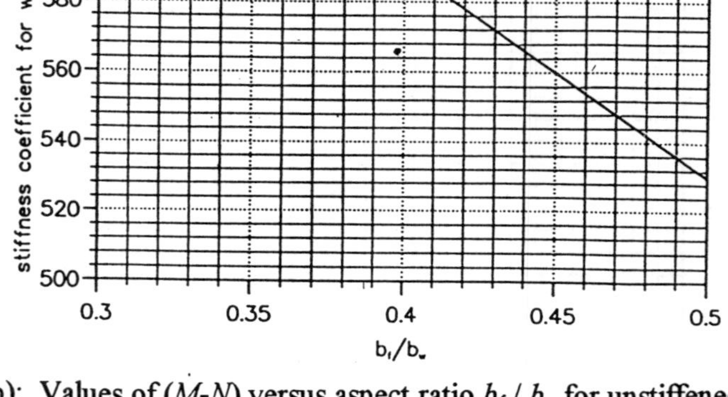

Figure 6(b): Values of (M-N) versus aspect ratio b_f / b_w for unstiffened-mitre joint.

Table 1. Comparison of torsional and bending moments for different finite element models

FE Model	Bending Moment (kNmm)	Torsional Moment (kNmm)
shell	519.3	159.2
beam	503.5	161.3
difference	3.0%	1.3%

4 Conclusion

A model for the approximate analysis of axially discontinuous thin-walled structures subjected to torsion has been developed. The model is applicable to plain channel section structural members with 90 degree axial discontinuity for three different joint types. The model is simple and can represent warping as well as bending and twisting behaviour. The model illustrates the behaviour of the thin walled structures with respect to torsion and moment about minor axis at the section where the joint is connected to the nonloaded member. The results obtained from the simple model were found to agree well with results obtained from detailed model analysed using full finite element method. Discrepancies between results from the beam model and those from a shell element model should not exceed 10 per cent.

References

1. Hanizah, A.H. *Behaviour of Axially Discontinuous Thin-walled Structures when Subjected to Torsion*, D.Phil Thesis, University of Sussex, UK, 1994.
2. Hanizah, A.H. & Ibrahim, A. Torsional behaviour of axially-discontinuous thin-walled frame structures, *PSSC 95, Proceedings of the 4th. Pacific Structural Steel Conf.*, (ed N.E. Shanmugam & Y.S. Choo), pp. 311 to 318, Singapore, 1995.
3. Morrell, P.J.B. The influence of joint detail on the torsional behaviour of thin-walled structures having an axial discontinuity, *Proceedings of the Int. Conf. On Thin-walled Structures*, University of Strathclyde, 1979.

The analysis of riveted aluminium/steel combined members under bending action

M. Macdonald,[a] G.T. Taylor,[b] D. Dunne,[b] J. Rhodes[c]
[a]*Department of Engineering, Glasgow Caledonian University, Cowcaddens Road, Glasgow G4 0BA, UK*
[b]*Department of Energy and Environmental Technology, Glasgow Caledonian University, Cowcaddens Road, Glasgow G4 0BA, UK*
[c]*Department of Mechanical Engineering, University of Strathclyde, Glasgow G1 1XQ, UK*

Abstract

Extruded aluminium hollow box sections with steel section inserts are riveted together to form a combined member increasingly used as the basis for the production of skeletal structures ranging from shelters on transport routes to larger covered walkways at airports, hospitals and industrial compounds. The paper reports on a comprehensive study using the finite element method supported by experimentation, of the stress and deflection behaviour of typical combined members. Analytical and experimental results are also presented.
The results of the study are presented and the paper concludes on the engineering value of the use of aluminium/steel combined members including the implications for the design of light aluminium/steel structures.

1 Introduction

The last decade has seen an increasing market for light skeletal structures constructed from extruded aluminium sections with steel section inserts riveted together. Southside Engineering Ltd in Glasgow, Scotland have achieved considerable success with their products. Fig 1.1.

The primary strength of these structures is derived from inserting a thin, cold formed steel, plain channel into an extruded thin walled, aluminium box-section, each component then riveted together at particular intervals along their length resulting in a combined member.

Many of the structures manufactured in Glasgow now require a structural integrity check to British Standards criteria. This requires consideration of snow and wind loading Ref. (1) and the determination of stress response of the structure Refs. (2),(3).

The design and stress analysis of complete structures can only be effectively carried out by finite element analysis and only beam elements can be used to represent the combined members purely on economical grounds. The steel and aluminium beam elements are 'tied' together in the FE model at nodes located at rivet points. The aluminium and polycarbonate panels are represented by plate elements in FE models. Initial research work was undertaken where a typical aluminium/steel combined member was investigated as a cantilever modelled by beam elements and then more rigorously by plate elements. This early work was reported in Ref. (4).

The research work has continued with a more detailed and comprehensive study using rigorous finite element models. The development of a 'best' finite element model representation of the physical problem is briefly described and the significant results of the study are presented in this paper. Complementary to this work an experimental investigation was carried out initially focusing on a series of load capacity tests, followed by a stress investigation of a combined member where strain gauge technology was applied. The results of this are also presented herein.

2 Load Capacity Tests of Combined Member.

2.1 Representation of Structural Member A typical real world structure was represented by an encastre/free end cantilever of length 1.26m. The load was applied at the free end of the cantilever, acting about the major axes of the sections. The cross-section of the structural member is shown in Fig 2.1 and it should be noted that tests were carried out on riveted and unriveted sections to establish if riveting had any effect on the members performance.

2.2 Load Capacity Test Equipment The design and manufacture of a fixture that would provide the conditions of a built-in end was undertaken. This was fabricated from steel plate and is shown in Fig 2.2 with the cantilever fixed in place by 4 bolt connections which pass through the structural member.

A Tinius-Olsen Universal Testing Machine, which has a large 90 tonnes load capacity, was used to test the cantilever to failure. This machine system consists of 3 main sub-systems, i.e. the machine crosshead, machine table and control console. The crosshead allows the application of the downward load and this is hydraulically and accurately controlled allowing loading ranges to be set. The movement of the crosshead as the load was applied and hence the free end deflection of the cantilever was measured by means of a deflectometer connected to a Bryan's A3 X-Y plotter. The applied load was measured by means of a sensitive load cell within the machine table which was also connected to the plotter and hence a graph of load-deflection for the cantilever was obtained which highlighted the point at which failure of the cantilever occurred and hence the load capacity of the structural member was obtained. A schematic diagram of the experimental set-up used in the investigation is shown in Fig 2.3.

2.3 Test Procedure and Results A general test procedure was devised and is outlined below and should be read in conjunction with Figs 2.1, 2.2 and 2.3:

a. I-section placed on machine table and secured by bolts and G-clamps.
b. 90° adjustable angle bracket placed on I-section and secured by 4 bolts.
c. Built-in end fixture placed on I-section and secured to 90° adjustable angle bracket using 4 bolts.
d. Cantilever placed in built-in end fixture and secured in position using adjustable plate and 4 bolts.
e. Load applicator placed and tightened in grips within the machine crosshead and positioned at the free end of the cantilever.
f. Deflectometer placed on machine table and positioned below the machine crosshead and set at zero from the control console.
g. Load range estimated at control console and the load application rate set.
h. The load-deflection scales set at the plotter.
i. The load is applied until failure of the cantilever occurs as highlighted at the built-in end and on the graphical output.

The above general procedure was adopted for several cantilever test configurations, i.e. for riveted and unriveted structural members, single member tests (i.e. steel channel and aluminium hollow box section) and loading about X-X and Y-Y axes. The results are shown in Table 1. Other tests were carried out on combined members where the steel channel inside the aluminium hollow box section was a tight fit and on others were there was a loose fit to find out the effect that the fit had on the load capacity of the combined member. Ref (5) gives a more comprehensive description of the above experiments.

2.4 Design Code Approach To obtain a load capacity for comparison with that from the load tests, BS5950, Part 5, Section 3 [Ref (2)] was referred to so that the load capacity of the cold formed steel channel could be determined. To determine a load capacity for the aluminum hollow box section, BS8118, Part 1,Chapter 4 [Ref (3)] was referred to.The results obtained from the codes are shown in Table 1. Further details of the above analyses can be found in Ref (5) along with other theoretical approaches to the problem.

2.5 Member Load Capacity The load capacity tests of the combined members, riveted and unriveted showed that the failure was as expected at the built-in end where a plastic hinge mechanism formed in the webs and flanges of the aluminium box section, and the steel channel failed due to twisting and warping effects. The difference in load capacity of riveted and unriveted members was found to be of the order of 5%. It was observed although not specifically quantified during the tests that riveted combined members where a snug fit existed, an increase in load capacity of the order of 5% was obtained over loose fitting members. The design codes produced load capacities that were conservative as they provide elastic solutions only with no allowance for plasticity. Ref.(5) shows other theoretical approaches to obtain a load capacity which compare much more favourably with the test results.

3 Finite Element Investigation of Combined Member.

3.1 Introduction The beam was modelled using PAFEC-FE operating on Hewlet Packard workstations where element type 44210 Ref (6) was used.

3.2 Convergence Study A convergence study was run for each of the primary members wherein a series of models with increasing mesh density was generated. From this study a suitable mesh was generated that would ensure accuracy of stresses and deflections.
Having generated a suitable mesh for the aluminium box section the process was repeated for the steel channel. The FE models were then combined to represent the combined member.

3.3 Development of 'Best' Model Representation of Engineering Problem The two members must satisfactorily interact at their interface, a suitable representation of the rivet connections must be made and restraint and load application have to be considered.
Loads were applied at the cantilever free end as discrete point loads and a planer restraint applied at the encastre end. A single row of rivets, centrally located on the web is modelled by tying corresponding steel and aluminium nodes together at rivet locations by using a 'generalised contraint'. This allows specification of relationships between degrees of freedom at node pairs and the selection of appropriate freedoms reflecting the restraining effect of the rivet.
The interaction of the steel and aluminium sections at their interface provided a major modelling problem. Early models used discrete generalised constraints to specify the interactive behaviour. This, however, does not allow the model to reflect the independent displacement behaviour of each member. Differential displacement occurs as the steel channel twists resulting in it apparently 'ghosting' through the box section. PAFEC provides a GAPS module which serves to identify the physical gap and/or contact between the aluminium box and the steel channel. A number of different models were generated and solutions examined. The generalised constraint was confirmed as the most economical representation of rivets and the GAPS option applied selectively across the interface was used.

3.4 Analysis of Cantilevered Combined Member The physical problems considered were cantilevers of length 1.26m and one of length 1.84m with one end fixed and one end free, the first being consistent with test specimens of Section 2. A load of 1kN was applied at the free end giving bending about the major axis.

A series of cantilever problems was modelled where the web depth of the steel channel was varied from 92mm (loose fit) to 94mm (snug fit). This series of solutions was repeated without rivets to establish the effect of rivets and a snug fit.

3.4.1 Load Capacity of Combined Members The maximum stress profiles on the inside surface of the aluminium and steel sections are shown in Figs 3.4.1.1 -3.4.1.3. The Von Mises equivalent stress is shown versus the distance around the section in the fixed end region of the cantilever where the highest stresses occur. The very nature of the problem gives rise to a complex stress problem. Combined members where the steel channel, either fitting snugly into the aluminium box section or not, will behave very similarly as long as they are riveted together. It appears that a snug fit does not change stress patterns around the sections and magnitudes of stress are not significantly different. This suggests that for riveted no significant load capacity gain is made on fitting the steel section snugly into the box section. Rivets being excluded from the combined beam does have a clear effect on the stress patterns around each section. The magnitudes of peak stress are not significantly different but show interesting variation. Stresses in the aluminium reduce on riveting and in the steel show an increase. The steel channel is obviously brought into action sooner in a riveted combination. The maximum peak stress in steel increases by about 5% although other peak stresses increase by a larger margin. The aluminium maximum peak stress reduces by about 12%. It could be interpreted from this that the load capacity of the combined beam on riveting and considering the aluminium member as the critical component, will increase by about 12%. Here a higher yield value of steel over aluminium is assumed.
Deflections of the combined member are no less complex than stresses as the sections translate and rotate in two dimensions. For a riveted combined member where the steel is snug with the box section, the cantilever tip deflections vary from those where the steel has a loose fit by less than 5%. Again the loose fit and the snug fit apparently making no significant difference to load capacity. The deflections for a non riveted section versus a riveted combined section (snug fit) shows the deflections to differ by 3-4%.

3.4.2 Correlation of Plate Finite Elements and Beam Elements Since the design of the light skeletal structures focused in this study is based on beam element analysis it was an objective of the study to compare beam element results with those from plate finite elements. Tables 2 and 3 present the results obtained for combined section beams only although analyses of individual members were also done. The results of the analysis of individual members suggests that beam stresses should be increased by about 25% for the aluminium box section and

about 11% for the steel channel. This however does not apply to members combined by riveting. Examination of the results suggests that no adjustment of beam stresses are necessary. This is based on the plate results for the 1.84m cantilever since those for the 1.26m cantilever seem to be in error.
An experimental investigation of a combined member using strain gauge technology was carried out and the results for a 1.84m cantilever gave the results which show that finite element and gauge results are in general agreement. A more comprehensively gauged combined beam, however, is required to be tested for a more detailed correlation.

4 Conclusions

A significant number of problem parameters are involved in the investigation of the combined member studied here. Good quality control of the experimental and finite element work is essential. The testing of the combined member highlighted this point as the specimens supplied by the manufacturer were not always of consistent dimensions. Members riveted together, either with the steel fitting snugly inside the box or not behave almost identically which was also verified by the tests. The effect of a more significant size of gap could be further investigated. Riveted and unriveted combined members do appear to behave differently. The tests and the FE results show an increase in load capacity for riveted members of about 5-10% is gained. Beam finite element stresses for individual members should be factored by about 25% for the aluminium box and 11% for the steel channel. Beam stresses for a combined member would appear to be close to these of a more rigorous plate element solution and a factor on stresses is not necessary. This point needs further investigation. The strain gauge results show general agreement with finite element results but a more detailed comprehensive experimental analysis is required for better correlation.

References

1. BS CP3 *Data for the Design of Buildings,*Chap V, Parts 1(1967) and 2(1972),*Loading:Dead,Imposed and Wind Loads.* BSI
2. BS5950,Part 5:1990 *Code of Practice for the Design of Cold Formed Sections,* BSI, London 1990.
3. BS8118,Part 1 *The Structural Use of Aluminium,* BSI, London 1990.
4. Taylor,G.T. The Design by Finite Elements of Riveted Aluminium/Steel Composite Structures,*Proceedings of the 7th World Congress on Finite Element Methods.*Monte Carlo, November 1993 (pp330-335).
5. Macdonald,M. *Bending of a Thin-Walled Combined Section Beam,* MSc Thesis, University of Strathclyde, Glasgow, September 1993.
6. PAFEC Software, Strelley Hall, Nottingham, England.

Figure 1.1: **Simple Bus Shelter Product** (Courtesy: Southside Engineering, Glasgow)

Table 1
Experimental Investigation and Design Code Results

Specimen Configuration	**Experiment, P(kN)**	**Design Code P(kN)**
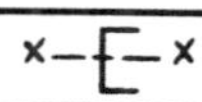	3.78	3.63 (BS5950)
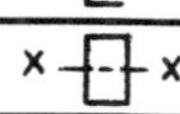	3.84	3.80 (BS8118)
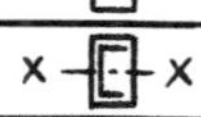 unriveted	10.25	7.43
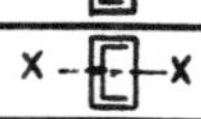 riveted	9.70	7.43
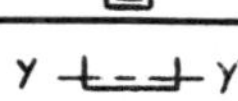	1.74	0.8
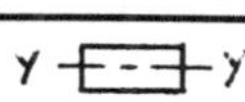	1.76	1.71
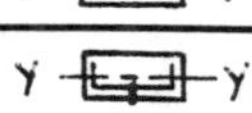 riveted	4.10	2.51

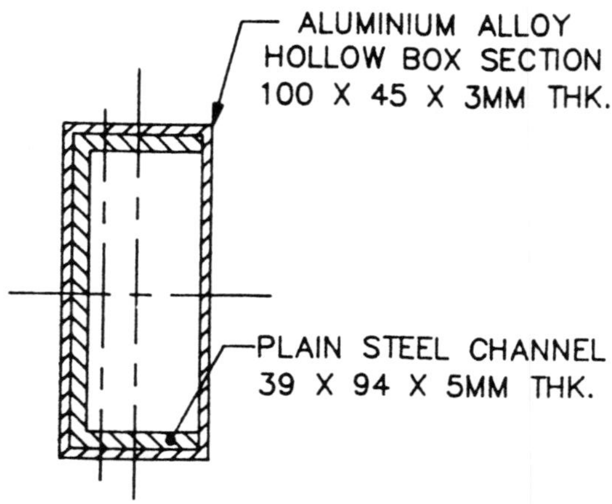

Figure 2.1: Structural Member Cross-Section.

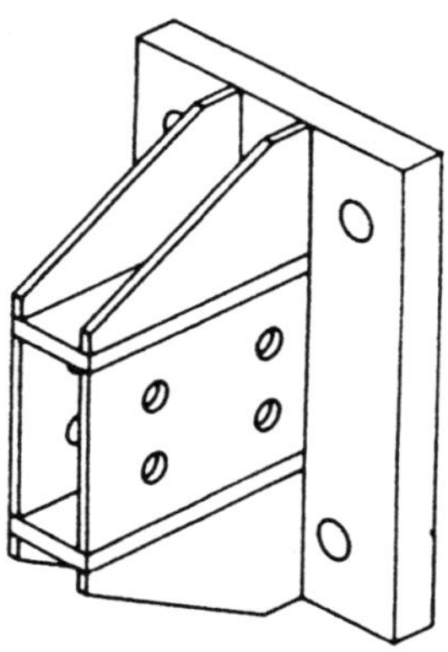

Figure 2.2: Built-in End Fixture.

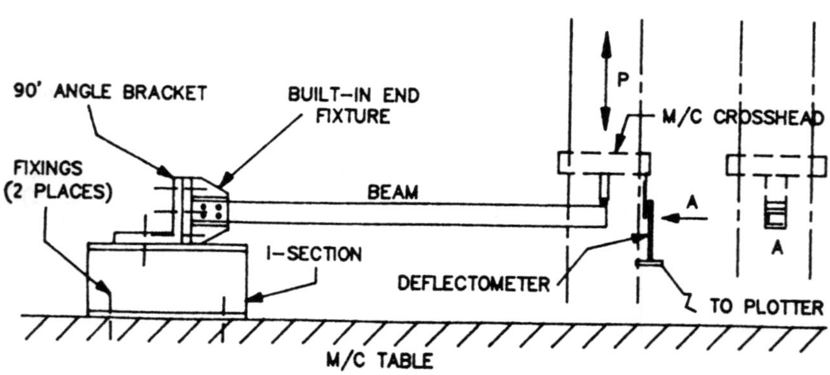

Figure 2.3: Experimental Investigation - Schematic Layout.

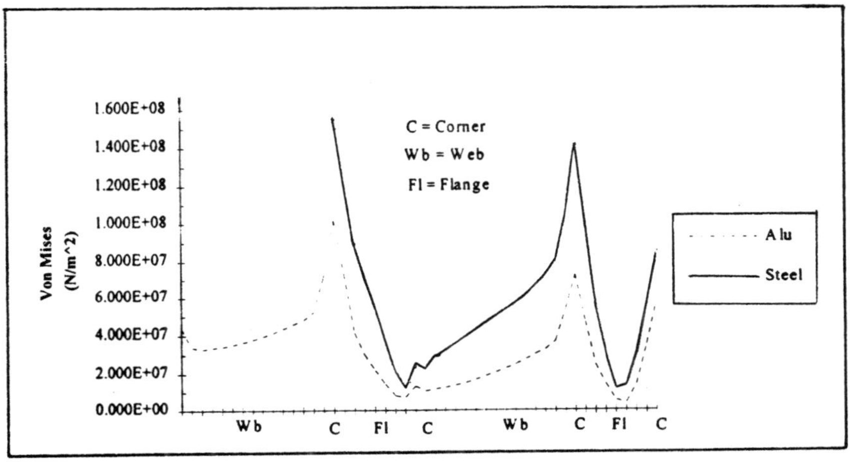

Figure 3.4.1.1 Steel Web 92mm (Loose Fit) - Riveted

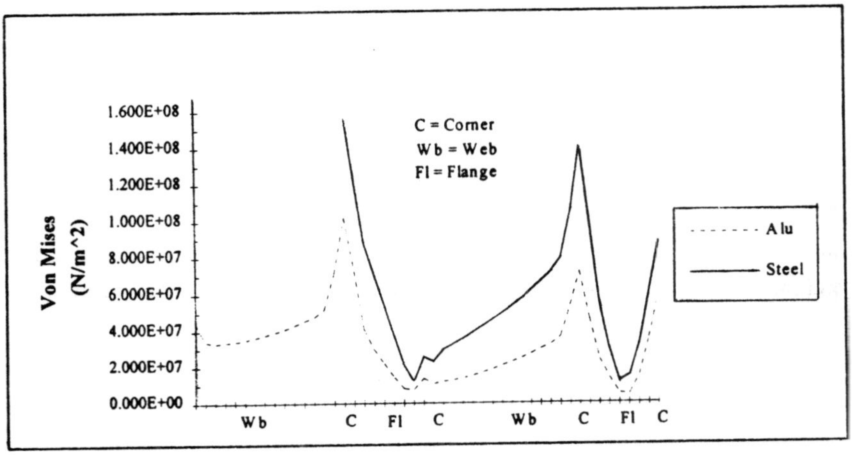

Figure 3.4.1.2 Steel Web 94mm (Snug Fit) - Riveted

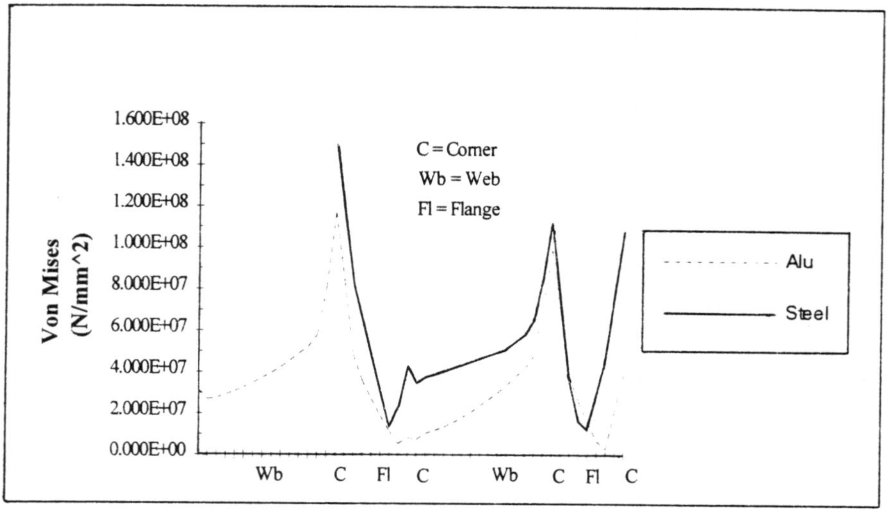

Figure 3.4.1.3 Steel Web 94mm (Snug Fit) - No Rivet

Table 2
Beam Element Results

Combined SectionLength	Free End Deflection	Max. Bending Stress(MN/m^2)
1.84m	4.9mm	Steel: ± 39.9; Al:± 15
1.26m	2.78mm	Steel: ± 51; Al: ± 19

Table 3
Plate Element Results

Combined Section Length	Free End Deflection (Averaged)	Max. Bending Stress(MN/m^2) (Averaged)
1.84m	St: 4.12mm; Al: 3.98mm	Steel: ± 40; Al:± 16
1.26m	St: 9.8mm; Al:10.29mm	Steel: ± 65; Al: ± 35

Effect of microshrinkages on the fatigue behaviours of cast stainless steel under rotating bending and reversed torsion

H. Nisitani, S. Tanaka, T. Teranishi
Faculty of Engineering, Kyushu Sangyo University,
2-3-1 Matsukadai, Higashi-ku, Fukuoka 813, Japan

Abstract

Considering the fatigue crack initiation site, all the materials can be classified into the following three types. That is, the first type is the material whose crack initiation site is a slipped part, the second type is the material whose crack initiation site is always a defect and the third type is the material whose crack initiation site is the slipped part or the defect depending on loading conditions. In this study, the characteristics of the third type is clarified through the successive observations of fatigue process in a stainless cast steel.

1 Introduction

The fatigue fracture of an annealed steel is usually caused by the propagation of a crack initiated from a slip band or a grain boundary on the specimen surface (Type Ⅰ -material in Table 1) [1,2]. On the other hand, in the case where the material includes the defects bigger than a definite size, the fatigue fracture is usually caused by the propagation of a crack initiated from the defect (Type Ⅱ -material)[2-4]. In some cases where the material includes defects, however, it depends on loading conditions whether the crack initiation site is the defect or not (Type Ⅲ -material).

In the former two cases where the crack initiation site is always a slipped part or always one of the defect, the fatigue processes have been studied until now [1-4]. The fatigue process in the last case is not clarified.

Table 1: Fatigue crack initiation site.

Type I material	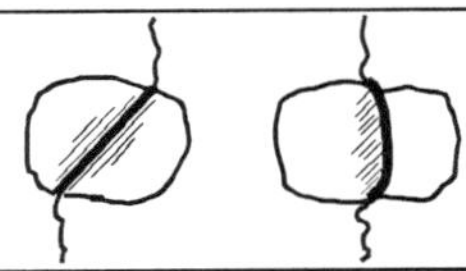	The crack initiation site is a slipped part
Type II material		The crack initiation site is a defect
Type III material	The crack initiation site depends on loading conditions	

In this paper, rotating bending and torsional fatigue tests were carried out on the specimens of a cast stainless steel, and the initiation and propagation behaviors of the microcrack were investigated through the successive observations by the plastic replica method. The observation of fracture surface was made using a SEM. The effects of microshrinkages on the fatigue behaviors under the above two types of loading were discussed based on the obtained results. Moreover, the characteristics of crack propagation were examined by the small-crack growth law [5] and the difference between the dependencies of Nf on the stress amplitude under the two types of loading was explained by the crack growth law.

2 Material, specimen and testing method

The material used was a cast stainless steel (SCS6). The chemical composition and the mechanical properties of the material are shown in Table 2. The microstructure of the material is shown in Figure 1. This material includes some microshrinkages (an arrow mark in Figure 1).

Figure 2 shows the dimensions of specimens in two types of loading. The stress concentration factors of the specimens are 1.06 under bending and 1.03 under torsion [6]. Before testing, all the specimens were polished by emery paper, and then were electro-polished to the depth of about 20 μ m to eliminate the work hardened layers. The Ono-type rotating bending fatigue testing machine (the capacity : 98Nm and the rate of stress repetitions : about 40Hz) and a torsional fatigue testing machine TB-10 (the capacity : 98Nm and the rate of stress repetitions : about 33Hz) were used. The value of stress in this paper means the maximum stress at the minimum cross section of the specimen.

In order to observe the fatigue process and to measure the crack length, the testing machines were stopped at the given cycle intervals during stress repetitions, plastic replicas were taken from the specimen surface, a gold was coated on them in vacuum, and finally they were observed by a metallurgical

Table 2: Chemical composition and mechanical properties.

Chemical composition (wt%)								Mechanical propaties		
C	Si	Mn	P	S	Ni	Cr	Mo	$\sigma_{0.2}$	σ_B	ϕ
								MPa		%
0.056	0.48	0.67	0.03	0.004	3.91	11.8	0.27	602	824	64

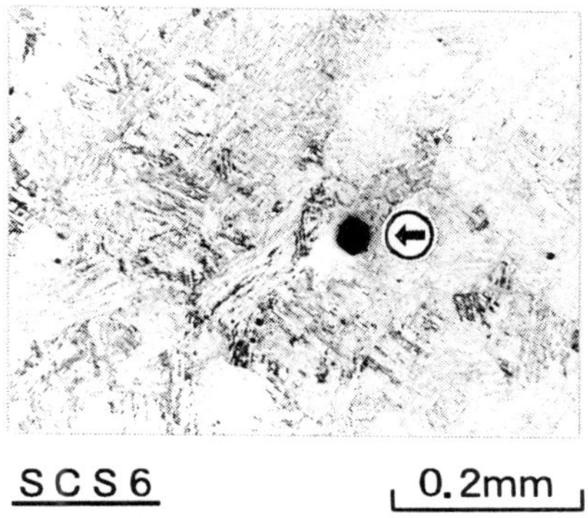

Figure 1: Microstructure of the material.

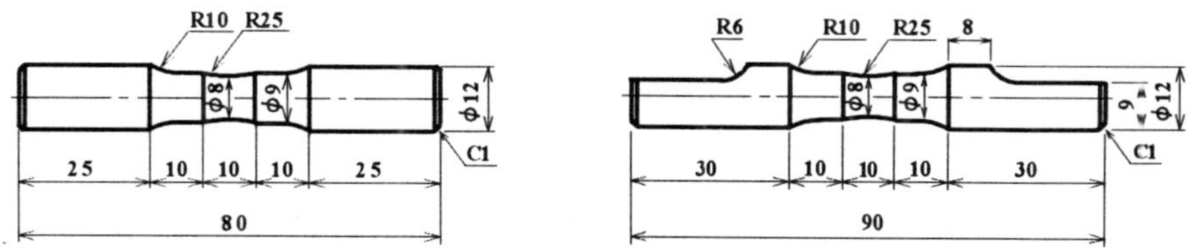

(a) Rotating bending. (b) Reversed torsion.

Figure 2: Fatigue specimens.

microscope (×400). The crack length means the projected crack length in the direction of the axis in rotating bending, and the projected crack length in the direction of 45° against the axis in reversed torsion. Moreover, the observation of fracture surface was made using a SEM.

3 Experimental results and discussion

3.1 S-N curves and fatigue process

Figure 3 shows the S-N curves for rotating bending and reversed torsion.

The index numbers are kb=8.2 (under rotating bending) and kt=8.9 (under reversed torsion), respectively. Therefore, the following relations are obtained.

$$1/N_f \propto \sigma^{8.2} \tag{1}$$

$$1/N_f \propto \tau^{8.9} \tag{2}$$

The fatigue limits are σ_w=360MPa (under rotating bending) and τ_w =278MPa (under reversed torsion), respectively. The ratio of fatigue limits τ_w/σ_w is 0.77. This ratio in the cast stainless steel (τ_w/σ_w=0.77) is larger than that of a defect-free isotropic carbon steel (τ_w/σ_w=0.65) [7]. It is due to the following reason : The crack initiation sites under two types of loading are different in this material. That is, only the fatigue limit in rotating bending is decreased by the existence of defects.

In the case of rotating bending, if this material does not include the defects, the fatigue limit is about 420MPa which is estimated by the linear notch mechanics [8]. The ratio of fatigue limits τ_w/σ_w using σ_w=420MPa is about 0.63. This value is nearly equal to that of a defect-free isotropic carbon steel (τ_w/σ_w=0.65) [7].

The reason why crack initiation sites depend on loading conditions is as follows : Generally, the effect of a defect on the fatigue limit in reversed torsion is smaller than the effect of the same defect in rotating bending [4]. In the case where the material includes the defects smaller than a definite size, both the fatigue limits of two types of loading are not influenced by the defects [3]. Considering the circumstances mentioned above, it is concluded that a defect of a definite size becomes the origin of fracture in rotating bending but does not in reversed torsion.

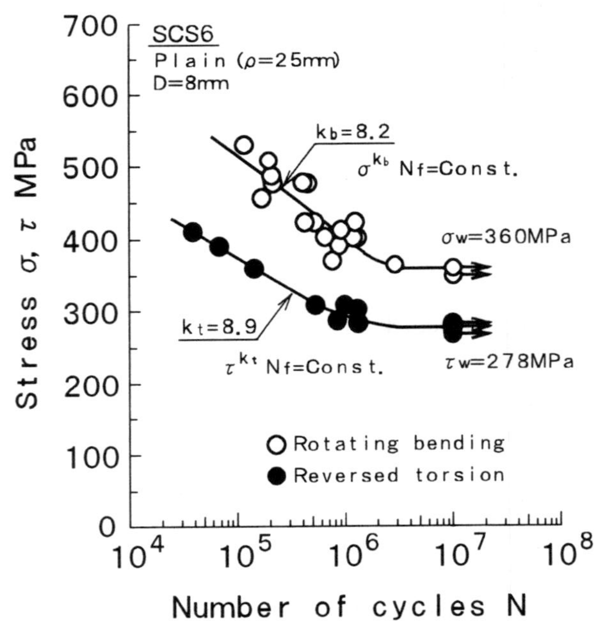

Figure 3: S-N curves.

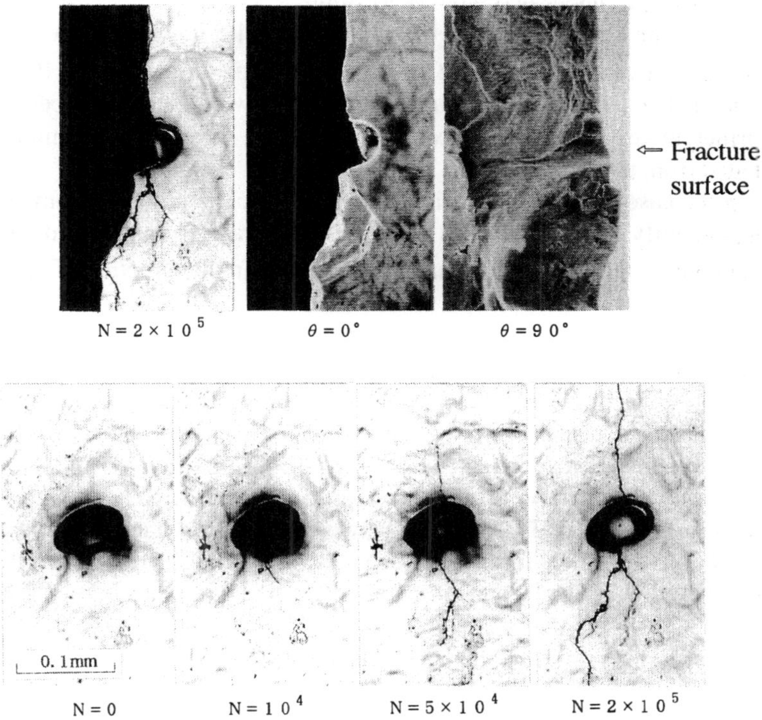

(a) Rotating bending (SCS6 σ =400MPa Nf=4.19×10^5 ⟷ Axis).

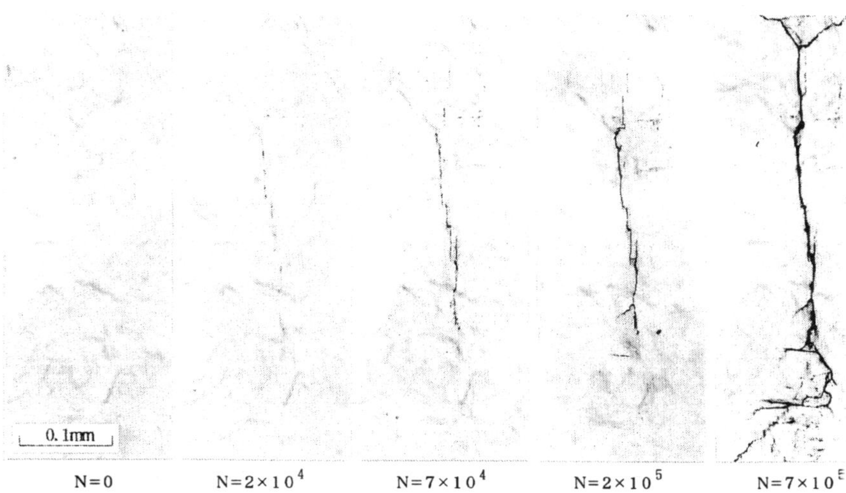

(b) Reversed torsion (SCS6 τ =295MPa Nf=1.30×10^6 ⟷ Axis).

Figure 4: Successive observations of specimen surfaces (fracture surface in bending is also shown).

Figure 4 shows the successive observations of specimen surface and fracture surface under two types of loading.

When the stress is higher than the fatigue limit, in the case of rotating bending, a microcrack appears from the defect (see the case of N=10^4 cycles in Fig.4(a)). The initiated microcrack propagates toward a direction perpendicular to principal stress and finally causes fracture. In most case, the microcrack is initiated from a microshrinkage.

In the case of reversed torsion, the microcrack is initiated from slip bands independently of defects (Fig.(b)). The microcrack propagates at first in the direction parallel to τ_{max} but finally propagates in the direction of 45° against the axis.

3.2 Characteristics of crack propagation

Figure 5 shows the relations between ln *l* and the relative number of cycles N/Nf. The crack propagation lives in two types of loading are controlled mainly by the life in which a crack propagates from 0.2 to 5mm; these lives are about 0.8Nf, indicating that evaluation of microcrack propagation process is also important for the cast stainless steel. The relations between ln *l* and N/Nf are nearly linear. This means that growth rate d*l*/dN approximately proportional to crack length *l*. From Fig.5, the growth laws of a small crack in the two types of loading are obtained

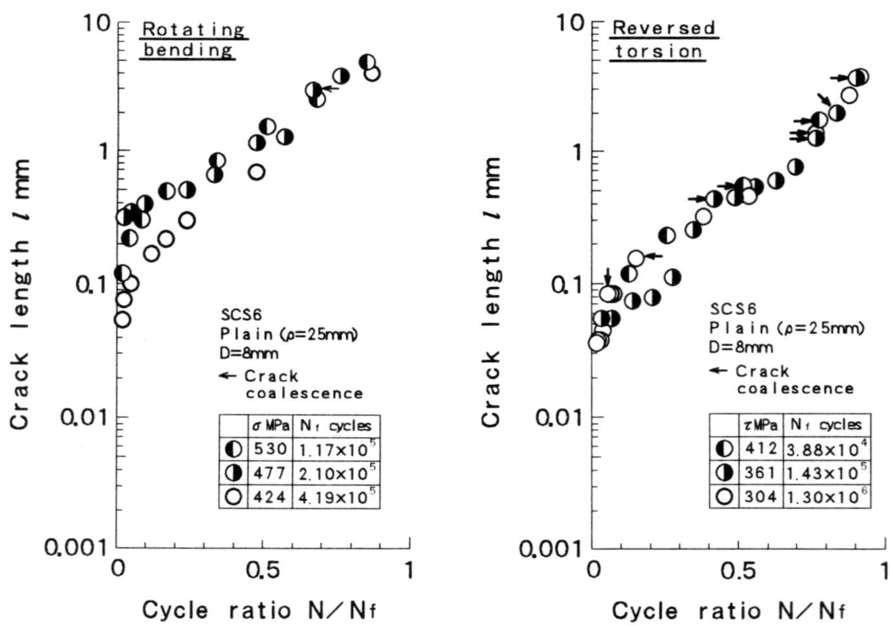

(a) Rotating bending. (b) Reversed torsion.

Figure 5: Relation between crack length and number of cycles normalized by the fatigue limit.

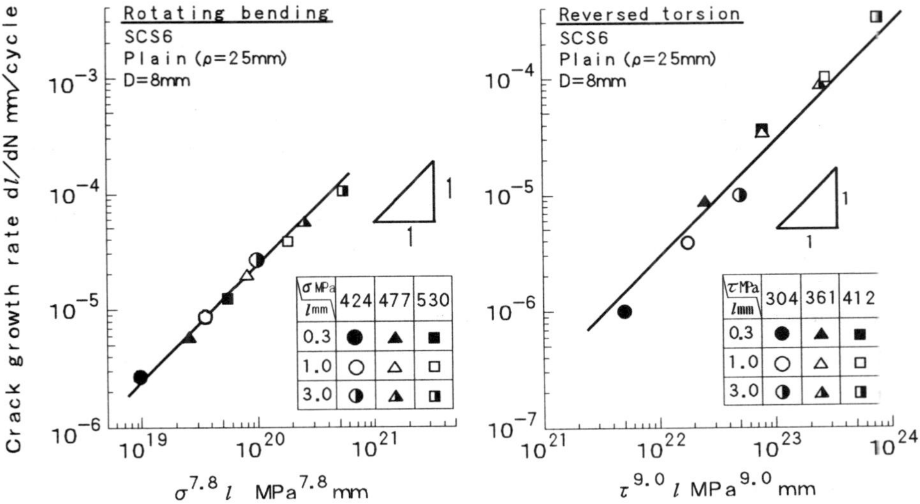

(a) Rotating bending. (b) Reversed torsion.

Figure 6: relation crack growth rate and $\sigma^{7.8}l$ or $\tau^{9.0}l$.

as follows :

$$dl/dN \propto \sigma^{7.8}l \quad (3)$$

$$dl/dN \propto \tau^{9.0}l \quad (4)$$

Figure 6 shows the relations between dl/dN and $\sigma^{7.8}l$, $\tau^{9.0}l$. As seen from this figure, dl/dN of small crack are determined uniquely by $\sigma^{7.8}l$ in rotating bending and by $\tau^{9.0}l$ in reversed torsion.

The index number of stress n_b in rotating bending is smaller than that n_t in reversed torsion (n_b=7.8 < n_t=9.0). This is due to the difference in relative stress levels ($\sigma / \sigma_{0.2}$ and $\tau / \tau_{0.2}$) of both loadings. $\sigma / \sigma_{0.2}$ is about 0.67～0.84 and $\tau / \tau_{0.2}$ is about 0.96～1.30, in the present experiment. When the relative stress is larger than 1, the reversible plastic zone size at a crack tip will increases greatly by the small amount of the increase in stress [9]. This means that the index number of stress is larger under high stress level than under low stress level.

Because the total fatigue lives of this material are occupied mostly by the crack propagation lives, the index number of stress in σ-Nf or τ-Nf relation (Figure 3) is nearly equal to that in the above crack growth law, respectively. Therefore, the difference between the dependencies of Nf on the stress amplitude in the two types of loading can be explained through the above crack growth laws [9].

4 Conclusions

Rotating bending and torsional fatigue tests were carried out on the plain specimens of a cast stainless steel. The initiation and propagation behaviors of microcracks were investigated using the plastic replica method. The main results are summarized as follows:

(1) In the present material, the critical cracks were initiated at microshrinkages in rotating bending, and the critical cracks in reversed torsion were initiated at slip bands.

(2) The ratio of fatigue limits in the cast stainless steel (τ_w/σ_w=0.77) is larger than that of a defect-free isotropic carbon steel (τ_w/σ_w=0.65) [7], because of the fact (1).

(3) The crack growth laws in rotating bending and reversed torsion are expressed by $dl/dN \propto \sigma^{7.8} l$ and $dl/dN \propto \tau^{9.0} l$, respectively. The difference between the dependencies of Nf on the stress amplitude under the two types of loading can be explained by the above crack growth law.

Reference

1.Nisitani, H., *Theory of Fatigue Strength* (in Japanese), Oumu-sha , 1985.

2.Murakami, Y., *Metal Fatigue:Effects of Small Defects and Nonmetallic Inclusions* (in Japanese), Yokendo Ltd,1993.

3.Murakami, Y. & Endo, M. The Torsional Fatigue Strength of Annealed 0.46% Carbon Steel Specimens with Micro-Hole the Diameter of Which is from 40 μ m～500 μ m (Comparison with Bending Fatigue Strength), *Japan Soc. Mech. Engrs* (in Japanese), 1981, **47**, Series A, 249-256.

4.Nisitani, H. & Kawano, T. Correlation between the Fatigue Limit of a Material with Defects and Its Non-Propagating Crack : Some Considerations Based on the Bending or Torsional Fatigue of the Specimen with a Diametrical Hole, *Japan Soc. Mech. Engrs* (in Japanese), 1971, **37**, 1492-1496.

5.Nisitani, H. A Consideration Concerning Fatigue Crack Growth Laws, *Preprints Japan Soc. Mech. Engrs* (in Japanese), 1979, No.790-13, 63-65. : Nisitani, H. Unifying Treatment of Fatigue Crack Growth Laws in Small, Large and Non-Propagating Cracks, *Mechanics of Fatigue-AMD*, 1981, 47 (Ed.Mura T.) ASME, 151.

6.Nisitani, H. & Noda, N. Stress Concentration of a Fatigue Test Specimen Having a Partially-Circular Circumferential Notch under Torsion, Tension or Bending, *Japan Soc. Mech. Engrs* (in Japanese), 1985, **51** Series A, 775-783.

7.Nisitani, H., Fukuda, T. & Fujimura, K. Comparison of Fatigue Limits in Rotating Bending and Torsional Fatigue of Rolled Steel Plain Specimens Subjected to Diffusion Annealing, *Japan Soc. Mech. Engrs* (in Japanese), 1994, **60** Series A, 2175-2181.

8.Nisitani, H., Tanaka, S., Teranishi, T., Tanaka, Y. & Iwadate, T. Effect of Notch and Grain Size on the Fatigue Strength of Cast Stainless Steel, *Japan Soc. Mech. Engrs* (in Japanese), 1995, **61** Series A, 1725-1729.

9.Nisitani, H. & Fukuda, T. Relation Between S-N Curve and Small-Crack Growth Law in Torsional Fatigue of Annealed Carbon Steel Plain Specimen, *Japan Soc. Mech. Engrs* (in Japanese), 1989, **55** Series A, 710-717.

Load adjustment in real-time processing systems

W. Tarng,[a] T.-H. Lin[b]
[a]*Department of Mathematics and Science Education, National Hsin-Chu Teachers College, Hsin-Chu, Taiwan, Republic of China*
[b]*Department of Electrical and Computer Engineering, State University of New York at Buffalo, Buffalo, NY 14260, USA*

Abstract

Modern process control systems are usually incorporated with high-speed computers to satisfy the requirements of real-time response and growing complexity. In such systems, the whole mission is divided into a number of tasks which are scheduled to meet their real-time constraints. In this paper, an adaptive scheduling algorithm based on current task queue length is proposed to handle the problem of transient overload in real-time processing systems due to the change of task characteristics. Our simulation result shows that the queue-length-based (QLB) algorithm can achieve better performance than traditional static scheduling algorithms.

1 Introduction

The major difference between real-time systems and general-purpose computer systems is that the correctness of a real-time system depends on not only the logical results of computation but also the time when these results are produced. Modern process control systems are usually incorporated with high-speed computers to satisfy the requirements of real-time response and growing complexity. In such systems, the whole processes are divided into a number of tasks and then scheduled to meet their real-time constraints (or deadlines). A deadline is *hard* if missing the deadline may result in catastrophic results, and is *soft* if missing the deadline only cause some degree of performance degradation.

Applications of hard real-time systems like airplane or nuclear reactor control systems are life-critical and therefore tasks must be scheduled carefully to meet their deadlines. When designing a real-time system, tasks are often categorized as *periodic* and *aperiodic* to simplify the scheduling mechanism and to guarantee the hard deadlines. Periodic tasks are invoked repeatedly by fixed time intervals while aperiodic tasks may arrive randomly. A general assumption for real-time scheduling is that periodic tasks must meet the hard deadlines equal to their periods and aperiodic tasks have soft deadlines to be satisfied. Because only periodic tasks have hard deadlines, they will be scheduled first and thus given higher priority over aperiodic tasks. However, the response times of aperiodic tasks can also be reduced if they are scheduled by the time slots allocated to a high-priority periodic server [1,2].

In [3], the period and service time of periodic server can be determined according to the deterministic characteristics of periodic tasks (computation times and periods) and the stochastic characteristics of aperiodic tasks (inter-arrival-time and computation-time distributions) to further improve the system performance. In practice, the periods and computation times of periodic tasks or even tasks themselves may change during different stages of system operation, for example, the take-off, cruise, and landing stages in a flight control system. Therefore, static scheduling algorithms can not fulfill the requirements of such systems in spite of their simplicity and low scheduling overhead. An adaptive scheduling algorithm based on dynamic period transformation mechanism [4] is proposed to handle the problem of transient overload due to the change of task characteristics in real-time systems. The adaptive algorithm defines a number of static schedules optimized for different load levels and switches among these schedules during run time to improve system performance. A real-time mission is first divided into several operational phases according to their functional requirements and estimated load conditions. In each phase, a static algorithm is used to schedule periodic tasks including the periodic server of which the service rate is determined based on the arrival rates of aperiodic tasks. To accommodate the change of load conditions, modification of task periods are performed when the system moves from one phase to another.

Using the gradient method, the optimal periods in each phase can be derived *a priori* from the deterministic characteristics of periodic tasks and stochastic characteristics of aperiodic tasks. However, in certain environments the stochastic characteristics of aperiodic tasks can not be estimated clearly such that it is very difficult to determine the service rate of the periodic server. One possible solution is to measure these parameters on-line and perform period transformation accordingly. In this paper, a queue-length-based (QLB) version of dynamic period transformation algorithm is proposed for dynamic load adjustment in real-time systems, and it is suitable for the applications where the stochastic characteristics of aperiodic tasks are not available in advance. The operation of the QLB algorithm is similar to the original dynamic period transformation mechanism except that the phases are specified by the number of aperiodic tasks waiting for execution. Usually, the periodic server in a phase with more waiting aperiodic tasks has higher service rate than it does in a phase with less waiting aperiodic task. The QLB algorithm monitors the number of pending aperiodic tasks whenever the CPU is idle to decided whether a period transformation is necessary. If the queue length does not match the defined status of current phase, a period transformation is performed immediately to adapt to the load change by increasing or decreasing the service rate of the periodic server. Our simulation result shows that the QLB algorithm can provide better performance than traditional static scheduling algorithms.

The remainder of this paper is organized as follows. In Section 2, the dynamic period transformation mechanism is reviewed first and then the QLB algorithm is presented. The performance measure to determine the optimal periods of periodic tasks and the service rate for aperiodic tasks in the QLB algorithm is described in Section 3. A queueing model is created to evaluated the performance of the QLB algorithm in Section 4. Section 5 is the conclusion of this paper.

2 The QLB Dynamic Period Transformation Algorithm

The dynamic period transformation mechanism [4] can achieve load adjustment by dividing a real-time mission into several operational phases, where the periods of periodic tasks and service rate for aperiodic tasks are determined by the gradient method in advance to optimize system performance. During system operation, a dynamic period transformation is performed when the mission moves from one phase to another to change the periods of periodic tasks and service rate for aperiodic tasks according to current load condition. In each phase, the rate-monotonic (RM) algorithm [5], the optimal static scheduling algorithm, is used to schedule periodic tasks including the periodic server. The RM algorithm schedules periodic tasks according their priorities, which are pre-determined according to the periods of tasks. The rule for assigning priorities is that tasks with shorter periods are given higher priorities. Usually, the period of the server is the shortest in order to receive the highest priority to provide better responsiveness for aperiodic tasks.

Because the computation resources such as CPU time and I/O channels are fixed, increasing the service rate of aperiodic tasks (by decreasing the period of server) will require increasing the periods of other tasks as compromise. An important part of the dynamic period transformation algorithm is to find the optimal periods in each phase, which are determined by the gradient method based on a pre-defined performance measure. For safety reason, the satisfaction of deadlines for periodic tasks in each phase must be examined off-line by the static checking algorithm proposed in [3]. Also, it is better to perform the dynamic period transformations when the CPU is idle, because idle times can separate the worst-case conditions in two different phases. Usually, the CPU idle times occur frequently and therefore the delay time for the request of period transformation is often very short.

Using the gradient method, the optimal periods of real-time tasks in each phase can be determined to improve the system performance. However, not all real-time missions can be clearly divided into a number of operational phases or provide stochastic characteristics of aperiodic tasks in each phase before their operation. In that case, it is very difficult to define the operational phases or to determine the service rate of periodic server. Therefore, we propose the QLB algorithm which performs period transformation based on the current workload of aperiodic tasks, i.e., the number of waiting aperiodic tasks in the task queue. In this approach, only the computation-time distribution of aperiodic tasks is required for the determination of task periods in each phase. For example, we can create a number of $n+1$ phases for a real-time mission and each phase is defined by the number of pending aperiodic tasks. The mission will stay in Phase j, $0 \leq j \leq n-1$, if the number of waiting aperiodic tasks equals j, and in Phase n if there are at least n aperiodic tasks pending. Usually, the service rate of the server is higher in the phase where the queue of waiting aperiodic tasks is longer. In Phase 0, because no aperiodic tasks are waiting for execution, the server can be suspended to reduce the overhead and CPU time utilization. A wider range of load conditions (queue length) can be covered by increasing the number of phases or states in each phase. The latter approach has the advantage of lower overhead since period transformations are performed less frequently, but it also has the disadvantage of longer reaction time to load changes. The operation of the QLB algorithm is very simple. It obtains the number of waiting aperiodic tasks at the beginning of idle times to decide whether a period transformation is necessary. If the current queue length does not match the specified condition of current phase, a period transformation is performed to adapt to the load change by increasing or decreasing the service rate of the server. The following example shows the operation of the QLB algorithm.

Example 1:

Three phases are created for a real-time mission by the QLB algorithm to schedule three periodic tasks and aperiodic tasks in Figure 1. The legitimate queue length of aperiodic tasks in each phase is defined as zero in Phase 0, one in Phase 1, and two or more in Phase 2, respectively. As described before, τ_1 is used as the server for aperiodic tasks. The mission starts with Phase 0 at $t=0$. Because no aperiodic tasks have arrived, only τ_2 and τ_3 are invoked and τ_2 is executed first. An aperiodic task τ_{a1} arrives at $t=3.5$, which requires five units of computation time. The first idle time appears at $t=5$ after the first iteration of τ_3 is completed. Because the number of aperiodic tasks has been changed to one, a period transformation is performed to move the system state from Phase 0 to Phase 1. The periodic server τ_1 with one unit of service time is invoked immediately to serve τ_{a1}. The second idle time occurs at $t=8$ after the second iteration of τ_3 is completed. Since the number of aperiodic task is not changed, the system stays in Phase 1 and the idle time can also be used to execute τ_{a1}. Another aperiodic task τ_{a2} arrives at $t=10.5$ with computation time equal to two.

The third idle time occurs at $t=11$ after the third iteration of τ_2 is completed. Since the number of pending aperiodic tasks has now become two, the system switches from Phase 1 to Phase 2 to increase the service rate of τ_1.

Task	Phase 0			Phase 1			Phase 2		
	C_i	T_i	Priority	C_i	T_i	Priority	C_i	T_i	Priority
τ_1	Suspended			1.0	4.0	1st	1.0	2.0	1st
τ_2	1.5	3.0	1st	1.5	5.5	2nd	1.5	7.5	2nd
τ_3	2.0	5.5	2nd	2.0	6.0	3rd	2.0	8.0	3rd

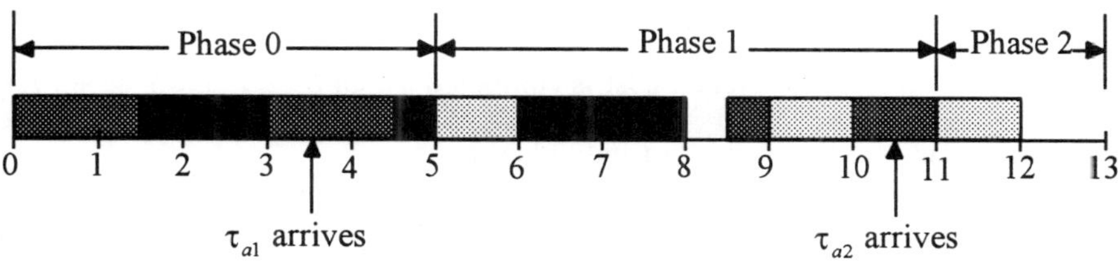

Figure 1: The operation of the QLB algorithm.

3 The Optimal Periods of The QLB algorithm

An important part for the development of the QLB algorithm is to find the optimal periods in each operational phase. The definition of operational phases in the QLB algorithm is based on current queue length of aperiodic tasks. When the computation-time distribution of aperiodic tasks and the service rate of periodic server are given, the mean response time of aperiodic tasks can be estimated [3]. In many applications, the performances of real-time systems such as accuracy and stability are strongly related to the periods or response times of tasks. Usually, a shorter period or response time can produce better results or higher stability. However, it is not possible to schedule all tasks with very short periods since the computation resources are fixed. A systematic approach for improving system performance is to determine the periods of tasks according to their importance and timing constraints. We develop the following performance measure to determine the optimal periods of tasks in the QLB algorithm based on the gradient method.

Our performance measure is a vector comprising a number of components to evaluate each task, where the performance of a periodic tasks is evaluated by its period, and the performance of aperiodic tasks is evaluated by their mean response time. We define the longest period that a periodic tasks can be scheduled to satisfy the worst-case requirements as its *boundary period*. To determine the optimal periods, the performance vector has to be converted into a scalar by the mapping of the performance vector to a weight vector specifying the importance of each individual task. For a real-time system containing m

periodic tasks $\tau_1,\ldots,\tau_m$ and the aperiodic task τ_a, the system cost is determined based on the following assumptions.

1. The computation time C_i of periodic task τ_i is a constant and the cost of running τ_i is a non-decreasing function Ψ_ι of its period T_i.
2. The computation times of aperiodic task τ_a are exponentially distributed with mean $1/\mu$, and the cost of running τ_a is a non-decreasing function Ψ_a of its mean response time.

According to the above assumptions, the mean response times of aperiodic tasks can be approximated by $T_1/\mu C_1$ because T_1 is usually very small, and the cost of running aperiodic tasks can then be expressed as the cost of running the server, which is also a non-decreasing function of its period T_1 because C_1 and μ are all constant. Finally, the system cost is computed by $\Phi(T_1,\ldots,T_m)=\sum_{i=1}^{m} w_i\Psi_\iota(T_i)$, where $w_1,\ldots,w_m$ are the weights of individual tasks.

The optimization process is performed to minimize the system cost $\Phi(T_1,\ldots T_m)$ using the gradient method subject to the timing constraints of periodic tasks. At the beginning of optimization process, the periods of all tasks are set to their boundary values $\overline{T_1},\ldots,\overline{T_m}$. Then, the minimization of system cost is performed step by step by moving the period vector $(T_1,\ldots,T_m)$ towards the opposite direction of the gradient vector $(\frac{\partial\Phi}{\partial T_1},\ldots,\frac{\partial\Phi}{\partial T_m})$. In each step, the FC algorithm [3] is used to check whether all periodic tasks can satisfy their timing constraints. If the deadlines of all periodic tasks are satisfied, we compute the gradient vector again and repeat the same procedure. Otherwise, instead of decreasing all periods at the same time, we start to reduce the periods one at a time to further minimize the system cost. Similarly, the FC algorithm has to be applied for the checking of timing constraints. The whole procedure stops when all periods can not be reduced any further. Since the optimal periods in each phase are determined by the gradient method, the QLB algorithm can always provide the best results no matter how the arrivals of aperiodic tasks change.

4 Performance Evaluation

In this section, an *M/M/n* queueing model is created to analyze the performance of the QLB algorithm. We will compare the performance of the QLB algorithm with that of the RM algorithm under different load conditions. For a mission containing *n+1* phases, the *M/M/n* queueing model can be used to compute the probability of staying in each phase and the long-term system performance can then be evaluated by the mean of system costs in all created phases. For simplicity, the *n+1* phases of the QLB algorithm are defined in such a way that the legitimate number of waiting aperiodic tasks in Phase *i* is equal to *i* when $i<n$. Also, the number of legitimate waiting aperiodic tasks can be greater than or equal to *n* in Phase *n*.

According to the operation of the QLB algorithm, any discrepancy in the queue length of aperiodic tasks will result in a period transformation to switch the system from current phase to one of the neighboring phases. The state transition diagram of the simplified model is shown in Figure 2, where λ is the arrival rate of aperiodic tasks and $B_i\mu$ is the service rate of the periodic tasks in Phase i. The service rate is computed by $B_i\mu = {}^{C_1\mu}/_{T_1^i}$ where T_1^i is the optimal period of the server in Phase i. A question may arise for the introduction of aperiodic tasks' arrival rate λ here because the QLB algorithm is designed specifically for the applications where the arrivals of aperiodic tasks can not be estimated. In fact, the parameter λ in our queueing model is only used to evaluate the performance of the QLB algorithm in general environments and is not needed in the determination of task periods.

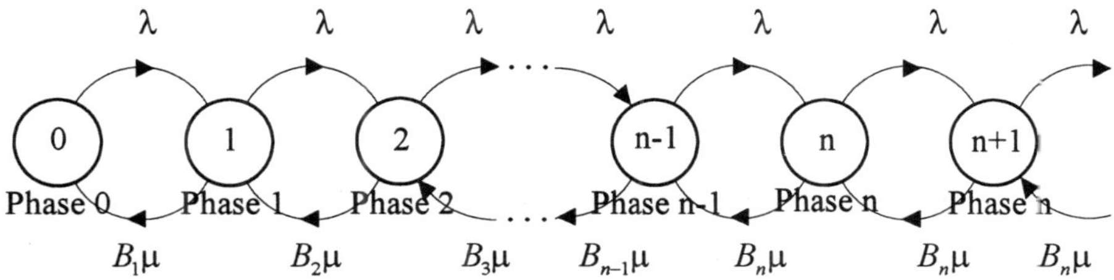

Figure 2: State transition diagram of the QLB algorithm.

If the probability that i aperiodic tasks are waiting in task queue is denoted by p_i, then we have

$$p_i = \begin{cases} \dfrac{p_0(\frac{\lambda}{\mu})^i}{\prod_{j=1}^{i} B_j}, & 0 \le i \le n-1 \\[2ex] \dfrac{p_0(\frac{\lambda}{\mu})^i}{B_n^{\,i-n}\prod_{j=1}^{n} B_j}, & i \ge n \end{cases} \tag{1}$$

Since $\sum_{i=0}^{\infty} p_i = 1$, we can derive p_0 from Equation (1) as

$$p_0 = \frac{1}{1 + \sum_{i=1}^{n-1} \dfrac{(\frac{\lambda}{\mu})^i}{\prod_{j=1}^{i} B_j} + \sum_{i=n}^{\infty} \dfrac{(\frac{\lambda}{\mu})^i}{B_n^{\,i-n}\prod_{j=1}^{n} B_j}} \tag{2}$$

All other p_i's can be obtained by the substitution of p_0 from Equation (2) into Equation (1). The probability for the system to stay in phase i, denoted by P_i, can now be computed as

$$P_i = \begin{cases} p_i, & 0 \le i \le n-1 \\ 1 - \sum_{j=0}^{n-1} p_j, & i = n \end{cases} \tag{3}$$

The following example shows the computation of probabilities and system performance for a real-time mission scheduled by the QLB algorithm.

Example 2:
A real-time mission containing five periodic tasks τ_1, τ_2, τ_3, τ_4, and τ_5 as listed in Table 1 are scheduled by the QLB algorithm and their optimal periods in each phase are computed by the gradient method and listed in Table 2. It is assumed that the arrival rate of aperiodic tasks $\lambda = 0.1$ and the mean computation time of aperiodic tasks $1/\mu = 1$. Four phases are created where the optimal periods of periodic tasks are determined by the gradient method and listed in Table 2.

Table 1. A real-time mission containing five periodic tasks.

Task	τ_1	τ_2	τ_3	τ_4	τ_5
C_i	1	2	3	4	5
$\overline{T_i}$	10	20	30	40	50
w_i	5	4	3	2	1
$\Psi_\iota(T_i)$	$0.1T_1/\mu C_1$	$e^{0.1T_2}$	$e^{0.1T_3}$	$e^{0.1T_4}$	$e^{0.1T_5}$

Table 2. Four phases created by the QLB algorithm.

Phase i	T_1^i	T_2^i	T_3^i	T_4^i	T_5^i	Φ	P_i
Phase 0	∞	9.6	9.6	19.1	19.1	38.5	0.35
Phase 1	8.8	13.1	13.1	13.1	26.1	51.4	0.31
Phase 2	7.4	11.1	11.1	22.1	22.1	56.0	0.23
Phase 3	3.4	16.6	16.6	16.6	33.1	87.3	0.11

The probability P_0, P_1, P_2, and P_3 are calculated as 0.35, 0.31, 0.23, and 0.11, respectively, and the system cost in phase *i* is computed by

$$\Phi(T_1^i, T_2^i, T_3^i, T_4^i, T_5^i) = 0.5T_1^i + 4e^{0.1T_2^i} + 3e^{0.1T_3^i} + 2e^{0.1T_4^i} + e^{0.1T_5^i} \tag{4}$$

and also listed in Table 2. Finally, the average system cost $E[\Phi]$ for the whole mission is computed as

$$E[\Phi]=\sum_{i=0}^{3}P_i\Phi(T_1^i,T_2^i,T_3^i,T_4^i,T_5^i)$$
$$=0.35\times 38.5+0.31\times 51.4+0.23\times 56.0+0.11\times 87.3 \quad (5)$$
$$=51.9$$

Under the same condition, we use the gradient method to compute the optimal periods for the RM algorithm as $T_1^R=4.9$, $T_2^R=14.6$, $T_3^R=14.6$, $T_4^R=14.6$, and $T_5^R=29.2$ and its system cost is about 61.9, which is higher than that of the QLB algorithm. It has to be noted that there is only one set of periods for the RM algorithm and no period transformations are performed.

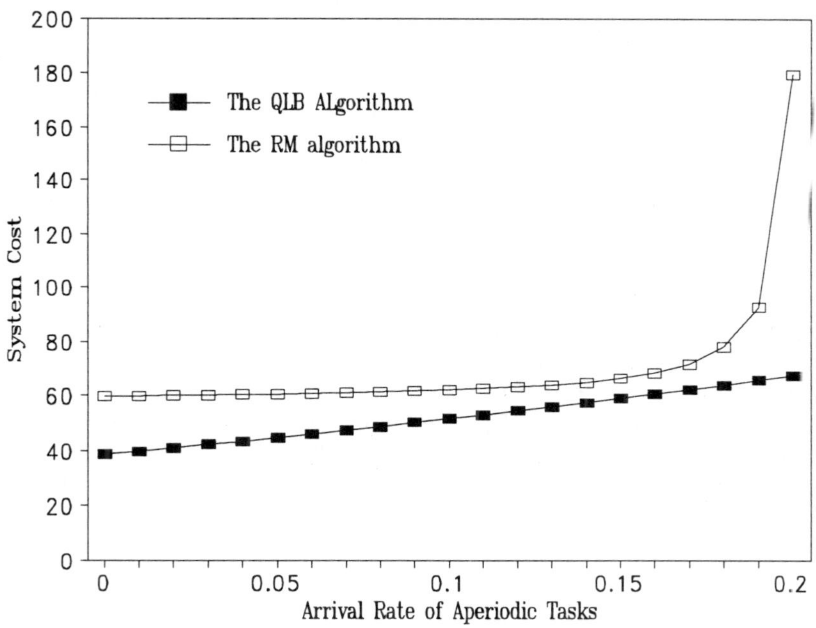

Figure 3: The QLB and RM algorithms under different task arrival rates.

Figure 3 shows the average system costs for the real-time mission in Example 3 scheduled by the QLB algorithm and the RM algorithm under different values of λ. The optimal periods for the RM algorithms are derived under the load condition λ=0.1 and then used for all λ ranging from 0 to 0.2. Similarly, we use the periods obtained in Table 2 for the QLB algorithm to operate under different values of λ. The result reveals the robustness of these two algorithms because they have to operate in the environments where the arrivals of aperiodic tasks may be different from the original estimation or even not estimated. According to this result, the QLB algorithm is always better than the RM algorithm for different values of λ. The QLB algorithm has a higher probability to stay in Phase 0 to reduce the system cost by suspending the server when λ is small. The QLB algorithm is also better than the RM algorithm when

λ is higher than estimated since the service rate for aperiodic tasks in the QLB algorithm can be extended to $C_1\mu / T_1^3 \approx 0.294$ in Phase 3, which is higher than that of the RM algorithm. This explains why the cost of the RM algorithm increases dramatically when λ is close to $C_1\mu / T_1^R \approx 0.204$ while that of the QLB algorithm dose not.

5 Conclusion

In many real-time environments, the characteristics of tasks may change during the operation of computing systems. Most static scheduling algorithms can not provide efficient mechanisms for dynamic load adjustment such that they are not suitable for such applications. The dynamic period transformation algorithm proposed in [4] can adapt to the load changes as long as the arrivals of aperiodic tasks can be estimated. In this paper, we have proposed a queue-length-based (QLB) version of the dynamic period transformation algorithm which can be specifically applied in the environments where the stochastic characteristics of aperiodic tasks are not provided beforehand. The QLB algorithm defines a number of operational phases according to the queue length of aperiodic tasks, and it performs a period transformation and switches to another phase if the number of waiting aperiodic tasks does not match the legitimate queue length. Since the optimal periods in each phase are determined by the gradient method, the QLB algorithm can always provide the best results no matter how the arrivals of aperiodic tasks change. The simulation result shows that the QLB algorithm has very high robustness in the environments where the arrivals of aperiodic tasks are different from the original estimation or even not estimated.

Reference

1. J. Lehoczky, L. Sha, and J. Strosnider, Enhanced aperiodic responsiveness in hard real-time environments, *Proceedings of Real-Time System Symposium*, pp. 261-270, San Jose, California, Dec. 1987.
2. B. Sprunt, J. Lehoczky, and L. Sha, Exploiting unused periodic time for aperiodic service using the extended priority exchange algorithm, *Proceedings of Real-Time System Symposium*, pp. 251-258, Huntsville, Alabama, Dec. 1988.
3. T.H. Lin and W. Tarng, Scheduling periodic and aperiodic tasks in hard real-time systems, *ACM Performance Evaluation Review*, vol. 19, no.1, pp. 31-38, May, 1991.
4. W. Tarng and T.H. Lin, Adaptive task scheduling in real-time systems, P*roceedings of International Conference on Parallel and Distributed Systems*, pp. 607-611, Taipei, Taiwan, Dec. 1993.
5. C. L. Liu and J. W. Layland, Scheduling algorithms for multiprogramming in a hard real-time environment, *J. ACM*, vol. 20, no. 1, pp. 46-61, Jan. 1973.

Experimental evaluation of environmentally assisted fracture: are there constant strain rate experiments?

J. Toribio

Department of Materials Science, University of La Coruña, ETSI Caminos, Campus de Elviña, 15192 La Coruña, Spain

Abstract

In constant strain rate (CSR) tests on pre-cracked or notched specimens, the local strain rate at the crack or notch tip —and not the global strain rate or externally applied testing displacement rate— is the variable governing the process. However, local strain rate changes with the time even for constant global strain rate. This paper suggests two possibilities to overcome this difficulty: (i) a real computer-controlled CSR test during which the local strain rate does keep constant by means of real time processing or (ii) a computer post-processing of results of a standard CSR test with variable local strain rate, so as to obtain universal results.

1 Introduction

It is well known [1-5] that one of the most promising experimental techniques to study environmentally assisted fracture phenomena is the performance of CSR tests on smooth, notched or pre-cracked specimens, the two latter being more convenient since they localize the environmental attack in the vicinity of the crack or notch tip. In these tests a constant displacement rate is externally applied at the sample ends by the testing machine.

When pre-cracked or notched specimens are used, it is the local strain rate at the crack or notch tip —and not the global strain rate or externally applied testing displacement rate— the variable which governs the process. However, this paper presents computational results from elastic-plastic finite element analysis demonstrating that the relationship between local and global strain rates depends on the load applied on the sample, which implies that local strain rate changes throughout the time, even for constant global strain rate externally applied, as in the case of CSR tests. As a consequence, local strain rate is not really constant during the experiment.

Further sections of the paper are devoted to the computation of the relationship between local and global strain rates in notched samples, including global and local analysis, and evolution of the plastic zone, fundamental in determining the local strain. Final section offers applications of numerical results to experimental evaluation of environmentally assisted fracture, which represents a kind of integration of computational methods and testing.

2 Problem statement

The material model used in the computations corresponds to a real high strength steel whose stress-strain curve can be represented by the following Ramberg-Osgood equation:

$$\varepsilon = \frac{\sigma}{E} + \left(\frac{\sigma}{P}\right)^n \tag{1}$$

in which E the elastic modulus, and P and n the Ramberg-Osgood parameters. For the chosen material E = 199.000 MPa, P = 2100 MPa, n = 4.9.

The geometry consists of an axisymmetric notched bar of length L. A set of four notched geometries of maximum and minimum notch depths and radii were chosen , as shown in Fig. 1.

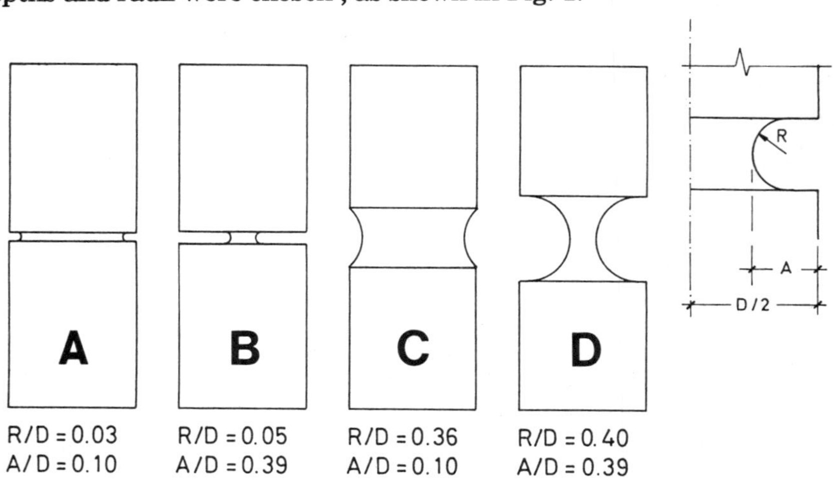

Figure 1: Notched geometries (of length L) used in the analysis.

The finite element method with an elastic-plastic code was used in the computations on the basis of a Newton-Raphson scheme. Load was introduced as displacement applied at the sample ends, step by step. Isoparametric eight node quadrilateral elements were used.

3 Global analysis

The first approach to the problem consists of a global analysis, with special emphasis on the evolution of the plastic zone, which has fundamental importance for the determination of local strain. Von Mises criterion was used to obtain the plastic zone:

$$\bar{\sigma} = \sigma_Y \tag{2}$$

where $\bar{\sigma}$ is the effective or equivalent stress and $\sigma_Y = \sigma_{0.1} = 500$ MPa, yield strength measured at 0.1% of strain. Properly speaking, it is the limit of material linear behaviour. Progressive plastic zones for different values of the displacement between sample ends distant a length L (global displacement u_G) appear in Fig. 2 and they reflect the progressive spreading of the plastic zone as the external load applied on the sample —in the form of displacement— increases. Yielding always initiates at the notch tip and spreads towards the inner part until the whole net section is under plastic regime.

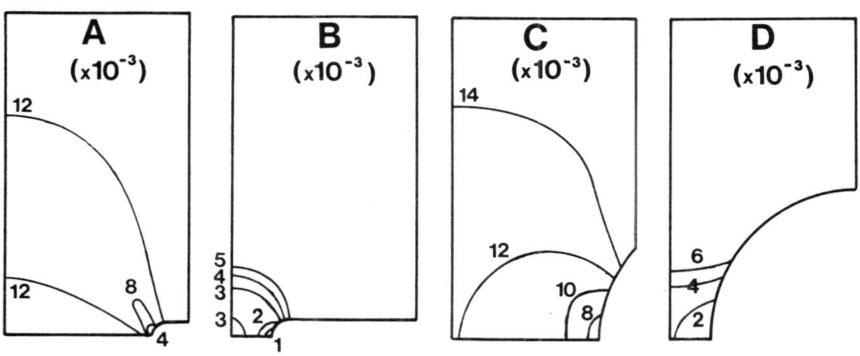

Figure 2: Progressive extension of the plastic zone in each geometry, for different values of u_G/D, where u_G is the global displacement applied at the sample ends (L=4D) and D the sample diameter.

As can be seen in Fig. 3, the four geometries present the same distributions of displacements in axial direction z along straight lines parallel to the sample axis at distance r. Displacements grow strongly in the notch vicinity ($0 < z < z_0$) with a typical nonlinear distribution and slightly and linearly in the rest of the sample, i.e.:

$$u_z = k_1 + k_2 z; \;\; 0 < r < a; \;\; z > z_0 \tag{3}$$

where a is the radius of the net section, and z_0 is the vertical coordinate for which a uniaxial stress state is established and thus the effect of the notch is negligible. From the computational results it can be said that $z_0/D = 2$ is an adequate length for that purpose.

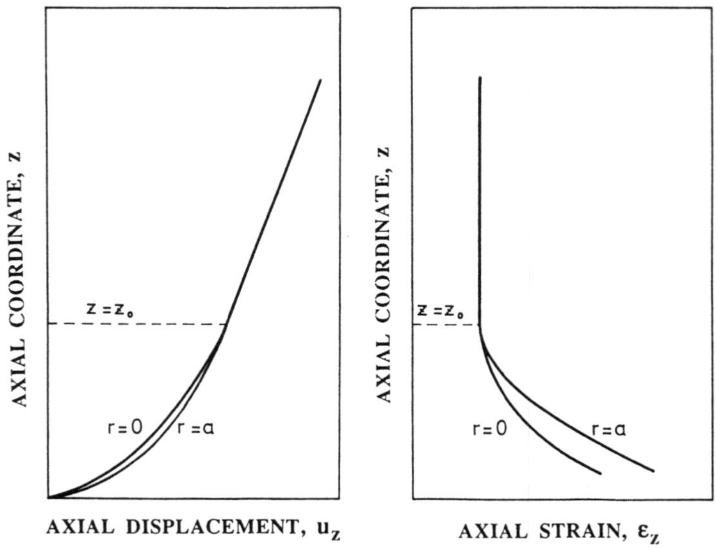

Figure 3: Axial displacement and axial strain distributions along the sample axis.

Strain tensor is defined from the displacement vector. In this work, focussed on notched specimens axially loaded, only the axial (z) component will be analyzed. On the basis of the hypothesis of small strains, the relationship between the cartesian components of the strain tensor and the displacement vector is the following:

$$\varepsilon_z = \frac{\partial u_z}{\partial z} \tag{4}$$

and ε_z-z distributions can be obtained from u_z-z curves through derivation. Fig. 3 shows these distributions for an intermediate step of loading. It is observed a strain concentration in the neck, higher next to the notch tip, and uniform strain in the rest of the sample ($\varepsilon_z = k_2$), starting from coordinate z_0, which marks the beginning of the part of the geometry under uniaxial stress state. On the basis of a Ramberg-Osgood curve of the material, it can be written as follows:

$$k_2 = \frac{4F}{\pi D^2 E} + \left(\frac{4F}{\pi D^2 P}\right)^n \tag{5}$$

in which F is the load, D the sample diameter, E the Young modulus and P, n the Ramberg-Osgood parameters, according to equation (1). The value k_2 of equation (5) has interesting experimental applications in research on environmentally assisted cracking, since it represents the uniaxial strain far from the notch, a variable which can be controlled by means of an extensometer applied outside the corrosion cell. It could be called, then, *remote strain*.

4 Local analysis

Local strain ε_L is defined as that associated with a local reference length B (small enough to guarantee the convergence of the method) parallel to the bar axis (z) and placed at the notch tip, i.e:

$$\varepsilon_L = \frac{\Delta B}{B} = \frac{u_L}{B} \tag{6}$$

where the enlargement ΔB of the local reference length B is the relative displacement between its ends (local displacement u_L).

Global strain ε_G is defined as the relative displacement between the ends of a sample of length L (global displacement u_G), divided by a characteristic length of the geometry (diameter D) to obtain a dimensionless variable:

$$\varepsilon_G = \frac{u_G}{D} \tag{7}$$

It must be noticed that the global strain is a dimensionless displacement and, contrarily to local strain, it is not a strain in the continuum mechanics sense, since the strain is non uniform along the axial direction in a notched geometry. Strictly speaking, therefore, it should be called *dimensionless global displacement.*

Local and global strain rates can be obtained on time-derivation of expressions (6) and (7).

$$\dot{\varepsilon}_L = \frac{u_L{}^{i+1} - u_L{}^{i}}{B\,\Delta t} \tag{8}$$

$$\dot{\varepsilon}_G = \frac{u_G{}^{i+1} - u_G{}^{i}}{D\,\Delta t} \tag{9}$$

where superscripts i and i+1 mean a loading step and the next one, and the value Δt represents the time internal between two loading steps, which can be constant or variable from one to another step. However, parameter Δt can be easily eliminated by calculating the relationship between local and global strain rate:

$$\frac{\dot{\varepsilon}_L}{\dot{\varepsilon}_G} = \frac{\dfrac{u_L{}^{i+1} - u_L{}^{i}}{B}}{\dfrac{u_G{}^{i+1} - u_G{}^{i}}{D}} \tag{10}$$

Local reference length must be short enough to guarantee the convergence of the mathematical method to the solution but, on the other hand, it must be greater than the relevant microstructural parameter (usually the material grain size), to assure the validity of the continuum mechanics approach, i.e., the agreement between the mechanical model and the microphysical reality of the material.

Mathematical convergence is guaranteed, since the displacement distribution is continuous, and therefore the limit define in (6) does exist. The objective is, then, to check from which local reference lengths the solution is accurate enough, however keeping the chosen reference length higher than the material grain size or similar characteristic microstructural size of the material.

Fig. 4 offers the relationship between local and global strain rates (for geometry A) as a function of global strain for different values of the local reference length B. The accuracy is excellent for B/D = 0.01. In this case the relative error of local strain rate is lower than 1%, even when B changes its magnitude order. Using D = 10 mm as a typical value for diameter of steel wires, it results B = 100 μm, one magnitude order higher than the first microstructural level of a standard high strength steel (average pearlite colony size: 10-15 μm), material for which this research has a special interest. The mathematical model, on the basis of continuum mechanics, is consistent with the microphysical reality of the material.

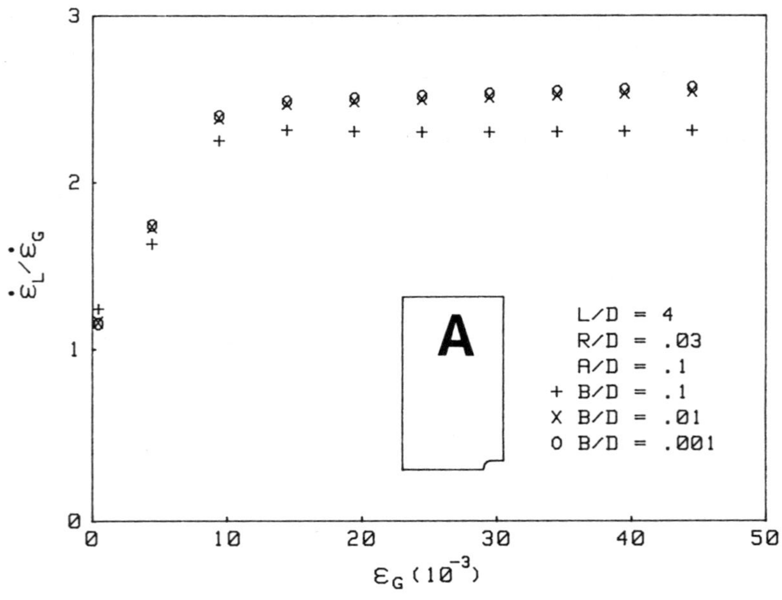

Figure 4: Relationship between local and global strain rates for different local reference lengths, showing the convergence of the numerical method (Geometry A).

Global reference length should be long enough to permit uniaxial stress state at its ends, so the influence of the notch on the stress state at the ends of the reference length becomes negligible, the distribution of axial displacement is linear (equation (3)) and the axial strain is uniform (equation(5)). A value L = 4D is enough to achieve this, according to the elastic-plastic numerical computations.

With these local and global reference lengths, the expressions (8) and (9) for the strain rates become:

$$\dot{\varepsilon}_L = \frac{u_L^{i+1}(0.01D) - u_L^{i}(0.01D)}{0.01\,D\,\Delta t} \tag{11}$$

$$\dot{\varepsilon}_G = \frac{u_G^{i+1}(4D) - u_G^{i}(4D)}{D\,\Delta t} \tag{12}$$

where the variables between parentheses are the local (0.01D) and global (4D) reference lengths, distances on which the relative displacements between the ends are measured to compute the local and global strains. It should be noticed that a limit is involved in the expression (11), which makes the value of local strain rate unique. It is, by definition, independent of the diameter D, although this variable appears to define the local reference length.

The choice of a global reference length does not represent a restriction, and the results do not loose general applicability if the choice is carried out with the condition of uniaxial stress state at its ends (far enough from the notch to neglect its influence). In this case there is a linear relationship between the relative displacements of two different couples of points placed symmetrically with respect to the net section of the sample, corresponding to the ends of two distinct global reference lengths. For the conventional global reference length adopted in this research (L=4D) and any other global reference length (L > 4D; any distance between the control points in an environmentally assisted cracking test) the relationship is:

$$u_G\,(4D) = u_G\,(L) - k_2\,(L - 4D) \tag{13}$$

where $u_G(4D)$ is the relative displacement between the ends of a reference length 4D, $u_G\,(L)$ the same corresponding to the ends of a any other reference length L > 4D, and k_2 the strain far enough from the notch, where the stress state is uniaxial. In the case of Ramberg-Osgood modelling of the stress-strain curve of the material, it yields:

$$u_G\,(4D) = u_G\,(L) - \left[\frac{4F}{\pi D^2 E} + \left(\frac{4F}{\pi D^2 P}\right)^{n}\right](L - 4D) \tag{14}$$

with the same meanings as before.

It should be noticed that the expression (13) is purely mechanical and independent of the material, while relation (14) involves the choice of a particular material, in this case a typical high strength steel modelled by a stress-strain relationship of Ramberg-Osgood type, with parameters P and n.

5 Relation between local and global strain rates

In Fig. 5 a plot is given of the relationship between local and global strain rates for the local and global reference lengths chosen in previous section, the local length being at the notch tip. Evolution of plastic zone is also presented (shaded area is the plastic zone adopting a yield strength $\sigma_Y = \sigma_{0.1}$, limit of material linear behaviour).

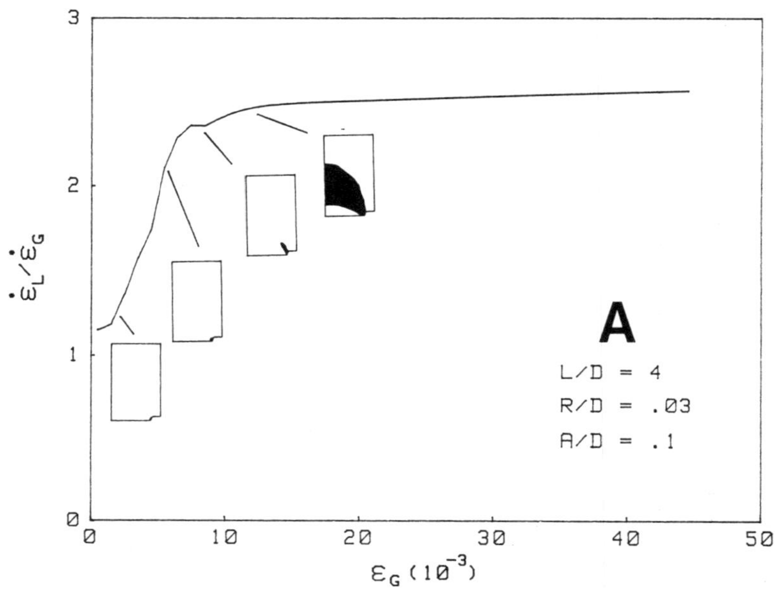

Figure 5: Relationship between local and global strain rates and evolution of the plastic zone for geometry A.

Relationship between local and global strain rates changes with the time, as the plastic zone spreads or the global strain (or more properly global displacement) increases. For notched geometries the curve has always the same general aspect, an three regions can be distinguished, each of them representing a phase of the process:

(i) Phase I (Elastic): Horizontal. The whole sample is under elastic regime. The relationship between local and global strain rate is constant or, in other words, a constant global strain rate (externally applied) produces a constant local strain rate at the notch tip.

(ii) Phase II (Transition): Large slope. Yielding starts at notch tip and spreads progressively towards the inner region. The shape of the plastic zone depends on the particular geometry, while its size increases as global strain increases.

(iii) Phase III (Plastic): Slightly increasing or quasi-horizontal. Plastic zone reaches the sample axis. In this case the relationship between local and global strain rate is more or less constant, with a high level of local strain rate due to the extended plastic zone.

6 Application to environmentally assisted fracture testing

Computational results of previous sections demonstrate that local strain rate at the notch tip —the key variable in environmentally assisted fracture of notched specimens— is not constant during a test in which externally-applied global strain rate (or testing displacement rate) is constant. Furthermore, it is impossible to perform a strictly *real* constant strain rate test in environmentally assisted fracture, even when the global strain rate is constant *and* a smooth sample is tested, since cracking, multi-cracking, dissolution or any kind of localized damage appears at the critical steps of the test, before the final fracture, and such defects behave as stress and strain concentrators in the same manner as notches, thus generating again the problem of variability of local strain rate as the test proceeds.

To overcome this difficulty of notched specimens, two approaches may be suggested in the framework of integration of computational methods and testing in evaluation of environmentally assisted fracture. The first one consists of using real time processing of information to have a computer-assisted control of the experimental device. The second one deals with post-processing of experimental information to obtain universal results.

With regard to the first way, it is possible to design a real CSR test on notched samples during which the local strain rate does keep constant in spite of variations of global strain rate, introducing the computational relationship between local and global strain rates as input data into the computer which controls the testing machine. By means of *real time processing* of the global strain signal during the test, a feedback order can be sent to the mechanical device, thereby producing a really constant *local* strain rate test.

Another approach consists of performing a standard CSR test with constant global strain rate or testing displacement rate. As demonstrated throughout this paper, local strain rate at the notch tip changes as the test proceeds (as the global displacement increases or the plastic zone spreads). However, outcome of this non-constant strain rate test could be *post-processed* on the basis of the results of this paper (relationship between local and global strain rates as a function of global strain given in Fig. 5) to obtain *universal* results independent of the notched geometry. This approach was successfully developed in previous works [6].

7 Conclusions

When pre-cracked or notched specimens are used, it is the local strain rate at the crack or notch tip —and not the global strain rate or externally applied testing displacement rate— the variable which governs the process.

The relationship between local and global strain rates depends on the load applied on the sample, which implies that local strain rate changes throughout the time, even for constant global strain rate externally applied, as in the case of CSR tests.

It is possible to design a real CSR test on notched samples, introducing the relationship between local and global strain rates into the computer which controls the testing machine, and using real time processing of the global strain signal during the test.

Another approach consists of performing a standard CSR test and post-processing its outcome on the basis of the relationship between local and global strain rates, to obtain universal results independent of the notched geometry.

Previously described approaches represent ways to integrate computational methods and testing in environmentally assisted fracture, a branch of structural integrity of special interest in engineering and technology.

Acknowledgements

This work was funded by the Spanish DGICYT (Grant CE94-001) and *Xunta de Galicia* (Grant XUGA 11801A93)

References

1. Parkins, R.N. Development of strain-rate testing and its implications, *ASTM STP 665*, 1979, 5-25.
2. Kim, C.D. & Wilde, B.E. A review of the constant strain-rate stress corrosion cracking test, *ASTM STP 665*, 1979, 97-112.
3. Scully, J.C. Propagation of stress corrosion cracks under constant strain-rate conditions, *ASTM STP 665*, 1979, 237-253.
4. ISO 7539-7. *Slow strain rate testing*, 1989.
5. Parkins, R.N. Slow strain rate testing-25 years experience, *ASTM STP 1210*, 1993, 7-21.
6. Toribio, J. & Elices, M. The role of local strain rate in the hydrogen embrittlement of round-notched samples. *Corros. Sci.*, 1992, **33**, 1387-1409.

Experimental study of the performance of structural steel members subjected to multi-dimensional ground motion

H.-L. Hsu

Department of Civil Engineering, National Central University, Chung-Li, Taiwan, Republic of China

Abstract

This paper presents an experimental investigation of the inelastic behavior of wide-flange structural steel members under combined time-dependent bending and torsional loading. A quantitative evaluation of energy dissipation capacity after occurrence of local buckling was made. Test results show that a W6x15 section with a d/b_f ratio equal to one performed better than a W8x10 section, which has a larger d/b_f ratio, in resisting combined bending and torsion. Test results also show that the presence of torsion significantly affects a member's bending-moment carrying capacity.

Introduction

Current design philosophy for structural systems requires that structures must be sufficiently ductile to prevent collapse during major earthquakes. Steel structures have been widely used in building and bridge construction because of their high ductility performance during earthquake ground motion [1]. The success of a structure in absorbing severe earthquake ground motion without collapse depends largely on each member's ability to absorb excessive energy through inelastic deformation.

Loading induced by an earthquake is usually complicated and time-dependent. Therefore, members designed for safety under static loading must still be checked for instability-failure potential under cyclic and earthquake loading[2]. The most important design concept in earthquake engineering is recognizing that the buckling of a structural member does not end that member's service life. However, there will be a reduction in the member's strength and energy dissipation capacity[3]. Therefore, it is essential to investigate the post-buckling behavior of steel members subjected to multi-dimensional earthquake ground motion , qualitatively and quantitatively.

Theoretical analyses of structural responses to combined dynamic loading are usually complicated and difficult [4-10]. This paper is focused on the performance of steel members subjected to combined time-dependent bending and torsional loading induced by multi-dimensional earthquake ground motion. The performance of various steel members is compared according to their strength, and deterioration in energy dissipation capacity during the post-buckling stage.

Test set-up

The test set-up was designed as a closed loop with respect to force equilibrium during testing. Two strong base beams connected by three tie beams formed the frame base, which was then stiffened and bolted to the test floor. Test specimens were mounted on one of the base beams in the vertical direction. The configuration is shown in Fig. 1.

The reaction frames were composed of two W14x135 wide-flange columns with 1/2 inch-thick stiffeners added where the actuators were mounted, and two bracing members with 1/2 inch-thick end plates. Loading was transmitted to the test specimens through a strong loading beam from two actuators (one for bending and one for torsion). By varying the displacements of each actuator, a combination of bending, shear and torsional loading was generated. Two one-inch-thick end plates were welded to the specimens and attached to the frame base and the loading beams. The details are shown in Fig. 1.

The hydraulic actuators used were MTS models 244.41 and 244.52. Load capacities for the actuators were 100 kips and 200 kips, respectively, and strokes for both were 10 inches. Actuator piston rods movements were monitored and recorded by internally mounted Linear Variable Differential Transformers (LVDTs).

Experimental procedures and results

In order to determine the effect of torsion on the performance of steel members with various cross-sectional dimensions, two A-36-grade wide-flange steel members, W6x15 and W8x10, were selected for testing. Four loading combinations including bending, shear and torsion were chosen to study the effect of torsion on a member's bending-moment capacity. They were:

(1) cyclic bending and shear,
(2) cyclic torsion,
(3) cyclic bending and shear with minor constant twist,
(4) cyclic bending and shear with moderate constant twist.

Displacement at the top of the specimen was used as the control parameter in all tests.

W6x15 Series Tests

The first series of tests was conducted on W6x15 members. The sectional area of these members was 4.43 inch^2 and the moment of inertia about the strong axis (I_x) was 29.1 inch^4. The width/thickness ($b_f/2t_f$) ratio was equal to 11.5 and the d/b_f ratio was one.

Cyclic bending and shear

This specimen was subjected to periodically increasing bending and shear forces until local buckling occurred. Then, the member was subjected to repeated loading by keeping the displacement amplitude at the value at which local buckling occurred. Loading history was sequentially composed of three cycles of 0.25, 0.5, 0.75 and 1 inch displacements at the top of the specimen, followed by 10 cycles of 1.25 inches of displacement amplitude. Displacement history for both actuators is shown in Fig. 2. The relation between load and displacement is shown in Fig. 3.

The specimen buckled when displacement amplitude at its top reached 1.25 inches. Energy absorption was obtained by calculating the work bounded by the hysteresis loop. The energy dissipated during each cycle was divided by the energy absorbed during the first cycle after buckling occurred to obtain the normalized energy dissipation factor.

Normalized capacity to absorb energy during the cycles after buckling occurred was reduced from 1 to 0.992, 0.956, 0.932, 0.915, respectively. These values were linearly curve-fitted to obtain the deterioration rate, which represents the average drop per cycle in energy dissipation capacity. The deterioration rate for this test was 1.9%

Cyclic torsion

This specimen was subjected to periodically increasing torsion. Pure Cyclic torsion was applied by fixing the center actuator and cyclically moving the actuator at the end to generate torsion. The applied torsion was obtained by multiplying the load measured by the load transducer with the distance between the two actuators. The relation between torsion and twist angle is shown in Fig. 4. Buckling occurred, due to pure torsion, when twist angle reached 0.133 radians. At this stage, noticeable distortion of the section was observed during the test. The deterioration in energy dissipation capacity was small for this cyclic torsion test. Normalized energy dissipation factors for the cycles after buckling occurred were 1, 0.998, 0.984, 0.966, respectively. The deterioration rate for this test was 0.75%.

Cyclic bending and shear with minor constant twist
Combined bending, shear and torsion was achieved by first controlling the displacements of both actuators to maintain a constant twist angle, then cyclically moved both actuators at same amplitude. A 0.01 radians constant twist angle was applied to this specimen. Local buckling occurred at the same displacement amplitude as that in the bending and shear test, 1.25 inch. Normalized energy dissipation factors for the cycles after buckling occurred were 1, 0.956, 0.916, 0.89, 0.871, respectively, These values represent a deterioration rate of 2.76%.

Cyclic bending and shear with moderate constant twist
This specimen was subjected to periodically increasing bending and shear with a constant 0.03 radians twist angle. The loading sequence was the same as that used in the cyclic bending and shear test. Local buckling occurred when the displacement at the top of the specimen reached 1.25 inches. Normalized energy dissipation factors for the cycles after local buckling occurred were calculated as 1, 0.91, 0.87, 0.83, respectively. The deterioration rate in energy dissipation was 4.1%. Comparisons of initial buckling strength and normalized energy dissipation factors for W6x15 sections with various degrees of torsion applied are shown in Figs. 5 and 6.

W8x10 Series Tests

The second series of tests was conducted on W8x10 specimens. The moment of inertia about the strong axis (I_x) was 30.8 inch^4 and the sectional area of these member was 2.96 inch^2. The width/thickness ($b_f/2t_f$) ratio was equal to 9.6 and the d/b_f ratio was two.

Cyclic bending and shear
The loading sequence was composed of three cycles of 0.1, 0.25, 0.5, 0.75 inches and five cycles of 1 and 1.25 inches. Initial buckling occurred when the amplitude reached 1 inch. At this stage, noticeable deterioration in energy absorption and strength was observed. When the displacement amplitude was increased to 1.25 inches, a further reduction in strength was clearly detected. Detailed relation between load and displacement is shown in Fig. 7.

Initial buckling strength during the first cycle of 1 inch amplitude was 15.12 kips for the positive side and 14.95 kips for the negative side. Strength for the cycles after initial buckling was 14.57, 13.38, and 12.38 kips, respectively. Normalized energy dissipation factors for each cycle when the displacement at the top of the specimen equaled 1 inch were reduced from 1 to 0.966, 0.919, 0.862, respectively. A deterioration rate of 2.85% was calculated by linearly curve-fitting these values.

Cyclic torsion

This specimen was initially subjected to three cycles of 0.01 radians twist angle, followed by a series of increasing twist angles up to 0.125 radians. Local buckling occurred when the twist angle reached 0.136 radians. The buckling zone was small and only occurred at the ends of the specimen. The result of the interaction between torsion and twisting angle is shown in Fig. 8. Normalized energy dissipated in each cycle with respect to that of the first cycle after local bucking was reduced from 1 to 0.996, 0.989.0.979, respectively. The deterioration rate obtained by curve-fitting these values was 0.8%

Cyclic bending and shear with minor constant twist

This specimen was subjected to periodically increasing bending and shear with a constant 0.0085 radians twist angle. The loading sequence for the actuator that generated bending was the same as that used in the bending and shear test. Significant local buckling occurred when the displacement amplitude at the top of the specimen reached 1 inch. The energy dissipated in each cycle was normalized with respect to that of the first cycle when local buckling of the member occurred. The normalized energy dissipation factors were reduced from 1 to 0.929, and 0.866. The deterioration rate for this test specimen was 3.35%

Cyclic bending and shear with moderate constant twist

This specimen was subjected to periodically increasing bending and shear with a 0.03 radians twist angle. The twisting angle was kept constant throughout the loading history. Local buckling occurred when the displacement at the top of the specimen reached 1 inch. Due to the larger twist angle, the normalized energy dissipation capacity dropped from 1 to 0.896, 0.839 and 0.789 for the cycles after the member had locally buckled. The deterioration rate calculated from these values was 5.32%. Comparisons of initial buckling strength and normalized energy dissipation factors for W8x10 sections with various degree of torsion applied are shown in Figs. 9 and 10.

Comparison of test results

In order to identify the effect of torsion on the reduction of load-carrying capacity as well as rate of deterioration in load-carrying capacity after the member buckled, a comparison of test results between members with and without torsion was made.

For all W6x15 sections tested with bending, local buckling occurred at the same displacement amplitude regardless of whether the twist angle was included.

The same phenomena were observed in the W8x10 sections. However, the initial buckling strength and deterioration of energy dissipation capacity after the members had locally buckled showed significant differences. This can be clearly seen in Figs. 5, 6, 9 and 10. The W6x15 section test results show that members subjected to minor and moderate degrees of constant twist angles deteriorated in energy dissipation capacity 1.45 and 2.15 times faster than that without torsion. The W8x10 sections showed average deterioration rates of members with minor and moderate constant twist angles were 3.35% and 5.32%, respectively. These values are larger than those of the W6x15 sections with the same order of torsion.

Summary and conclusion

In this paper, two wide-flange structural steel members were tested to study their performance during multi-dimensional earthquake ground motion. Information obtained from this investigation is of two-fold benefit. It clearly shows, by comparing the performance of members with and without torsion, that the presence of torsion can significantly reduce a member's energy dissipation and bending-moment carrying capacity. Therefore, the torsional effect should be included in the design considerations for earthquake-resistant structural systems. The experimental study mentioned above also showed that the effect of torsion increases when degree of torsion and d/b_f ratio increase, which can be useful in the selection of member sections for design and construction.

References

1. Fukumoto, Y. and Lee, G. C. (1992), Stability and Ductility of Steel Structures under Cyclic Loading, CRC Press, Boca Raton, FL.
2. International Conference of Building Officials (1991), Uniform Building Code, Whittier, CA.
3. Lee,E.T. and and Lee,G.C.,(1992). "Locally Buckled Plastic Hinge Behavior Under Monotonic and Cyclic Loading Condition," Proceeding of 9th Engineering Mechanics Conference, ASCE, College Station, Texas, May, 1992, pp. 1047-1050.
4. Chang, K.C. and Lee, G.C., (1986). "Biaxial Properties of Structural Steel Under Nonproportional Loading," Journal of Engineering Mechanics, ASCE, Vol. 112, No.8, pp. 792-805.
5. Lamba, H.S. and Sidebottom, O.M., (1978). "Cyclic Plasticity for Nonproportional Paths: Part 1-Cyclic Hardening, Erasure of Memory, and

Subsequent Strain Hardening Experiments." Journal of engineering Materials and Technology, Transaction of the ASME, Vol. 100, pp. 96-103.

6. McDowell, D.L.,(1985). "A Two Surface Model of Transient Nonproportional Cyclic Plasticity: Part II - Comparison o Theory with Experiments," Journal of Applied Mechanics, ASME, Vol. 52, pp.303-308.

7. Phillips, A., Tang, J.L., (1972). "The Effect of Loading Path one the Yield Surface at Elevated Temperatures," Int. J. of Solids and Structures, pp. 463-474.

8. Phillips, A., Liu, C.S., and Justusson, J.W., (1972). "An Experimental Investigation of Yield Surfaces at Elevated Temperatures," Acta Mechanica, pp. 119-146.

9. Phillips, A., and Tang, J.L., and Ricciuti, M., (1974). "Some New Observations on Yield Surfaces, " Acta Mechanica, Vol. 20, pp.23-39.

10. Tseng, N.T. and Lee, G.C., (1983). "Simple Plasticity Model of Two-Surface Type," Journal of Engineering Mechanics, ASCE, Vol. 109, No. 3, pp. 795.

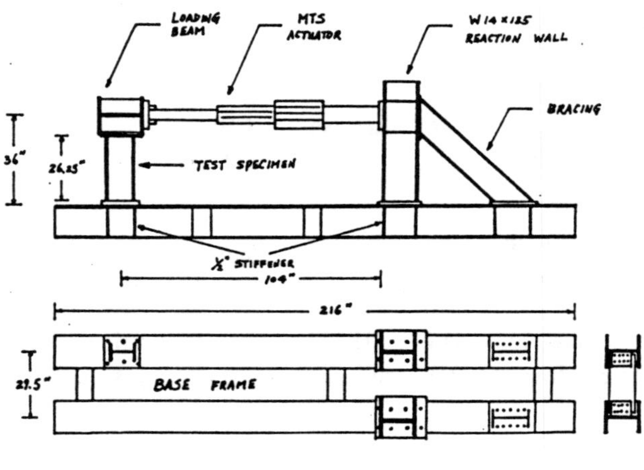

Figure 1: Configuration of test frame

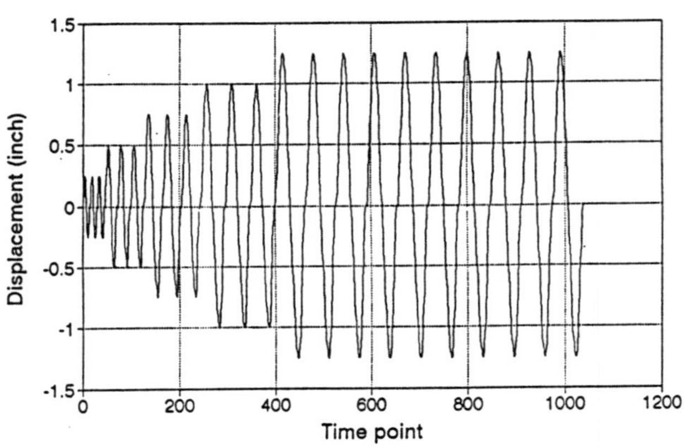

Figure 2: Loading history for W6x15 with cyclic bending and shear

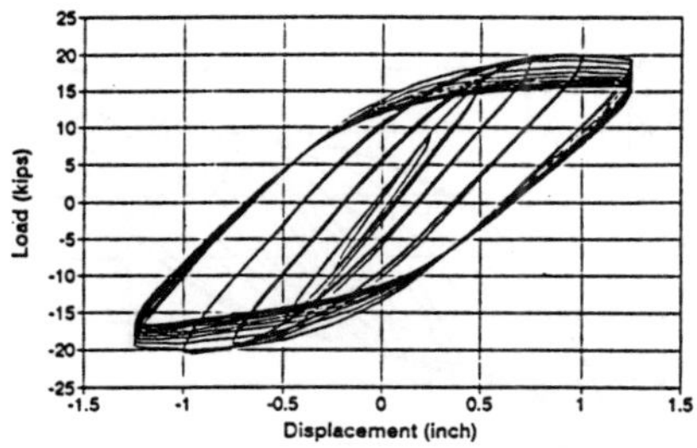

Figure 3: Load-displacement relation for W6x15 with cyclic bending and shear

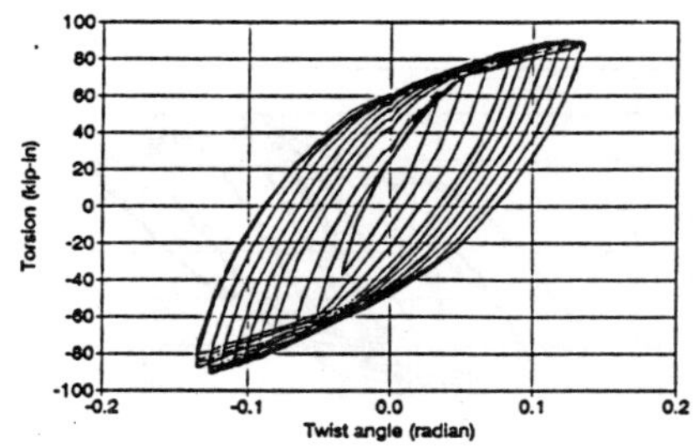

Figure 4: Torsion-twist angle relation for W6x15 with cyclic torsion

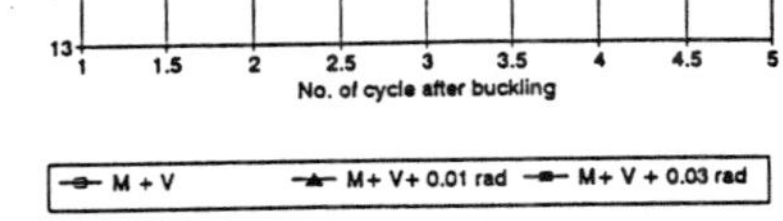

Figure 5: Comparison of strength for W6x15 sections with various degrees of torsion

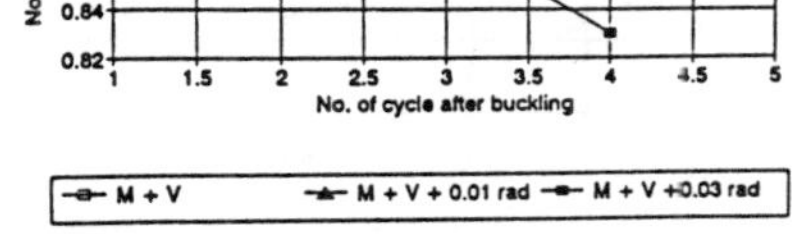

Figure 6: Comparison of normalized energy dissipation factor for W6x15 sections with various degrees of torsion

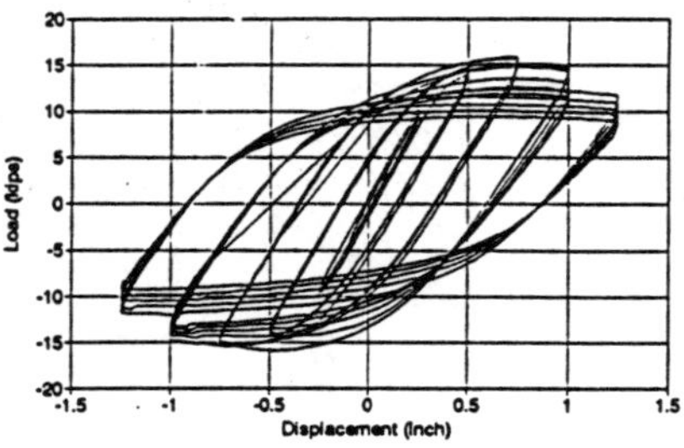

Figure 7: Load-displacement relation for W8x10 with cyclic bending and shear

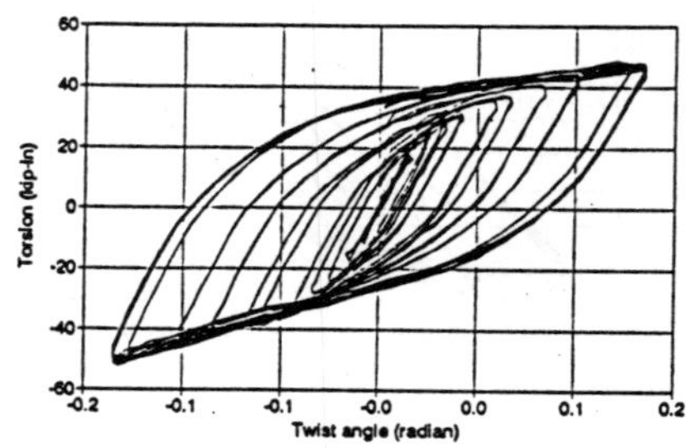

Figure 8: Torsion-twist angle relation for W8x10 with cyclic torsion

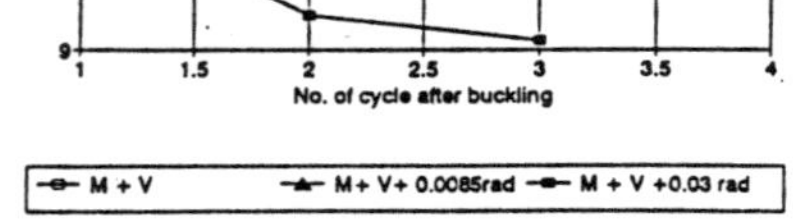

Figure 9: Comparison of strength for W8x10 sections with various degrees of torsion

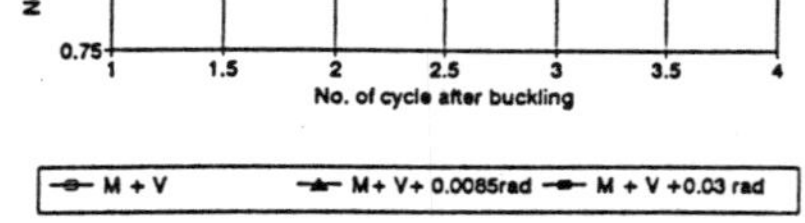

Figure 10: Comparison of normalized energy dissipation factor for W8x10 sections with various degrees of torsion

Predicting fatigue crack growth in complex components

A.C. Neves, S.M. Niku, R.A. Adey
Computational Mechanics, BEASY, Ashurst Lodge, Ashurst, Southampton, Hampshire SO40 7AA, UK

Abstract

Catastrophic failure of engineering structures is caused by cracks that extend beyond a safe size. Cracks are present to some extent in all structures and can lead to failure or decrease in structural strength if they grow. This paper described an automatic procedure for predicting the fatigue crack growth in two and three dimensional structures and the key data for fracture mechanics based design: the stress intensity factors.

The procedure is implemented in a general purpose engineering analysis system BEASY. The system uses a dual boundary element technique to simplify the modelling of the cracks and to provide highly accurate stress intensity factors. Applications are presented for single and multiple cracks.

1.0 Introduction

The cost to industry of fracture was recently estimated in a report of the US Department of Commerce entitled "The economic effect of fracture in the United States". In this report it estimated that the cost of fracture was $119 billion dollars per year (4% of gross national product).

It further estimated that approximately one third could be saved through the use of current fracture control technology and a further 25% could be saved through fracture related research. Therefore the annual cost of fracture could be halved by the application of better design tools based on fracture mechanics technology.

The combination of the easy to use and apply boundary element approach in BEASY and the new fracture growth tools incorporated in the system are aimed to satisfy this goal.

Catastrophic fracture failure of engineering structures is caused by cracks that extend beyond a safe size. Cracks, present to some extent in all structures, either as a result of manufacturing fabrication defects or localised damage in service, may grow by mechanisms such as fatigue, stress-corrosion or creep. The crack growth leads to a decrease in the structural strength. Thus, when the service loading cannot be sustained by the current residual strength, fracture occurs leading to the failure of the structure. Fracture, the final catastrophic event that takes place very rapidly, is preceded by crack growth which develops slowly during normal service conditions, mainly by fatigue due to cyclic loading.

Damage tolerance assessment is a procedure that defines whether a crack can be sustained safely during the projected service life of the structure. Damage tolerance assessment is, therefore, required as a basis for any fracture control plan, generating the following information, upon which fracture control decisions can be made:

The effect of cracks on the structural residual strength, leading to the evaluation of their maximum permissible size.

The cracks growth as a function of time, leading to the evaluation of the life of the cracks to reach their maximum permissible size, from which the safe operational life of the structure is defined.

Linear elastic fracture mechanics can be used in damage tolerance analyses to describe the behaviour of cracks. The fundamental assumption of linear elastic fracture mechanics is that the crack behaviour is determined solely by the values of the stress intensity factors which are a function of the applied load and the geometry of the cracked structure. The stress intensity factors, thus play a fundamental role in linear elastic fracture mechanics applications.

Crack-growth processes are simulated with an incremental crack-extension analysis. For each increment of the crack extension, a stress analysis is carried out and the stress intensity factors are evaluated. The crack path, predicted on an incremental basis, is computed by a criterion defined in terms of the stress intensity factors.

The boundary element method is well established as a powerful solution tool for fracture mechanics (see Aliabadi & Rooke [1]). The reason for its success is the boundary only representation, the high accuracy and the methods ability to represent the high stress fields near the crack front.

The theoretical foundation of the dual boundary elements and crack growth algorithm follows closely the work of Portela and Aliabadi [2] for 2D; and Mi and Aliabadi [3] for 3D. The authors will not repeat the theoretical basis but would refer the reader to references [2] and [3] for the detailed basis of the crack growth and dual boundary elements and to Brebbia and Dominguez [4] for a description of the boundary element method. The fatigue life predictions are evaluated using the generalizes formula presented in NASA/FLAGRO 2.0 [5].

In this paper the implementation within the BEASY boundary element system is described and application presented showing the wide range of applications possible.

2.0 Simulation Strategy

A crack can be represented in a boundary element model using two main approaches. The traditional approach requires the user to define a zone boundary along the crack surface and continue this through the body of the components being studied. This can be clearly seen in Fig. 1b where the problem is split into two zones and the edge crack is extended by a zone interface (shown as a dotted line) across to another external boundary. This technique while not requiring any special theoretical development places on extra burden on the user.

The second approach recently developed is to use dual boundary elements to represent the crack. Fig. 1a shows the model of the edge crack using this approach. In this case the modeling is extremely simple and economical. The crack is represented by two elements occupying the same physical location, each element representing a face of the crack.

Multiple cracks can be easily defined by simply specifying the initial crack size. Crack merging can also be automatically modelled.

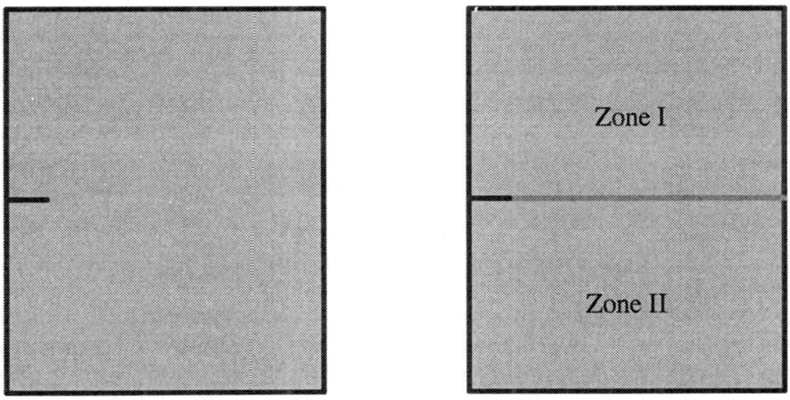

a) b)

FIGURE 1. Boundary element models of edge crack. a)Shows the crack modeled using dual boundary elements. b) Shows the crack modeled using two zones.

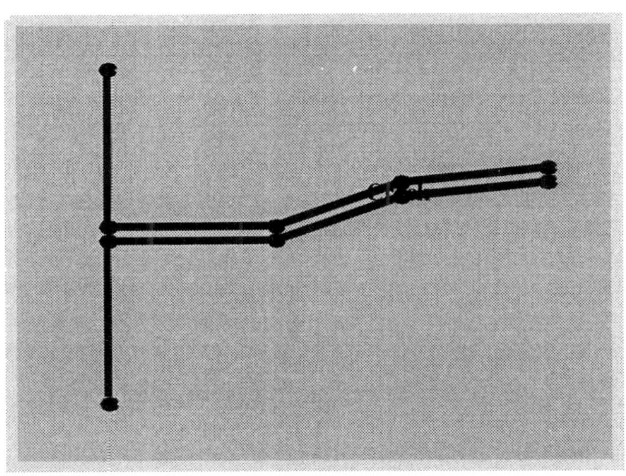

FIGURE 2. Dual boundary element representation of the crack.

2.1 Stress Intensity Factors

The system automatically computes the stress intensity factor using both J integral (2D only) and crack openning displacement (COD)formula. These methods can be used for computing Modes I, II and III intensity factors for edged and embedded cracks.

The J integral path is automatically computed by the software and follows the crack tip as it grows into the material.

2.2 Incremental Crack-extension Analysis

The incremental crack-extension analysis assumes a piece-wise linear discretisation of the unknown crack path. For each increment of the crack extension, the dual boundary element method is applied to carry out a stress analysis of the cracked structure and either the J-integral or the COD formula is the technique used for the evaluation of the stress intensity factors. The steps of this basic computational cycle, repeatedly executed for any number of crack-extension increments, are summarised as follows:

- Carry out a stress analysis of the structure
- Compute the stress intensity factors
- Compute the direction of the crack-extension increment
- Extend the crack one increment along the direction computed in the previous step
- Repeat all the above steps sequentially until a specified number of crack-extension increments is reached

For the sake of simplicity, the increment of crack extension is discretised with a fixed number of new boundary elements. In order to avoid numerical problems, concerned with the relative size of neighbouring elements, the crack increment length is kept between convenient limiting bounds, defined in terms of the size of the crack-tip element. Apart from this constraint, the length of the crack extension increment may be defined as the result of a compromise between accuracy and computational cost; the smaller the crack increment the more accurate and expensive is the analysis.

The results obtained from an incremental crack-extension analysis are a crack path diagram and diagrams of the stress intensity factor variation along the crack path.

2.3 Crack-extension Criterion

Several criteria have been proposed to describe the mixed-mode crack growth. Among them, the most commonly used are the maximum principal stress and the minimum strain energy density.

The maximum principal stress criterion formulated by Erdogan & Sih [7] postulates that the growth of the crack will occur in a direction perpendicular to the maximum principal stress. Thus, the local crack-growth direction is determined by the condition that the local shear stress is zero.In practice this requirement gives a unique direction irrespective of the length of the crack extension increment. Therefore the procedure adopted in the system is to use a predictor corrector technique to ensure the crack path is unique and independent of the crack extension increment used.

The minimum strain energy density formulated by Sih [8] is based on the hypotheses:

- The direction of crack propagation at any point along the crack front is toward the region with the minimum value of strain energy density factor S as compared with other regions on the same spherical surface surrounding the point.
- Crack extension occurs when the strain energy density factor in the region $S = S_{min}$ reaches a critical value S_{cr}.
- The length r_0 of the initial crack extension is assumed to be proportional to S_{min} such that S_{min}/r_0 remains constant along the new crack front.

In 2D applications, the system uses as default the maximum principal stress while in 3D cases only the minimum strain energy density is adopted.

2.4 Crack Growth Relationship For Fatigue

A number of crack growth laws have been developed relating the rate of growth of the crack to the stress intensity factor. BEASY evaluates the fatigue life prediction using the generalized NASGRO 2.0 equation [5]:

$$\frac{da}{dN} = \frac{C \cdot (1-f)^n \cdot \Delta K^n \cdot \left(1 - \frac{\Delta K_{th}}{\Delta K}\right)^p}{(1-R)^n \cdot \left(1 - \frac{\Delta K}{(1-R)K_c}\right)^q} \quad (1)$$

It should be noted that equation (1) may be reduced to the Paris equation [6] by setting the parameters p and q equal to zero and not considering the effect of crack closure, *i.e.* $f=R$ for $0<R<1$. The Paris equation is given by:

$$\frac{da}{dN} = C \cdot \Delta K^n \quad (2)$$

The database of fatigue material constants supplied by NASA and ESA has been integrated with the crack growth algorithm and this can be used to select standard data; or the user can define alternative database containing the material constants.

3.0 Numerical Validation

The accurate prediction of the crack growth is highly dependent upon the accuracy of the stress intensity factors. Any minor innaccuracy in these values is accumulated during crack growth process which can lead to a completely unrealistic prediction.

Therefore considerable care has been taken in selecting and implementing the techniques used to compute the SIF values. A large number of tests have been carried out to verify the accuracy and reliability of the software for simple modes and mixed mode cracks.

Comparisons of the stress intensity factors have been made with a number of published test solutions and the accuracy found to be very high. For single mode and mixed mode cracks the solution has been found to be almost exact.

3.1 Gear Tooth

The growth of a crack from the root of a gear is investigated. The problem was originally studied by Flasker and Pehan [9]. In this example, the NASGRO equation is used with stress ratio $R=0.5$ and total number of cycles $N=10^6$. The curve fitting constants of the titanium alloy (code P1CA13WA1) was taken from the database file NASMFM. Crack closure effect was not taken into consideration. The crack growth until failure,i.e.,K_{max} exceeds the fracture toughness (K_c) of the material.

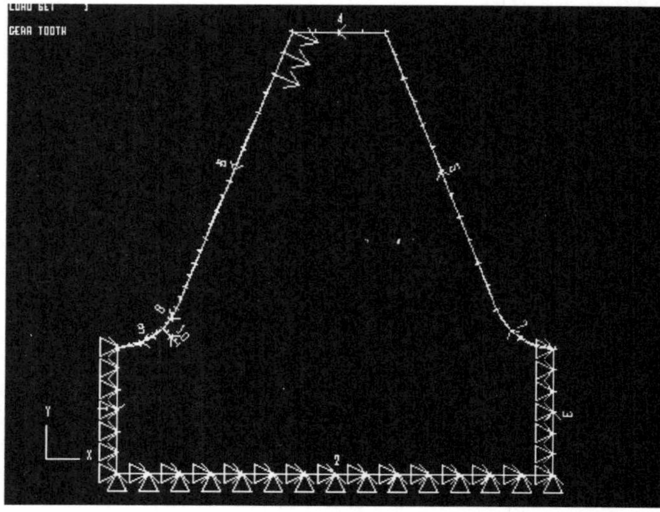

FIGURE 3. Complete model geometry of the gear tooth with the applied loading.

Figure 4shows the deformed shape of the gear tooth and the crack path. This compares well with experimental results and manual finite element calculations reported by [9]

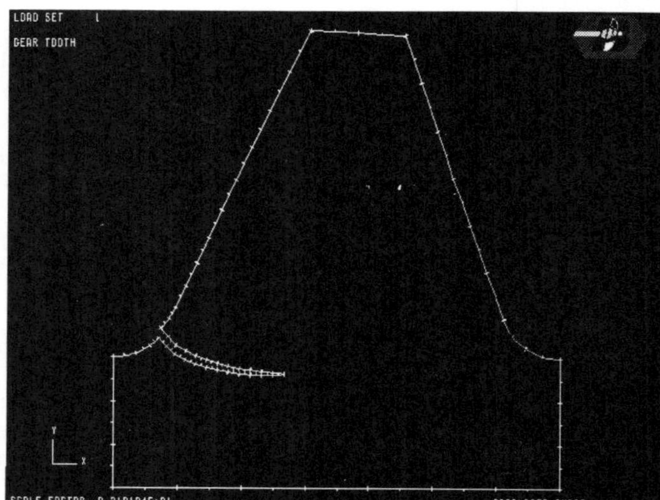

FIGURE 4. Deformed shape of the gear tooth.

Figure 5 shows the stress intensity factors varying for each crack growth increment. One can notice that at the final increment, the effective stress intensity factor is greater than the fracture toughness. The extension of the crack per number of cycle loading is given in Figure 6.

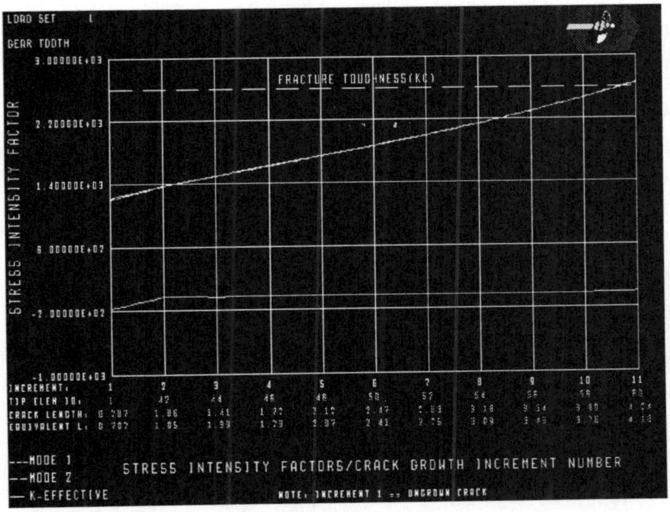

FIGURE 5. Stress intensity factors x crack growth for the gear tooth.

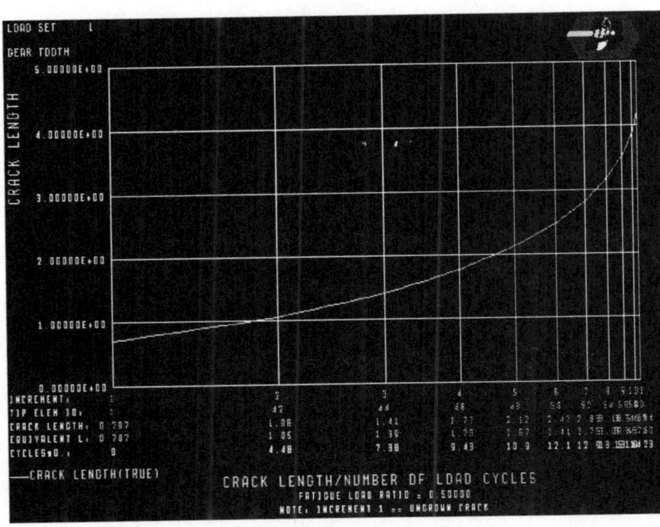

FIGURE 6. Crack length x Number of load cycle .

3.2 Multiple crack growth

In this application, the effect of cracks initiating from the rivet holes of an aircraft panel is investigated. The plate is fixed at the bottom side and a normal traction applied at the top. The complete crack path and history can be predicted for multiple site damage studies. The successful completion of this analysis provides the data required to predict the total fatigue life *before* the component goes into service. The safe design life can be stablished by use of suitable safaty factors.

Compared with other techniques, this computer simulation provides much more information much quicker and much cheaper. With any technique, a single prediction does not represent the final solution but is merely the starting point for sensitivity studies. The flexibility, accuracy and ability of this computer based techique to model multiple site damage in complex strucures is the authors believe a major advance.

As the cracks grow the software automatically adjust the rate of crack growth for each crack, based on the crack growth law. This can be seen in this example where the cracks near the edge grow much slower than the others.

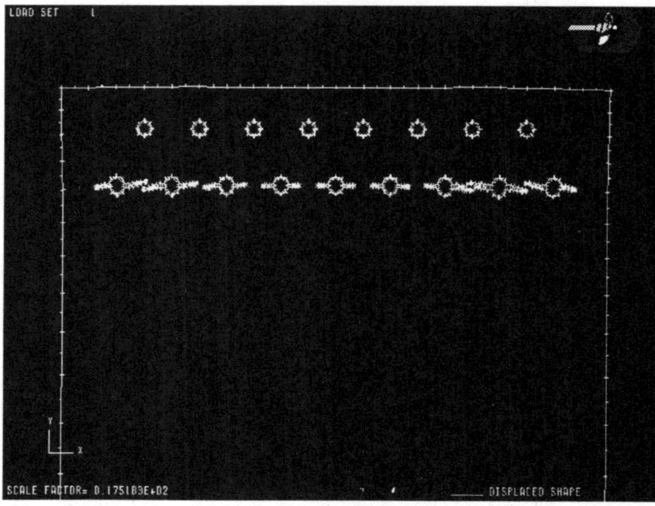

FIGURE 7. Aircraft panel with rivet holes.

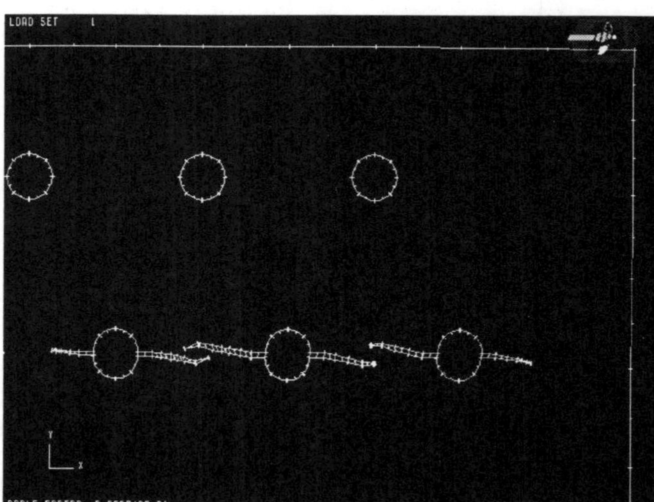

FIGURE 8. Detail of the cracks in the rivet holes.

3.3 Embedded Elliptical Crack

A mixed mode 3D crack growth problem is considered. A 45° inclined elliptical crack of major semi-axis a and minor semi-axis b is embedded in a cylindrical bar of radius R and height h as shown in Figure 9. The ratios of these sizes are $R/a=10$, $h/R=6$ and $b/a=0.5$. Material constants are Young's modulus $E=100\ kN.mm^{-2}$ and Poisson's ratio $\nu=0.3$. A fatigue cracking process is generating by a constant amplitude cyclic tensile loading applied on the ends of the bar with stress ratio $R=0$ and $\sigma_{max}=100\ N.mm^{-2}$. Material constants used in the Paris law are $C=1.5463x10^{-16}$ and $m=3.88$.

A total of four steps of increment were performed. The crack path after the last increment are shown in Figure 10 and it is in good agreement with reference [3].

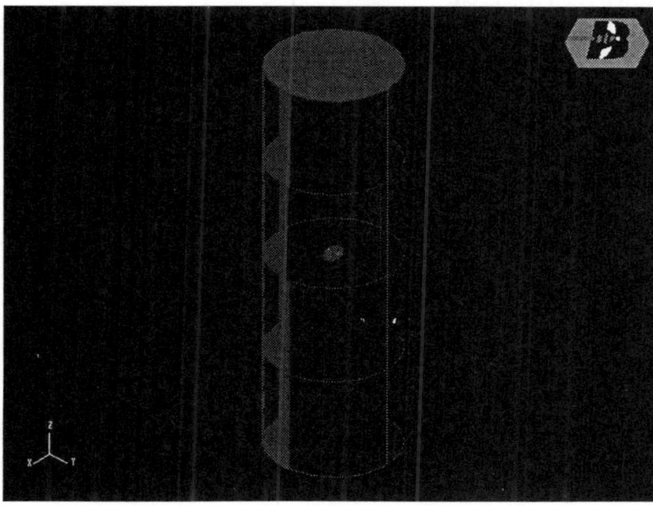

FIGURE 9. Cylinder with embedded elliptical crack.

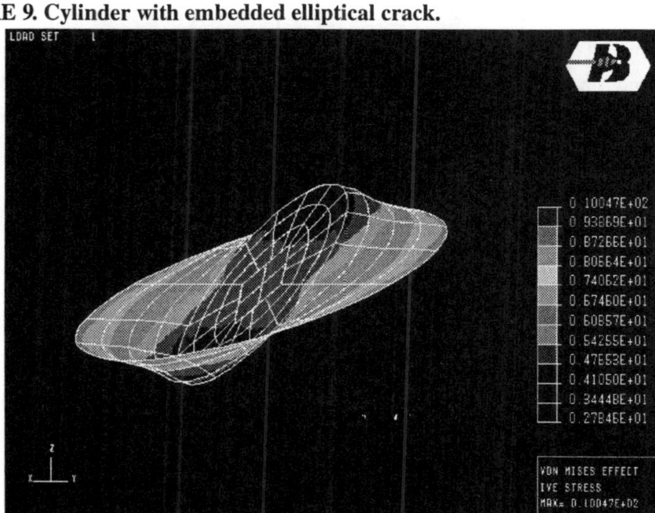

FIGURE 10. Von Mises stresses and crack growth path of the inclined elliptical crack

4.0 Conclusion

The paper describes a powerful system for the prediction of fracture data and fatigue crack growth in two and three dimensional structures. The techniques have shown high accuracy and the ability to automatically predict the growth of cracks. The system is particularly powerful because of the very simple modeling and complete lack of meshing problems associated with FEM solution of these problems.

Acknowledgement

The authors acknowledge the support of Dr Aliabadi during this work.

This work was partly funded under Project MASS of the European Community. Partners in the project include Computational Mechanics BEASY, Microparts, Technical University of Berlin and Wessex Institute of Technology.

Additional support was provided by the European Space Agency.

References

1. Aliabadi M.H., Rooke D.R., Numerical Fracture Mechanics, Computational Mechanics Publications, SOUTHAMPTON, U.K. 1991.

2. Portela A., Aliabadi M.H. & Rooke D.P. "Dual Boundary Element incremental analysis of crack propagation", Int Journal Computers and Structures, Vol. 46 , pp 237-247, (1993).

3. Mi, Y.; Aliabadi M.H. "Three-dimensional crack growth simulation using BEM", Computers & Structures, Vol 52, No 5, pp 871-878, (1994).

4. Brebbia C.A., Dominguez J., Boundary Elements - "An Introductory Course", Computational Mechanics Publications, Southampton, U.K. (1989).

5. Forman, R.G.; Shivakumar, V.; Newman, J.C., "Fatigue Crack Growth Computer Program "NASA/FLAGRO" Version 2.0", National Aeronautics and Space Administration Lyndon B. Johson Space Center, Houston, Texas, (1993).

6. Paris, P.C., "The Crack Growth of Fatigue Cracks Due to Variation in Load", PhD. Thesis, Lehigh University, Bethlehen (1962).

7. Erdogan, F and Sih, G.C. "On the crack extension in plates under plane loading and transverse shear", J. Basic Eng., 85,pp 519-527, (1963).

8. Sih, G.C. "Mechanics of Fracture Initiation and Propagation", Kluwer Academic, Dordrecht (1991).

9. Flasker J., Pehan S., "Crack propagation in tooth root with variable loading". Comm in Num Meth in Eng., Vol. 9,pp 103-110 (1993).

Design of partially prestressed concrete beam-column connection with singly prestressed sections under cyclic loadings

Ir.B. Budiono
Civil Engineering Department, Institut Teknologi Bandung, Jl. Ganesha 10, Bandung 40132, Indonesia

Abstract

Prestressed Concrete framed structure are commonly designed to provide presstressing steel to resist gravity loads. Therefore most of the prestressing tendons are located at the top of the beam at beam-column connections. Experimental and numerical non-linear finite element studies developed by the author show that these types of structures are capable to resist severe cyclic loading provided sufficient care is given. Special attention must be considered in the limitation of ductility demand, partial prestressing ratio and detailings in the potential plastic hinge regions. In this paper a simple and semi-rational design method for partially-prestressed interior beam-column connection in proposed. As the behavior of connections are non-flexural, the design method developed is based on major and minor strut modeling using plasticity theory.

1 Introduction

ACI 318M-89 requires that beam-column joints of frames in designated earthquake regions must the designed for seismic conditions. The code specifies that the shear strength of the beam-column connections are to be designed for inelastic response to severe seismic loadings with flexural plastic hinges to be developed in the beams adjacent to the column faces. Flexural actions in the beams generate joint shear forces. In addition, the joint forces are multiplied by an overstrength factor (α) to allow for the higher mean yield strength of the flexural reinforcement than the specified characteristic strength, fy. ACI 318M-89 gives α is equal to 1.25. The requirements in the code indicate that the joint shear strength is a function of only the shear strength of the concrete. Thus, only minimum transverse reinforcement is required in the joint.

In generating shear forces acting on the beam-column connections, the New Zealand code of practice (NZS 3101-82) uses a similar approach to that of ACI 318M-89. Flexural actions of the plastic hinges of the beams develop the horizontal shear forces in the joint, while the vertical shear forces are generated by flexural actions of the upper and lower columns of the joint. However, the shear resisting mechanisms specified by NZS 3101-82 are different than those developed in ACI 318M-89. Two major shear resisting mechanisms are considered; they are: (i) diagonal concrete strut and, (ii) panel truss actions. The diagonal strut action resists the forces resulting from the compressive and shear forces in the beam and the column. The shear forces are transmitted by the flexural reinforcement to the joint via bond. These bond forces produce diagonal tension force in the joint which are resisted by vertical and horizontal reinforcement to gave equilibrium in an equivalent panel truss action. For a low axial compressive force in the column (representing the worst condition in the beam-column connection under severe seismic actions) the portion of the joint shear forces resisted by the panel truss action is assumed to be no more than 50 % of the total shear force acting in the joint, as recommended by Paulay et al. (1978).

In this paper, a simple and semi-rational design method for partially-prestressed interior beam-column connection is proposed. As the behaviour of connections are non-flexural (ACI 318 RM-89 Commentary, Section R12.6.3), the design method developed is based on major and minor strut modeling using plasticity theory.

2 Forces Acting on Beam-Column Connections

Under large lateral loads the forces acting on an internal joint using a truss model are illustrated in Figure 1.

The diagonal compression force in the joint, Dj (as shown in Figure 1), is calculated from equilibrium. At point A (as shown in Figure 1), diagonal compressive force of the left beam and the upper column of the joint, Dv_1 and Dv_3, respectively, act on the joint with an angle θ. If the angle θ is taken as 45°, then the horizontal and vertical component of Dv_1 are equal to V_1, and V_1 can be taken as shear force of the left beam. A similar expression can be derived for Dv_3 that yields V_3 which is equal to upper column shear force. Therefore the horizontal and vertical joint shears, V_{jh} and V_{jv}, respectively, are

$$\left.\begin{aligned} V_{jh} &= C_1 + T_2 + V_1 - V_3 \\ V_{jv} &= C_3 + T_4 + V_3 - V_1 \end{aligned}\right\} \qquad \text{(A.1)}$$

For the horizontal shear in the joint (V_{jh}), C_1 combined with V_1 form the flexural compressive force at the top of the left beam, T_2 is tension force in the top of the joint, $V_1$and V_3 are the shear forces acting on the left beam and the upper column of the joint, respectively. A similar derivation can be obtained for the vertical shear in the joint (Vjv).

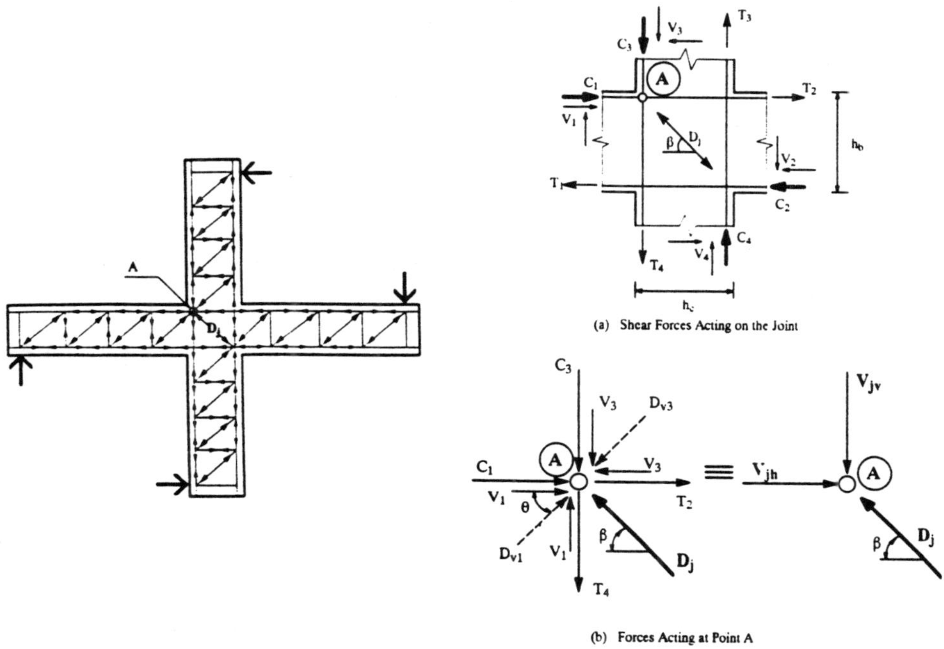

Figure 1. Truss Analogy in Beam-Column Connection

Horizontal equilibrium of the left beam and vertical equilibrium of the upper column (shown in Figure 1) give

$$\left.\begin{aligned} T_1 &= C_1 + V_1 \\ T_3 &= C_3 + V_3 \end{aligned}\right\} \qquad \text{(A.2)}$$

Substituting Eq. A.2 into Eq. A.1 yields

$$\left.\begin{aligned} V_{jh} &= T_1 + T_2 - V_3 \\ V_{jv} &= T_3 + T_4 - V_1 \end{aligned}\right\} \qquad \text{(A.3)}$$

The tension forces Ti (i=1,2) calculated when the beams reach their ultimate moments are given by

$$T_i = \alpha \left[A_{si} f_{yi} + A_{pi} f_{pui} \right] \quad (i = 1,2) \qquad \text{(A.4)}$$

where Asi is the reinforcement area; Api is the area of the prestressing steel; fyi is yield stress of the reinforcement; fpui the ultimate stress of the prestressing steel; and α is an overstrength factor.

As the column is proportional to be flexurally stronger than the beams, the elastic vertical shear of the joint (Vjv) becomes difficult to asses when plastic hinges develop in the beams adjacent to the column faces. However, the worst case occurs when the axial force acting on the column is small. In this case, Vjv can be taken as

$$V_{jv} = V_{jv} \tan \beta \qquad \text{(A.5)}$$

Vjh and Vjv are the horizontal and vertical components of the diagonal force in the concrete compressive strut (Dj) in the joint as shown in Figure 2 b, and

$$D_j = V_{jh} / \cos \beta \qquad \text{(A.6)}$$

where the angle β, for all practical purposes, can be taken as

$$\beta = \tan^{-1} \left(h_b / h_c \right) \qquad \text{(A.7)}$$

where h_b and he are the total depth of the beam and the column sections, respectively.

3 Shear Resisting Mechanisms of Beam-Column Connections

Shear across a joint core is transferred by major and minor diagonal concrete strut actions. The major strut action is shown in Figure 2a. The major diagonal strut is capable of transmitting a significant fraction of shear forces across the joint. If all of the shear forces resisted by the major strut (Dc), then Dc is equal to Dj, as given by Eq. A.6.

If the shear forces acting on the joint is greater than the major strut capacity then the excess forces (Ds) are transmitted by the minor strut. In this case, vertical and horizontal hanger reinforcement are required to carry the diagonal tension forces induced by bond forces. The bond forces are developed from the flexural reinforcement of the beams transferred across the joint core. The minor strut action is shown in Figure 2b.

In the general formulation, the shear resisting mechanism of a beam-column connection is given by

$$D_j = D_c + D_s \qquad \text{(A.8)}$$

where Dc and Ds are the forces carried by the major and minor struts, respectively. Equilibrium in the horizontal and vertical directions are given by :

$$\left.\begin{aligned} V_{jh} &= V_{ch} + V_{sh} \\ V_{jv} &= V_{cv} + V_{sv} \end{aligned}\right\} \qquad \text{(A.9)}$$

where Vc and Vs are the components of the forces in the major and minor struts, respectively; the subscripts h and v are representing the horizontal and vertical directions, respectively.

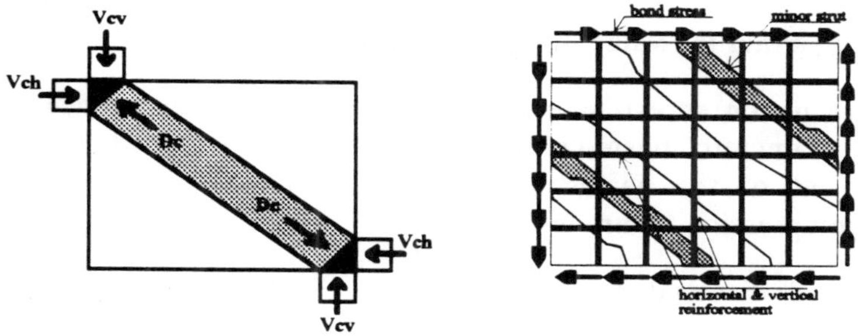

(a) Major Strut Action (b) Minor Strut Action

Figure 2. Plastic Truss Model of An of Interior Beam-Column Connection

3.1 Major Diagonal Compressive Strut (Dc)

A significant portion of the shear forces acting on the joint are resisted by a diagonal compression strut. As the concrete does not behave as a perfectly plastic material, particularly when higher strength concrete is used, the effective concrete strength is lower than the characteristic compressive strength, fc'. Some parts of the failure surface may be softening before other parts have mobilized their full strength (Foster, 1992). In addition, Vecchio and Collin (1986) showed that large transverse strains reduce the peak compressive strength. To accommodate these reductions of the peak strength, the effective concrete strength is given by

$$f_c^* = \nu f_c' \qquad \text{(A.10)}$$

where ν is an efficiency factor ($\nu \leq 1$). Warwick and Foster (1993) proposed an efficiency factor for concrete with characteristic strength ranging from 20 to 100 MPa. Noting that the shear span to effective depth ratio (a/d) = tan β, the Warwick and Foster relationship can be written in the form

$$\nu = 1.25 - \frac{f_c'}{500} - 0.72 \tan\beta + 0.18 \tan\beta^2 \leq 0.85 \ldots \text{ for } \tan\beta < 2$$

$$\nu = 0.53 - \frac{f_c'}{500} \ldots \text{ for } \tan\beta \geq 2 \qquad \text{(A.11)}$$

The maximum forces that can be taken by the compression strut is

$$Dc = \phi c\, f_c^*\, dc\, b \qquad \text{(A.12)}$$

where ϕ_c is the material strength reduction factor for compressive failure, f_c^* is the maximum average compressive stress in the compressive strut at failure (given by Eq. A.10), d_c is the depth of the compression strut, and b is the width of the joint. Foster (1994) recommended that the value of ϕ_c be taken as 0.6.

In order to obtain an estimate for d_c, a finite element experiment on a reinforced concrete beam-column connection was undertaken. The reinforcement of the beams and the column was proportioned to produce compression failure in the joint.

The idealized diagonal compression strut, together with the bearing support in each corner of the joint, is illustrated in Figure 3. The horizontal and vertical dimensions of the bearing supports are given by

$$\left.\begin{aligned} W_h &= d_c \cos \beta \\ W_v &= d_c \sin \beta \end{aligned}\right\} \qquad \text{(A.13)}$$

where W_h and W_v are the horizontal and vertical bearing areas of the joint, respectively, and d_c is the width of the strut.

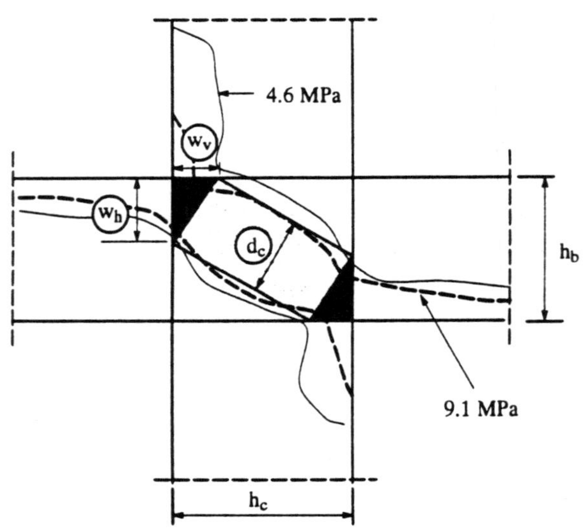

Figure 3. Bearing Support and the Width of the Major Diagonal Strut

Based on the parameters discussed earlier, it is found that the width of horizontal bearing area is approximately 0.5 h_b. The equations for the horizontal and vertical components of the major strut are

$$V_{ch} = \phi_c f_c^* \, 0.5 \, h_b \, b \qquad \text{(A.14a)}$$

$$V_{cv} = V_{ch} \tan\beta \qquad \text{(A.14b)}$$

where Vch and Vcv are the horizontal and vertical components of the major strut, respectively; ϕc is the strength reduction factor for compression failure; f_c* is determined using Eq. A.10; h_b and h_c are the overall depths of the beam and column, respectively; and b is the width of the joint.

3.2 Minor Compressive Struts

If the capacity of the major strut is exceeded then minor struts develop to balance the excess shear flow on the boundary of the joint. This shear flow is generated by bond stresses transferred from the flexural reinforcement of the beams across the joint zone (shown in Figure 2b). To maintain equilibrium, the forces in the minor compression struts must be balanced by additional reinforcement in the horizontal and vertical directions trough the joint. The shear forces resisted by minor struts in the horizontal and vertical directions are

$$\left.\begin{aligned} V_{sh} &= V_{jh} - V_{ch} \\ V_{sv} &= V_{jv} - V_{cv} \end{aligned}\right\} \qquad \text{(A.15)}$$

For practical purpose, Vsv is given by

$$V_{sv} = V_{sh} \tan\beta \qquad \text{(A.16)}$$

When both horizontal and vertical reinforcement are provided, the areas of the reinforcement required

$$\left.\begin{aligned} A_{sh} &= \frac{1}{\phi_s} \frac{V_{sh}}{f_{jy}} \\ V_{sv} &= A_{sh} \tan\beta \end{aligned}\right\} \qquad \text{(A.17)}$$

where Ash and Asv are the areas of horizontal and vertical reinforcement in the joint, respectively, ϕs is the strength reduction factor for tension failure; fjy is the yield strength of the joint reinforcement.

ϕ_s can be taken equal to 0.7 as recommended by Foster (1992), and vertical joint reinforcement may be provided by the column reinforcement distributed around the perimeter of the column.

3.3 Diagonal Tension Failure

The finite element result, showed that the joints behaved elastically under the cyclic tests with non-uniform stress contours in the joint.

The non-uniform stress distribution in the strut develops bursting forces, T_b, and leads to the possibility of a diagonal splitting failure, as shown in Figure 4. It may be observed in Figure 4 that the bursting force, T_b is given by

$$T_b = \frac{n}{4m} D_c \qquad \text{(A.18)}$$

Finite element experiments (Budiono, 1995) gave n approximately equal to 0.5 m. Taking n = 0.5 m and substituting into Eq. A.18 gives

$$\left.\begin{array}{l} T_b = 0.125 \ D \\ \text{where} \\ D_c = V_{ch} / \cos \beta \end{array}\right\} \qquad \text{(A.19)}$$

In addition, finite element studies undertaken by Warwick and Foster (1993) suggest that the percentage of web reinforcement in the vertical and horizontal directions should not be less than 0.2 % when $f_c' \leq 50$ MPa, and 0.4 % when $f_c' > 50$ MPa.

To allow the full strength of the compression strut to be develop, diagonal splitting failure should be avoided. Vertical and horizontal reinforcement may be added to the joint to increase the capacity of the joint against the diagonal splitting mode of failure.

$$\left.\begin{array}{l} A_{wh} = \dfrac{1}{\phi_s} \dfrac{T_b \sin \beta}{f_{jy}} \\ A_{wv} = \dfrac{1}{\phi_s} \dfrac{T_b \sin \beta}{f_{jy}} = A_{wh} / \tan \beta \end{array}\right\} \qquad \text{(A.20)}$$

This web reinforcement required is added to the shear reinforcement required by Eq. A.17.

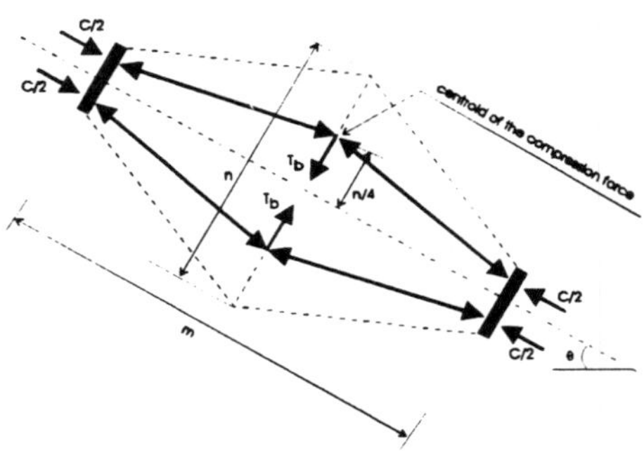

Figure 4 Result Forces in Compression Strut Due To the Changing Stress Trajectories in the Web

3.4 Provisions for Ductile Beam-Column Connections

(a) Partial Prestressing Ratio and Ductility Factor

In order to obtain a suitably ductile response in a partially-prestressed concrete beam-column connection, the partial prestressing ratio (PPR) and the ductility factor should satisfy the limitations given as follow (Budiono, 1995) :

(i) For PPR less than 0.2 the displacement ductility factor (μ) of 4.0 should be applied.

(ii) For $0.2 \leq$ PPR < 0.6 limited ductility approach should be considered ($\mu = 2.0$), while for PPR ≥ 0.6 no inelastic ductility is permitted.

(b) Plastic Hinge Region

At failure, finite element analysis showed that the concrete in the hinge zones was crushing and the effectiveness of the prestress was greatly reduced. To ensure that shear failure is prevented during the flexural hinge formation, it is recommended that in the hinge zone the contribution of concrete in resisting the factored design shear force is to be neglected.

(c) Strong Column Weak Beam

To ensure that column is stronger than the beam during the formation of the beam plastic hinges, it is recommended that in the vicinity of the joint, the flexural strength of each column should be at least greater than 1.5 times the maximum beam flexural strength.

(d) Bond Stress in the Joint

To ensure the breakdown of bond between the longitudinal reinforcement across the joint and the concrete is prevented, the column depth (h_c) should be taken larger than 20 times the largest diameter of the longitudinal reinforcement of the beams passing through the joint, as recommended by ACI-ASCE Committee 352 (1985).

4 Summary and Conclusion

A design for detailing joints partially-prestressed, beam-column connection is proposed. The design considers the strength of joint when plastic hinges develop in the beams adjacent to the column faces. Two mechanisms in resisting shear force in the joint are considered; that is the shear strength contributed by the major and minor compressive struts.

Horizontal and vertical joint reinforcement are required to carry tension forces generated by bursting action of the major struts and for hanger reinforcement to carry tension force derived from minor strut action.

In the beam hinge regions, the contribution of concrete to the shear strength is neglected, as part of the concrete in these regions become ineffective due to crushing. This allows for the formation of ductile hinges by flextural yielding of the longitudinal reinforcement in the beams. To ensure that the columns are stronger than the beams and that the breakdown of the bond stress in the joint is prevented, the special provisions for column strength recommended by ACI-ASCE Committe 352 (1985) are maintained.

REFERENCES

1) ACI-ASCE Committee 352, *Recommendations for Design of Beam-Column Joint in Monolithic Reinforced Concrete Structures*, ACI Journal, Proceedings, Vol. 82, No. 23, May - June, 1985, pp. 266 - 283.
2) ACI 318-89 (Revised 1992) and Commentary-ACI 318 R-89 (Revised 1992), American Concrete Institute, Detroit, Michigan.
3) Budiono, B., *Hysteretic Behaviour of Partially-Prestressed Concrete Beam - Column Connections*, Ph.D Thesis Department of Structural Engineering, School of Civil Engineering, University of New South Wales, 1995.
4) Foster, S. J., *The Structural Behaviour of Reinforced Concrete Deep Beams*, Ph.D Thesis, Department of Structural Engineering, School of Civil Engineering, University of New South Wales, August 1992.
5) Paulay, T., Park. R., and Priestley, M. J. N., *Reinforced Concrete Beam-Column Joints Under Seismic Actions*, ACI Journal, Proceedings, Vol. 75, No. 60, November 1978.
6) New Zealand Standard Association, NZS 3101, *Code of Practice for the Design of Concrete Structures*, 1982.
7) Vecchio, F. J. and Collins, M.P., *The Modified Compression Field Theory for Reinforced Concrete Elements Subjected to Shear*, ACI Journal Proceedings V 83, No. 22, March - April 1986
8) Warwick, W.B., and Foster, S.J., *Investigation Into the Efficiency Factor Used in Non - Flexural Reinforced Concrete Member Design*, UNICIV Report No. R-320, July 1993, University of New South Wales, Australia.

The effects of column stirrup and longitudinal beam reinforcement on exterior beam-column joint under cyclic loading

W. Merati,[a] P. Widagdo[b]
[a]*Civil Engineering Department, Institut Teknologi Bandung, Jl. Ganesha 10, Bandung 40132, Indonesia*
[b]*Civil Engineering Department, Atmajaya University Yogyakarta, Indonesia*

The aim of this research is to investigate the effect of ratio of longitudinal beam reinforcement in the joints under cyclic loading. As the measure of evaluation, the hysteretic diagram for joint response subjected to a certain cyclic loading pattern has been used. Three 1: 1 scale exterior beam-column joint specimens have been used to represent 3 variations of longitudinal beam reinforcement ratio, which affect the number of joint stirrup required.

The test results have been presented in the form of crack pattern, energy dissipation and accumulation for various loading cycles, and hysteretic loop graphs. It is concluded that the ratio of longitudinal reinforcement does not affect the ductility of the beam-column joint, provided that the number of joint stirrup is sufficiently installed. Furthermore, it is also concluded that higher ratio of longitudinal beam reinforcement requires higher number of stirrups to reach a constant cumulative energy dissipation capacity.

1 Introduction

Cyclic loading as occurs during earthquake, inflicts complex behaviour on the structural strength. One important structural component of a building of which the behaviour under cyclic loading is important to know is the beam-column joint. At the beam-column joint no failure can occur since it can decrease the overall structural performance. Present reinforced concrete building design concept uses the strong column each beam philosophy, which means that the plastic hinge should occur at the beam rather than at the column or beam-column joint.

The aim of this research is to observe the beam-column behaviour with varying compressive to tensile reinforcing steel ratio which affects the amount of stirrups at the joint under cyclic loading.

In the Indonesian Concrete Code, SKSNI-91(1991) it was stated that the compressive to tensile reinforcing steel ratio was not taken into account for determining the amount of required stirrups at the joint since in this case it was assumed that the longitudinal reinforcing steel bars bonding ability does not increase the shear strength of the joint. In the later literature, T. Paulay, et al. (1992) suggested to take the bonding ability of the reinforcing bar since analytically it can be proved that it helps increase the joint shear strength. One way to determine the shear strength at the joint is by measuring joint ductility that occurs. An interesting question can be raised on how much bond between the reinforcing steel and the concrete contributes to the joint shear strength when the joint reaches the yielding condition.

Analytically the contribution of the reinforcing steel bonding ability to the shear strength of the joint suggested by T. Paulay and Priestley is as follows.

2 Joint Horizontal Shear Mechanism

Forces acting on beam-column joint due to vertical and horizontal loads generally can be described as shown in Fig. 1. Shear force acting on the joint is V_{jh} and V_{jv}, which are the horizontal and vertical shear forces respectively. The horizontal shear force can be subdivided into the following components:

$$V_{jh} = V_{ch} + V_{sh} \qquad (1)$$

where V_{ch} is the shear force contribution from the concrete (concrete strut mechanism) and V_{sh} is the shear force contribution of the reinforcing steel (truss mechanism).

Horizontal shear force contribution by the concrete strut mechanism can be subdivided into several components:

$$V_{ch} = \Delta T_c - V_{col} \qquad (2)$$

where ΔT_c is the fraction of steel compression force C_s developed in the bottom beam reinforcement, introduced to the strut by means of bond over the length of bar subjected to transverse compression from the lower column, C_c is the compressive force component on the beam, and V_{col} is the column shear force.

The compressive force component on the beam C_c can be calculated by using the following equilibrium equation on the beam:

$$C_c = T - C_s = (1-\beta/\lambda_0)\, T \qquad (3)$$

where T is the tensile force of the upper steel reinforcement, C_s is the compressive force of the bottom steel reinforcement, β is the ratio of the upper steel reinforcement to the bottom steel reinforcement areas, and λ_0 is the overstrength factor of the steel reinforcement.

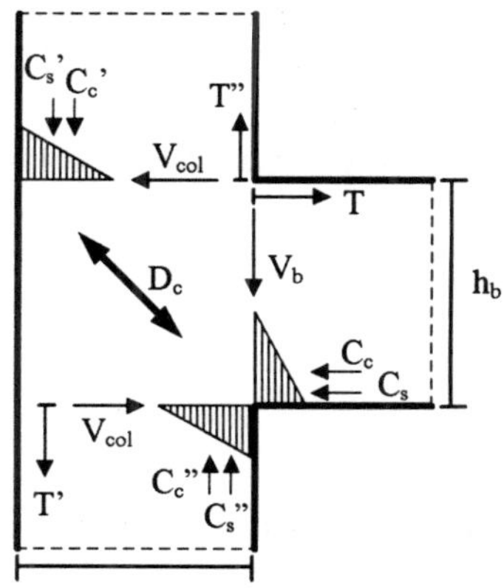

Figure 1: Forces on the exterior beam-column joint

The ΔT_c component is computed by using the following assumptions

1. Bonding force transfer of the bottom beam reinforcement which is bent inside the joint core is assumed to be effective within 0.70 h_k. Therefore, the bonding force magnitude per unit length can be calculated by using the following formula:

$$u_o = \frac{C_s}{0.7.h_k} \approx \frac{1.4.C_s}{h_k} \tag{4}$$

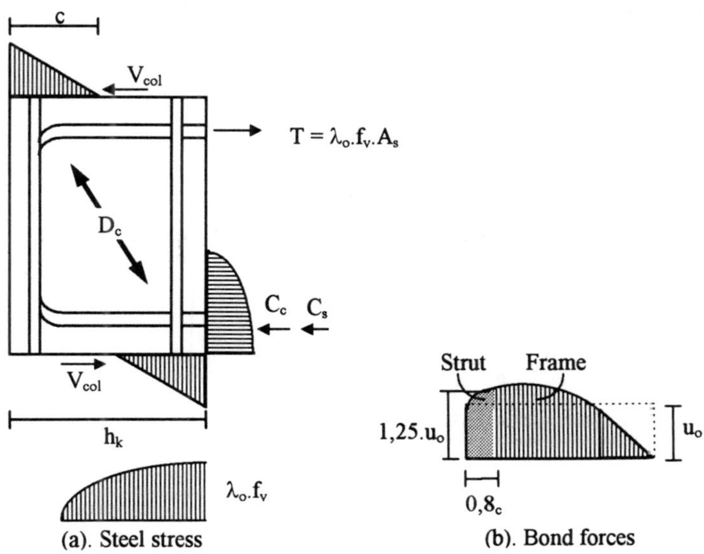

Figure 2: Development of shear forces within exterior beam-column joint

2. The bonding force per unit length u_0 which contributes to ΔT_c is assumed to be effective in the compressive part within the length of 0.80 c as shown in Fig. 2. Therefore, ΔT_c can be calculated by using the following formulas

$$\Delta T_c = (1.25.u_o).(0.80.c) = u_o.c \qquad (5)$$

or

$$\Delta T_c = \frac{1.4.c.\beta}{\lambda_o.h_k}T \qquad (6)$$

By substituting the values of C_c from Eq. (3) and ΔT_c from Eq. (5), the value of V_{ch} is obtained as follows:

$$V_{ch} = \left(1 - \frac{\beta}{\lambda_o}\right).T + \frac{1.4.c.\beta}{\lambda_o.h_k}T - V_{col} \qquad (7)$$

The shear force V_{sh} which is contributed by the frame mechanism can be obtained by substituting equation

$$V_{jh} = T - V_{col} \qquad (8)$$

into Eq. (1) to obtain the following equation:

$$V_{sh} = T - V_{col} - V_{ch} \qquad (9)$$

The above equation can be simplified by substituting the value of V_{ch} from Eq. (7) into Eq. (9) to obtain the following equation:

$$V_{sh} = \left[1 - \frac{1.4.c}{h_k}\right]\frac{\beta}{\lambda_o}T \qquad (10)$$

The value of c which is the depth of the flexure compression zone of the elastic column can be approximated by

$$c = \left(0.25 + 0.85\frac{P_u}{f_c^{'}.A}\right).h_k \qquad (11)$$

By substituting Eq. (11) into Eq. (10), the following equation is obtained

$$V_{sh} = \left[0.65 - 1.19 \frac{P_u}{f_c^{'} . A_g}) \right] \frac{\beta}{\lambda_o} T \qquad (12)$$

and this equation is simplified into :

$$V_{sh} \approx \left[0.70 - \frac{P_u}{f_c^{'} . A_g}) \right] \frac{\beta}{\lambda_o} T \qquad (13)$$

where P_u is the compressive force acting on the column, β compressive to tensile steel reinforcement ratios at the beam, T is the tensile force on the beam, λ_0 is the overstrength factor of the steel, and h_k is the column depth.
Based on the above equations the joint horizontal stirrups required to carry horizontal shear force can be calculated by equation

$$A_{jh} = \frac{V_{sh}}{f_{yh}} \qquad (14)$$

or in different form

$$A_{jh} = \beta \left(0.70 - \frac{P_u}{f_c^{'} . A_g} \right) \frac{f_y}{f_{yh}} A_s \qquad (15)$$

where A_s is the tensile reinforcing steel area, A_g is the gross column area, f_y is the longitudinal reinforcing steel yield stress, and f_{yh} is the stirrup yield stress.

3 Experimental Tests

To evaluate the hysteretic behaviour and stability in the joint area where reinforcement calculation was carried out as described above, experiments on three samples in the form of exterior beam-column joint (one to one scale) varying the longitudinal beam reinforcement ratios (β = 0.5, 0.75, and 1.0) under cyclic loading with a certain pattern have been carried out.

Reinforcement of beam and column members were based on capacity design according to the Indonesian Concrete Code and Standard Practice SKSNI T-15-1991-03. Shear reinforcement in the joint was designed based on the formula suggested by T. Paulay et. al. [10].

Reinforcement design data on the specimens are shown in table 1, with drawings on Figs. 3, 4, and 5.

Table 1. Reinforcement amount and characteristics on the specimens

Reinforcement and Characteristic	Specimen BU-1	Specimen BU-2	Specimen BU-3
Beam			
Longitudinal Reinforcement			
Top	4 ϕ16 mm	4 ϕ16 mm	4 ϕ16 mm
Bottom	2 ϕ16 mm	3 ϕ16 mm	4 ϕ16 mm
f_y (MPa)	518	525	550
f_{su} (MPa)	624	632	652
Column			
Longitudinal Reinforcement	8 ϕ16 mm	8 ϕ16 mm	8 ϕ16 mm
f_y	512	521	574
f_{su}	618	658	667
Stirrup	3 ϕ10 mm	4 ϕ10 mm	6 ϕ10 mm
f_y	385	385	385
f_{su}	524	524	524

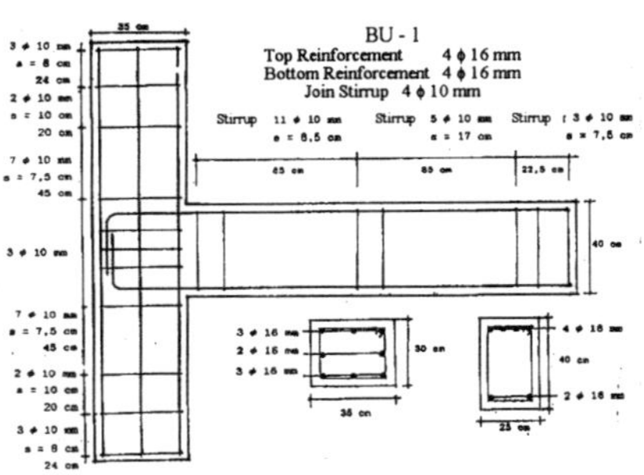

Figure 3: Reinforcement on BU-1

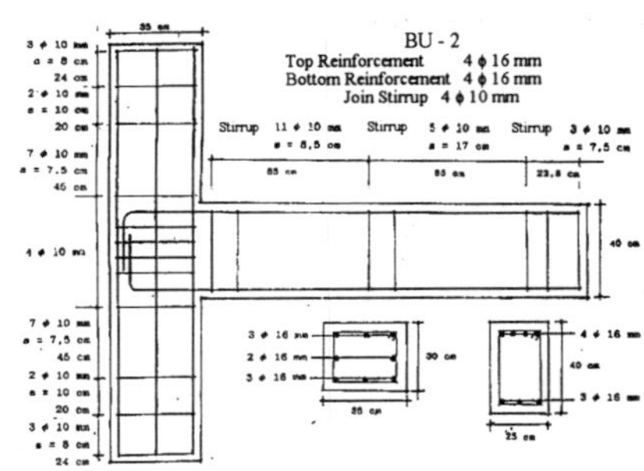

Figure 4: Reinforcement on BU-2

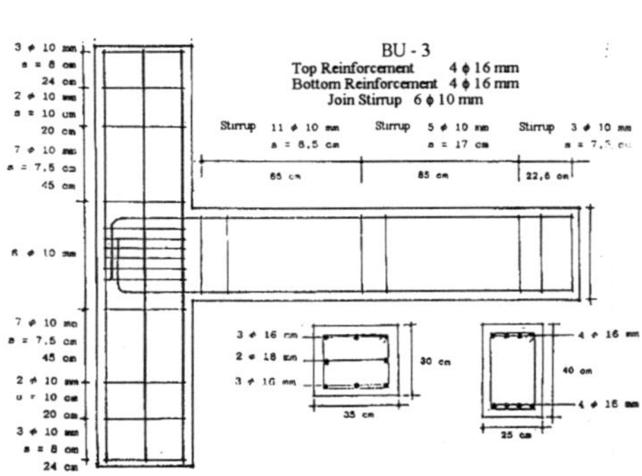

Figure 5: Reinforcement on BU-3

Concrete grade used for the specimens were of f_c' = 35 MPa. The setup of the testing is shown in Fig. 6, with cyclic loading pattern for each sample given in Fig. 7.

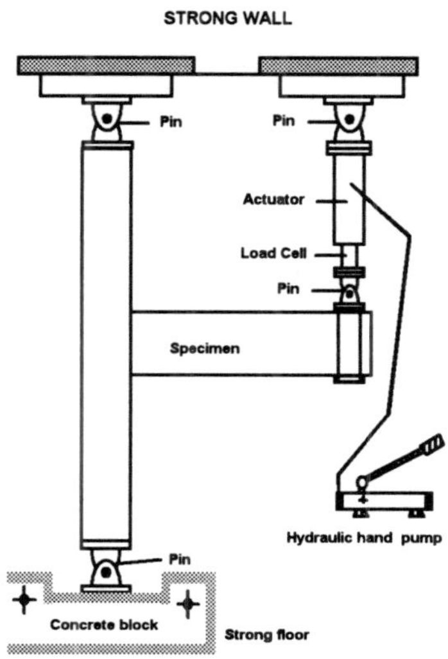

Figure 6: Specimen setup

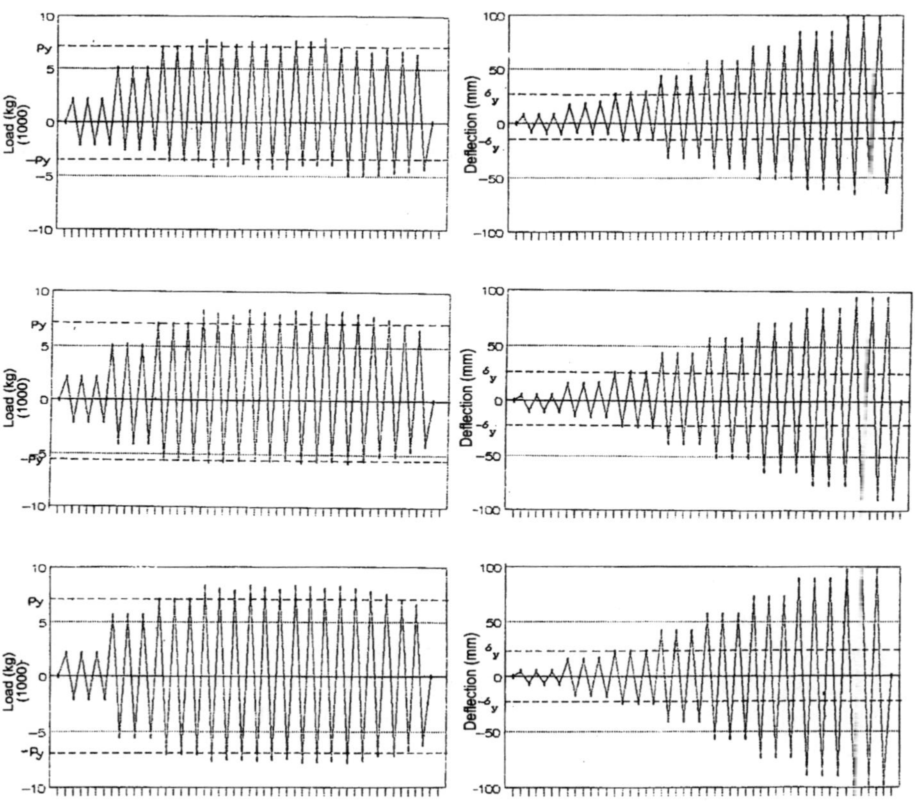

Figure 7: Loading pattern for BU-1 (a), BU-2 (b), and BU-3 (c)

4 Test Results

Test results are presented in the figures in the form of drawing (for crack pattern) and curves (for hysteretic loop and energy dissipation). Figure 8 shows crack patterns for specimens 1, 2, and 3 whereas Fig. 9 shows load-deformation measurements in the form of hysteretic loop curve.

Energy dissipation calculated from hysteretic loop area per cycle are shown on Figs. 10 and 11 for absolute and relative values respectively throughout the loading history.

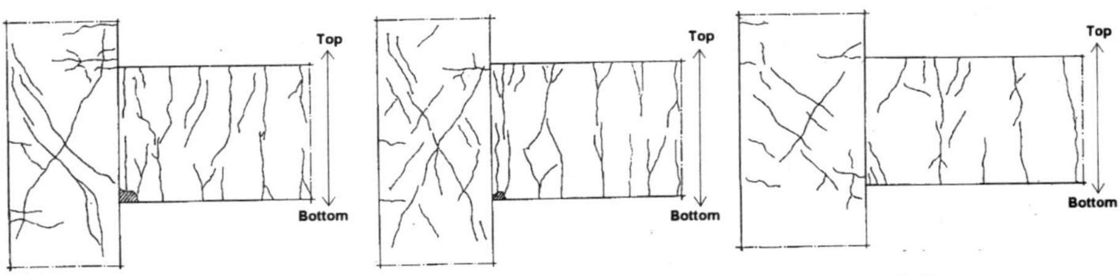

(a) (b) (c)

Fig. 8: Crack pattern lines of the specimens BU-1 (a), BU-2 (b), and BU-3 (c)

5 Discussions

From Fig. 8 it is observed that crack pattern lines on the beam perpendicular to its axis shows a moment dominated failure. In the joint area the crack lines are in the diagonal direction which shows the existence of shear failure.

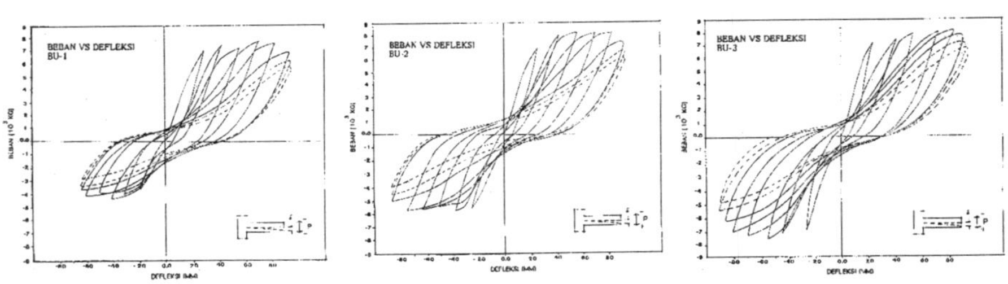

(a) (b) (c)

Fig. 9: Hysteretic loop of BU-1 (a), BU-2 (b), and BU-3 (c)

The applied load pattern was based on the stroke control consecutively giving loading stage with peaks of 3 cycles at elastic level, 3 cycles at 75% yield deflection level, and more than 3 cycles at larger than yield deflection level. With such loading pattern beam-column joint strength and ductility can be observed.

Straingages installed at the beam and column longitudinal reinforcing steel give information on the stress condition of the reinforcing steel.

To eliminate the differences among maximum loads that can be carried by each specimen in the dissipation energy calculation, in Fig. 11 a relative energy dissipation with respect to the yield energy dissipation ($P_y\delta_y$) was used. Figures 10 and 11 show energy dissipation increase every time as the load increases, however at the same load level a decrease of energy dissipation is encountered at later cycles. This shows the existence of stiffness and ductility degradation.

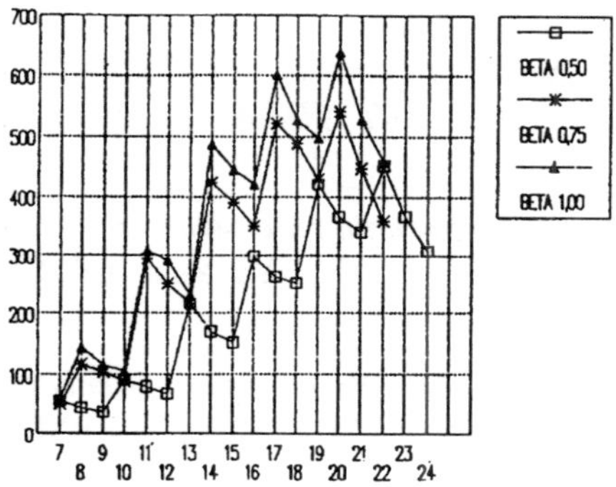

Fig. 10: Absolute energy dissipation

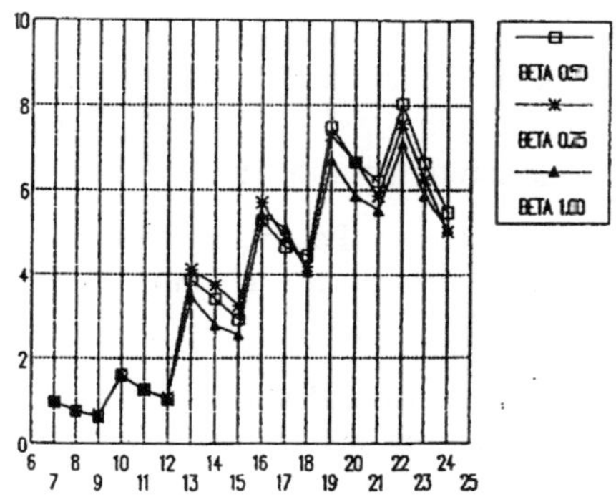

Fig. 11: Relative energy dissipation

6 Conclusion

From the tests and observations in this research, the following conclusions can be deducted:

1. Crack pattern that occurs on the beam shows the occurrence of bending failure whereas in the joint area the occurrence of shear failure which is the aim of this research.
2. The joints which are designed using specifications proposed by T. Paulay et al. [10], with compressive to tensile reinforcing steel ratio β = 0.50, 0.75, and 1.0 have almost the same energy dissipation (Fig. 10. b). This means that the reduction of stirrup proportional to the β ratio level can be done without reducing dissipation capacity, which also means the ductility of this exterior beam-column joint. The increase of β will increase the amount of joint stirrup. This also proves that bonding ability of longitudinal reinforcing steel contributes to the joint shear strength.
3. Longitudinal beam reinforcement ratio affects energy dissipation decrease at the n-th cycle compared to the first cycle at the same deflection. The increase of beam reinforcement ratio generally reduces energy dissipation decrease.

Acknowledgement

The authors would like to thank Directorate General of Higher Education of the Ministry of Education of the Republic of Indonesia for funding the research. The authors also wish to extend their gratitudes to their colleagues, Drs. Jody Firmansjah, Binsar Hariandja and Adang Surahman for their discussions, critics, and corrections given during the research.

References

1. ACI Committee 318., "*Building Code Requirements for Reinforced Concrete (ACI 318M-89) and Commetary-ACI RM-89*", American Concrete Institute 1989.
2. Departemen Pekerjaan Umum, "*Tata Cara Perhitungan Struktur Beton untuk Bangunan Gedung SK SNI T-15-1991-03*", Yayasan LPMB, Bandung, 1991.
3. Departemen Pekerjaan Umum, "*Tata Cara Perhitungan Rencana Campuran Beton Normal SK SNI T-15-1990-03*", Yayasan LPMB, Bandung, 1990.
4. Fintel, R., "*Design Recommendations for Type 2 High-Stretch Reinforced Concrete Connections*", American Concrete Institute, May-June 1991.
5. Fu, H.C., Erki, M.A. and Seckin, M., "*Review of Effects of Loading Rate on Reinforced Concrete*", Journal of Structural Engineering, ASCE, December 1991.
6. Ma, S.Y.M., "*Experimental and Analytical studies on the Hysteretic Behaviour of Rectangular and T-Beams*", Earthquake Engineering Research Center - Univ. of California at Berkeley, Report No. 76 - 2, May 1976.
7. Merati, W., "*Cyclic Behaviour of Structural Elements*", Research Contract No. 003/P4M/DPPM/L3311/PAU/1993, ITB, Oct., 1994.
8. Naem, F., "*The Seismic Design Handbook*", Van Nostrand Reinhold, New York, 1989.
9. New Zealand Standard Association, NZS 3101, "*Code of Practice for the Design of Concrete Structures*", 1982.
10. Paulay, T. and Priestly, M.J.N., "*Seismic Design of Reinforced Concrete and Masonry Building*", John Wiley & Son, Inc. 1992.

Experimental modelling for the investigation of shear in post-tensioned brickwork cantilever fin walls

F. Shafii,[a] B. Hobbs[b]

[a]Faculty of Civil Engineering, Universiti Teknologi Malaysia, Jonor, Malaysia

[b]Division of Civil Engineering and Building, Faculty of Science and Technology, University of Teesside, North Yorkshire, UK

Abstract

This paper describes a model scale investigation of shear behaviour in post-tensioned brickwork cantilever fin walls subjected to lateral load. Experimental testing of walls based on reduced height models to represent the critical zone under shear are described. The results of test walls based on this modelling technique are presented. The predictions of critical stresses by the finite element analysis correlate well with the values obtained by the Engineering Bending theory except for very high stress levels. It is concluded that the shear behaviour of prestressed walls can be studied by modelling at reduced scale.

1 Introduction

In post-tensioned walls the resistance to horizontal loads is provided by both the bending and shear strength of the wall. However, although the presence of prestress force improves the shear strength, it is often shear rather than bending which governs the load-carrying capacity, especially for earth retaining walls with geometric cross-sections. In geometrical sections, where the webs are relatively thin, these may lead to the development of diagonal cracking of the web. So far, only a small amount of research has been done to explore this particular problem in post-tensioned brickwork (Roumani and Phipps [1], Curtin, Shaw and Howard [2], Hobbs and Shafii [3], Shafii and Hobbs[4], Shafii [5, 6]). Additionally, the general understanding of shear behaviour in brickwork is made complicated by its directional properties, where failure may be regarded as being affected by both magnitude and direction of the principal stresses relative to the bed joint.

The majority of existing experimental research on post-tensioned walls has been based on testing full scale brickwork. This requires considerable time and effort in the preparation and construction of test specimens and consequently restricts the number of tests that can be undertaken during the period of research. Additionally, full scale testing may also be restricted by the availability of laboratory facilities and space to accommodate the brickwork structure. Previous research by Hobbs and Daou [1988, 1993] investigated the behaviour of free cantilever fin walls subjected to lateral loading using model scale studies. The work conducted by Hobbs and Daou [7, 8] identified shear as the critical mode of failure requiring further investigation.

A model scale investigation into the shear behaviour of post-tensioned brickwork cantilever fin walls is described in this paper. The structural performance and cracking behaviour of prestressed walls are examined. Comparisons are made between the calculated critical stresses at shear failure by the Engineering Bending theory and Finite Element analysis.

2 Reduced height models

Hobbs and Daou [8] reported that shear failures in prestressed walls are confined to within the shear span of the brickwork under loading. Based on these findings Shafii [5] developed a modelling technique for representing the critical region under shear in prestressed brickwork. The actual forces in a full height wall were reproduced by the modelling system using short wall and with an appropriate loading arrangement, Figure 1.

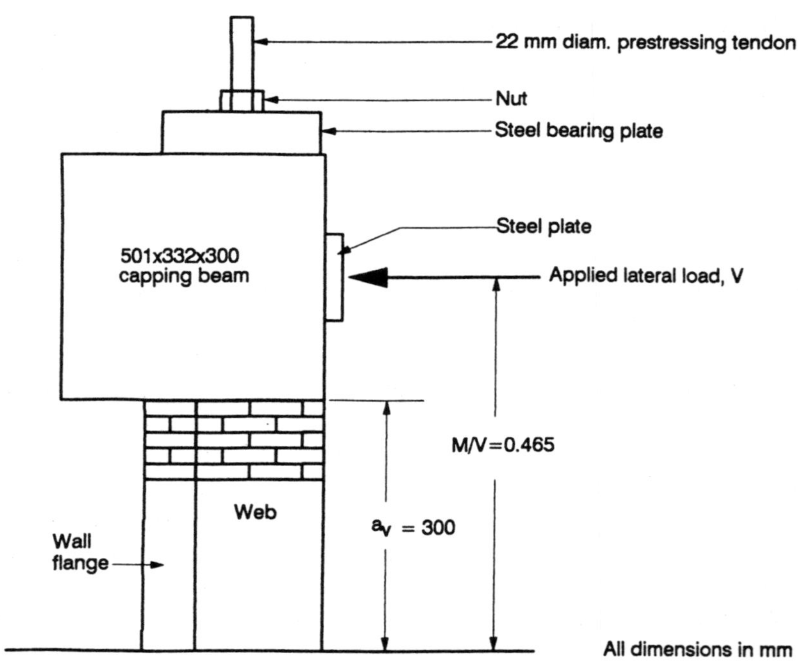

Figure 1: Test arrangement of reduced height model.

T-shaped sections with relatively thin webs were chosen for the cross sections giving the potential for shear failure in the web. A detailed description of the modelling technique is given by Shafii [5,6] and Shafii and Hobbs [4]. Tests conducted by Shafii [5] on standard sections used by Hobbs and Daou [8], showed that the reduced height models behaved similarly to the full height wall. It was concluded that with a proper modelling system, the shear behaviour of prestressed walls can be studied by modelling at reduced scale. It offers an alternative to full scale test and allowing wider range of tests to be undertaken at much reduced cost and time. The advantages gained from model testing include the ability to test up to the ultimate stage of failure, while full scale tests may be restricted to lower levels of applied load due to the risk of a dangerous collapse.

3 Main experimental programme

3.1 Wall sections and materials

This research was part of an investigation into the shear behaviour of post-tensioned brickwork retaining walls. Following the developments of the reduced height model, other wall sections were tested using the arrangements shown in Figure 1. The studies reported in this paper were based on tests undertaken for section 4 and section 7, shown in Figure 2. Similar to the section tested by Hobbs and Daou [7,8], the prestressing steel was accommodated within the void of the section at a distance of 120 mm from the extreme compression face of the wall.

The brickwork walls were built using 0.293 scale model bricks with dimensions 64x30x19 mm, manufactured in accordance to the specifications of high quality solid clay engineering bricks. The bricks were bonded with 3 mm mortar joints. Prestressing tendons made from high tensile grade EN26 steel bars with nominal diameter of 22 mm were used in the constructions of the walls. The properties of the material used in the construction of the walls are presented in Table 1. Wall specimens were constructed on a reinforced concrete base which was secured to a loading rig. A steel jig was used to guide the brickwork construction in order to maintain the joints and alignment at the required position. The specimens were cured for 28 days prior to testing.

3.2 Experimental methods

General details of the wall construction and prestressing arrangement are given in Figure 3. After 28 days of curing, the prestressing steel was accommodated at the required position within the void and anchored at the bottom end. Prestressing operations were undertaken from the top of the wall, by means of the prestressing steel block designed specially for this research. The prestressing force in the tendon was monitored using strain gauges installed at positions along the length of the steel. After prestressing to the required stress level the nut at the top anchorage was tightened. Crack detectors were assembled on the web, prior to the prestressing operation, to detect the occurrence of cracks during the test.

The lateral load was applied on the capping beam by means of a 500 kN capacity, computer controlled hydraulic jack. On applying the load, the capping beam will transfer the required applied moment on to the top of the wall. The hydraulic jack was operated using displacement control for applying the load on to the wall however, towards the final stage the displacements need to be adjusted in order to cope with the larger movements caused by the disintegration of the wall. The walls were tested until failure or when a state where very little or no change in the applied load occurs with further displacements.

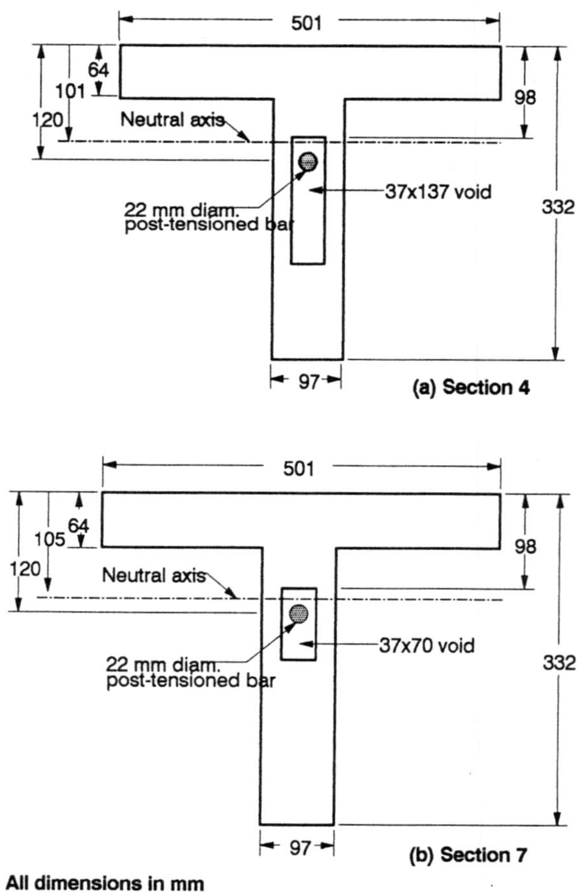

Figure 2: Cross section of test walls.

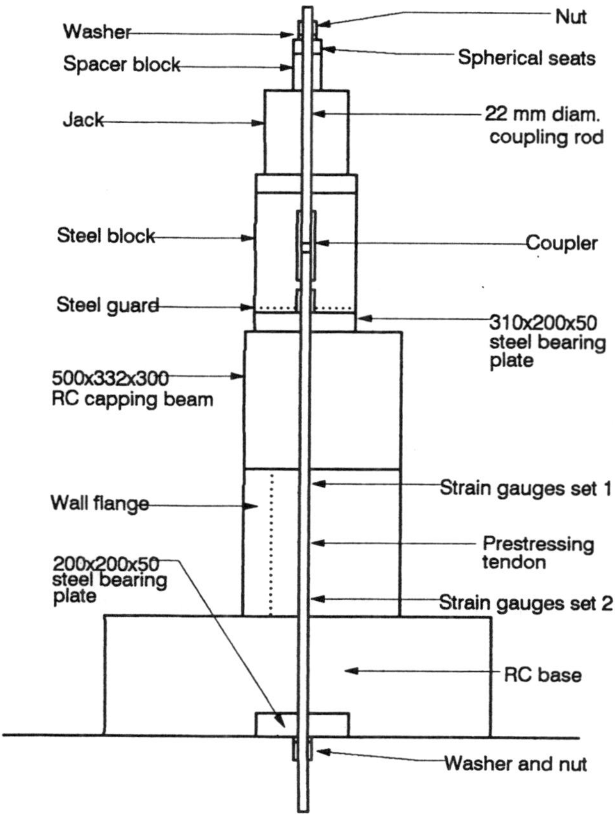

Figure 3: Details of wall construction and prestressing assembly.

Table 1. Material properties

<u>Bricks</u>	
Initial rate of suction ($kg/m^2/min$)	0.38
Water absorption (%)	3.71
Compressive strength on bed face (N/mm^2)	127.00
Axial tensile strength (N/mm^2)	8.40
Elastic modulus when tested on bed face (kN/mm^2)	22.32
<u>Mortar</u>	
Compressive strength (N/mm^2)	17.82
Elastic modulus (kN/mm^2)	14.92
<u>Prestressing steel</u>	
Tensile strength (N/mm^2)	1032.00
Elastic modulus (kN/mm^2)	212.00

4.0 Theoretical analyses of test walls

The stresses in the brickwork at the onset of diagonal cracking was calculated using a linear elastic analysis based on the Engineering Bending theory. A derivation of the formula for the compressive stress on the section, f_n, at the level on the wall where diagonal cracking occurs is given by Shafii [5]. The shear stress was assumed non-uniformly distributed along the depth of the brickwork section and the principal tensile strength of the brickwork calculated using the Mohr circle relationship as given in [5].

The Finite Element analysis, based on linear stress-strain relationship, was used to assess the validity of the elastic section analysis. In addition it provides a means of studying the overall distribution of stresses within the brickwork. A three-dimensional model was constructed using FEMGEN program module whilst the stresses and displacements calculations carried out by FINEL. The presentation of analysis results was performed by FEMVIEW program module. A detail developments of the Finite Element analysis is described by Shafii [5].

5.0 Results and discussions

The walls tested in this investigation failed by shear cracking of the web preceded by flexural cracking which started at the base of the wall when the flexural tensile strength of the brickwork was reached. As the lateral load increased the crack propagated through the web towards the compression flange.

The first diagonal crack occurred at some point in the web, independently and away from the flexural crack, when the critical value of the principal tensile stress was reached. As the load increased, the diagonal crack propagated through bricks and along mortar joints and extended diagonally towards the load point and then downwards to the compression zone. The subsequent propagation of cracks resulted in extensive damage to the web. Hobbs and Daou [8] reported similar failure mode for full height walls. The shear crack patterns for walls constructed from section 4 and 7 are shown in Figure 4 and Figure 5, respectively.

The load-deflection behaviour of the test walls can be categorised into three principal stages of behaviour. The first stage represents the stage of the wall behaviour before flexural cracking. At this elastic stage the deformations of the wall are more or less linearly proportional to the applied moment, i.e linear elastic behaviour. The end of this stage was marked by flexural cracking.

Afterwards, a short transition stage occurs in which the wall stiffness reduced rapidly as the flexural crack at the base of the wall penetrated towards the compression face. The crack propagated rapidly with only a small increase in the lateral load, thus causing a marked change in gradient of the curve.
The second stage is characterised by an increased rate of deflection with applied load and represents the behaviour of the wall after the flexural crack

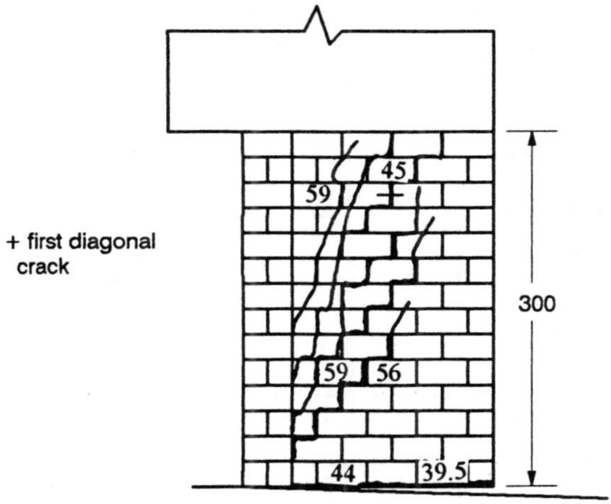

Figure 4: Shear crack pattern of wall 4.

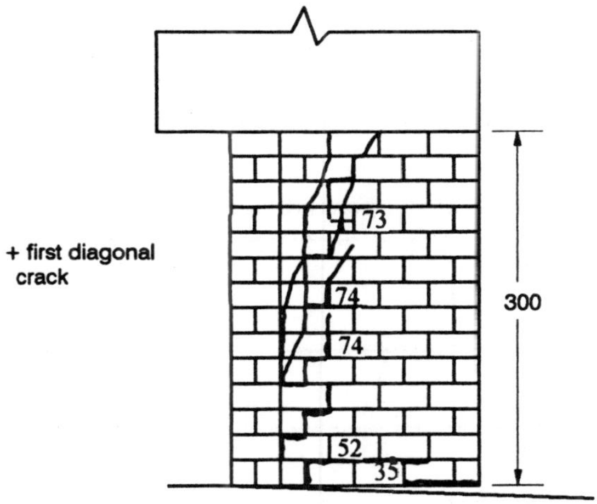

Figure 5: Shear crack pattern of wall 7.

occurs and while the steel stress is still in the elastic range. With increase in applied load, the flexural crack at the bottom of the wall progressed towards the flange and the wall rotate at the bottom of the wall around the compression zone. The end of the second stage was marked by the appearance of shear failure in the form of diagonal cracking. The third stage of the load-deflection curve is characterised by a very slow change in slope prior to the maximum applied load on the wall at ultimate.

The prestress force, P, shear force, V and moment at the base of the walls, M, at the critical stages of testing are shown in Table 2. It can be seen that the forces at both diagonal cracking and ultimate in wall 7 are comparatively greater than wall 4. This is expected because the cross sectional area of brickwork in wall 7 is larger and therefore, would resist greater applied load than in wall 4.

Table 2. Forces at the critical stages of testing

Wall	Flexural cracking			Diagonal cracking			Ultimate failure		
	P_c kN	V_c kN	M_c kNm	P_{dc} kN	V_{dc} kN	M_{dc} kNm	P_u kN	V_u kN	M_u kNm
4	52.5	39.5	18.4	158.3	45.0	30.9	213.3	62.1	28.9
7	152.6	35.0	16.3	214.8	73.0	34.0	226.3	78.6	35.2

The forces shown in Table 2 were used as input for the loading in the elastic section analysis and Finite Element analysis. The calculated shear strength, τ, principal tensile strength, f_{pt} and principal compressive stress, f_{pc}, of the walls at first diagonal cracking are presented in Table 3. In general the stresses at first diagonal cracking corresponding to wall 7 are relatively higher than in wall 4.

Table. 3 Brickwork stresses at the position of first diagonal cracking

Wall	Finite Element analysis		Elastic section analysis			
	f_n N/mm^2	τ N/mm^2	f_n N/mm^2	τ N/mm^2	f_{pc} N/mm^2	f_{pt} N/mm^2
4	1.28	2.64	1.85	2.76	3.84	1.98
7	2.97	3.05	3.54	5.25	7.31	3.77

As indicated in Table 3, there is a good correlation between the compressive stress obtained by the Finite Element analysis and elastic section analysis. The stresses produced by both methods of analysis correlates well for wall 4 however, slight discrepancies between the solutions were evident for wall 7. The discrepancies must be attributed to the theoretical assumptions made when modelling the walls in the respective analysis. Although both methods assumed a linear elastic relationship for the stress-strain behaviour of brickwork, the Finite Element method analysed the walls in three dimensions in comparison to two dimensions as in the elastic section analysis. Although the stresses produced by the finite element analysis are only approximate values but nevertheless this comparison indicates that the stresses calculated for prestressed walls based on elastic section analysis are similar to the values obtained by other established form of analysis particularly at low stress levels.

6.0 Conclusions

The shear behaviour of prestressed walls can be studied by modelling at reduced scale, the lower portion of the wall subjected to the critical stresses.The load-deflection relationships of the tested prestressed walls exhibit three phase behaviour namely, (i) initial uncracked response (ii) second stage following flexural cracking of the base section with the steel stress still in the elastic range and (iii) cracked section with shear failure characterised by the development of diagonal crack in the shear span.

The shear strength and principal tensile strength of brickwork at first diagonal cracking increases with the cross sectional area of brickwork.

The elastic section analysis based on the Engineering Bending theory provides reasonable values of stresses at first diagonal cracking, particularly at low stress levels, as verified by the Finite Element analysis. The elastic section analysis was carried out at the level on the wall where the first diagonal crack occurred, as observed in the test walls.

Acknowledgements

The authors wish to express their thanks to Messrs Marshalls Clay Products for the supply of the model bricks and technical staff of the University of Sheffield, Department of Civil and Structural Engineering for their willing assistance.The financial support afforded to the first author by The Association of Commonwealth Universities / British Council is gratefully acknowledged. Finally, thanks are also due to Universiti Teknologi Malaysia for granting study leave to carry out the research.

References

1. Roumani, N. and Phipps, M.E., The ultimate shear strength of unbonded prestressed I and T section simply supported beams, pp. 82-84, No 2, Proceedings of the British Ceramic Society, Masonry (2) Stoke-on-Trent, 1988.

2. Curtin, W.G., Shaw, G. and Howard, J., Structural testing of a post-tensioned brick fin wall, pp 333-343, Vol. 1, Proceedings of the 9th Int. Brick/Block Masonry Conference, Berlin, 1991.

3. Hobbs, B. and Shafii, F., Application of post-tensioning to brickwork retaining walls, pp r.1-r.13, Proceedings of the Asia-Pacific Structural Engineering Conference, Malaysia, 1992.

4. Shafii, F. and Hobbs, B., The structural modelling of post-tensioned brickwork cantilever fin walls, Proceedings of the 4th. International Masonry Conference, London, October, 1995 (in press)

5. Shafii, F., Shear behaviour of post-tensioned brickwork cantilever fin walls, PhD Thesis, University of Sheffield, 1994.

6. Shafii, F., Modelling of post-tensioned brickwork cantilever fin walls, Malaysian Science and Technology Congress, 1995 (in press)

7. Hobbs, B. and Daou, Y.A., Post tensioned T-section brickwork retaining walls, pp. 665-675, Proceedings of the 8th Int. Brick/ Block Masonry Conference, Dublin, 1988.

8. Hobbs, B. and Daou, Y.A., An experimental investigation of post-tensioned brickwork fin walls, pp. 120-130, Proc. Inst. of Civil Engineers, Building and Structures, 1993.

Structural vibrations calculation and testing of the heavy lifting appliance

D. Dinevski, M. Oblak
University of Maribor, Faculty of Mechanical Engineering,
Smetanova 17, 62000 Maribor, Slovenia

Abstract

Mechanical model of a hoisting appliance for standard ship containers (container crane) is made on the basis of finite elements method (FEM) and its vibration analysis is performed. The properties of the beams are modelled as line beam elements with the formulation for stretching, bending, torsion and transverse shear deformation effects. The actual hoisting appliance, for which the results are presented is the container crane with the load carrying capacity of 325/380 kN, product of Metalna Slovenia, stationed in Bangkok port, Thailand (Fig.1). Results of dynamic analysis of a box beam (reinforced with inner ribs) modelled with the mentioned line and 3D shell elements are compared. On account of a rather large dead weight (4700 kN) of the crane, the effect of the second order theory (geometric nonlinearity) on the eigenvalues and eigenmodes is estimated. To define the damping values and validate the modeling strategy an experiment on the similar structure was performed. The test and numerical results of the mentioned structure resonance frequencies and vibration mode shapes are compared.

1. Introduction

The aim of the conducted research was to obtain a relevant dynamic model of a container crane structure. The guiding reason is optimal design of the carrying structure with respect to its dynamic features, which have to be within certain limits. The wider purpose was to verify the method of the hoisting appliance structures dynamic model design with Finite Element Method (FEM), which means to assess:

- the type of the beam elements (result: Timoshenko beam is appropriate)
- the minimal density of the finite elements for particular parts of the structure
- whether to apply the 1st or the 2nd (geometric nonlinearity) order theory (result: 1st is good enough)
- the load function for the simulation of a general working cycle

On the other hand it is of significant importance in the design process to know in advance following dynamic properties of the structure:

- eigenvalues and eigenmodes,
- stiffness of the whole structure in the optional directions,
- damping characteristics,
- behaviour of the structure during working cycle,
- behaviour of the structure on the impact load,
- stress and deformation time history on the critical places during common operation procedure.

The dynamic model of container crane will be used for other types of numerical analysis (working cycle, ship transport simulation, fatigue analysis, dynamic optimisation). Results of the research will be applied to related structures through gained knowledge of modelling the dynamic features.

Figure 1. Bangkok port container cranes

The construction is under dynamic loading at following working operations and events:

- working cycle; acceleration or deceleration of the hoisting motion (working load on the spreader), horizontal acceleration (starting and braking) of the sway together with the load, swinging of the load, impact due to collision between the load and the base etc.,
- crane travelling along rail tracks (relatively seldom event),
- crane transport from the manufacturer to the buyer (ship transport of completed structure), where severe (but of short duration) dynamic loading is performed,
- climatic effects, seismic effects.

2. Finite element selection and testing

For the finite element dynamic analysis the NISA II program and a 2-node prismatic 3-D beam element with the formulation for stretching, bending, torsion and transverse shear deformation effects was applied. The element is analogous with the Timoshenko beam model (Equation (1), Fig.2.). Euler-Bernoulli model (without shear deformations and cross-sectional moment of inertia) was less accurate on the reference tests. All NAFEMS Benchmark reference tests for dynamic analysis were performed. Two examples are described.

$$
\begin{aligned}
&EI\frac{\partial^4 u}{\partial x^4} - \left(p - m\frac{\partial^2 u}{\partial t^2}\right) - J\frac{\partial^4 u}{\partial x^2 \partial t^2} + \\
&+ \frac{EI}{GA}\frac{\partial^2}{\partial x^2}\left(p - m\frac{\partial^2 u}{\partial t^2}\right) - \frac{J}{GA}\frac{\partial^2}{\partial t^2}\left(p - m\frac{\partial^2 u}{\partial t^2}\right) = 0
\end{aligned}
\tag{1}
$$

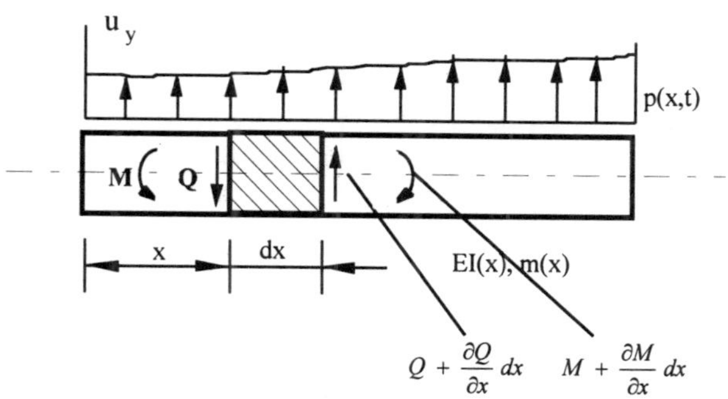

Figure 2. Timoshenko beam vibration formulation

2.1. Two examples of NAFEMS benchmark reference tests

The test beam is on Fig.3;
- Boundary conditions: x=y=z=Rx=0 at A,y=z=0 at B

The forcing function is a uniformly (along x axis) distributed force. The response is measured at the mid-span node (x=5m). Damping is ζ=2%

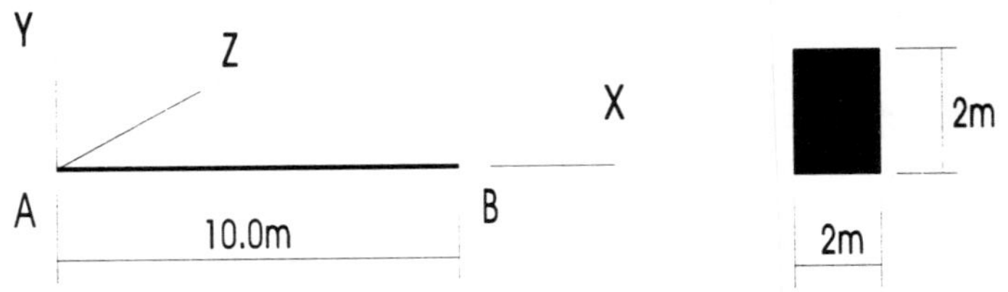

Figure 3. Geometry of the test beam

2.1.1. Harmonic forced vibration response test

Forcing function is:

$$F = F_0 \sin \omega t$$
$$F_0 = 10^6 N / mm$$
$$f = 0 to 60$$

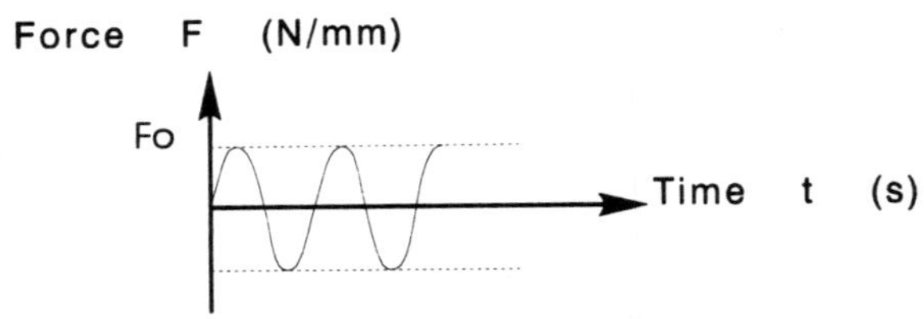

Figure 4. Forcing Function

Response: reference solution (analytically):
Peak displac.: d=13.45 mm at frequency f=42.65 Hz
Our solution (5 elements):
Peak displac.: d=13.42 mm at frequency f=42.71 Hz
The solution is in perfect agreement also with reference numerical solution (difference 0.1%).

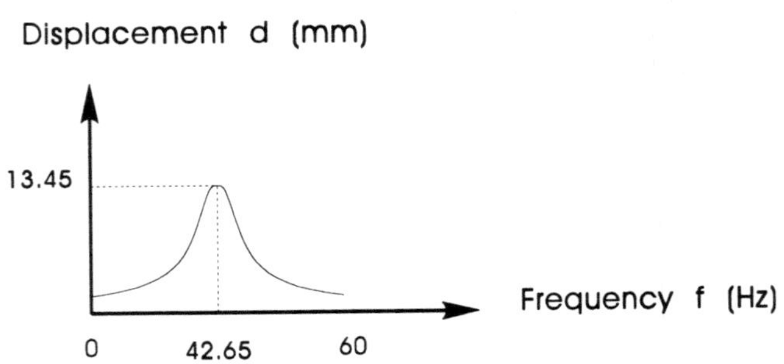

Figure 5. Response

2.1.2. Impact load response test (transient response)

Forcing function (temporal) is suddenly applied step load: $F_0=10^6$ N/mm

Response: reference solution (analytically):

Peak displac.: d=1.043 mm at time t=0.0117 s

Our solution:

Peak displac d=1.042mm at time t=0.0117 s

The solution is in perfect agreement also with reference numerical solution (diff. 0.2%).

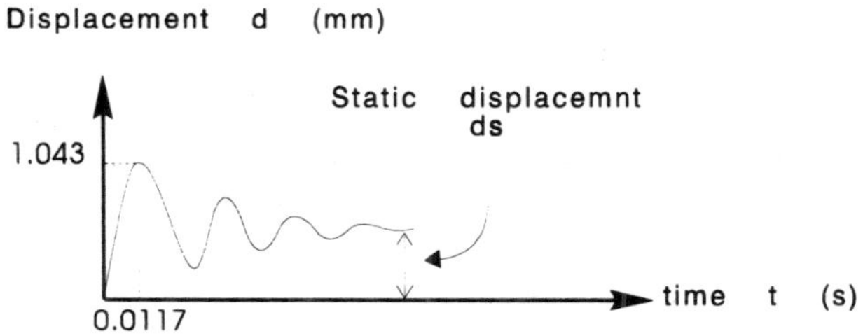

Figure 6. Response

2.2. Line and shell elements comparison test

Steel box beam with inner ribs (of rather large cross-section area) is a basic component of the structure and its mechanical properties would be in a best way modelled with shell finite elements. On the other hand, shell element model of the whole structure would be too complex to prepare and inefficient for the computation. Therefore the suitability of the line beam element was investigated in a way of comparing both (shell and line elements) results (eigenfrequencies and eigenmodes) for a typical beam (cantilever beam, cross-section on a Fig. 7, length is 10 m).

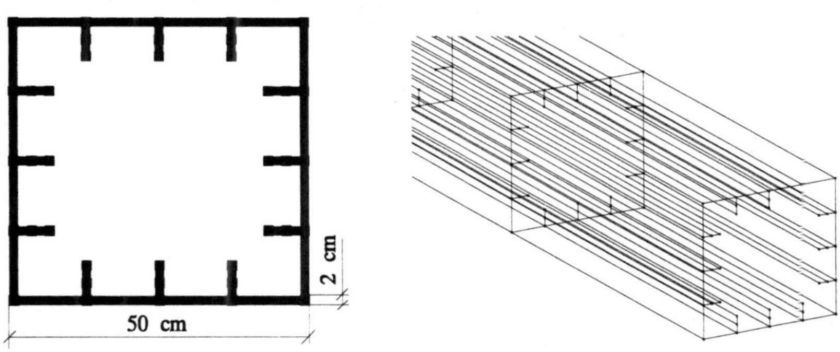

Figure 7. Beam cross-section and shell model detail

The line element is already described, the shell element is 3D, quadrilateral four node, six degrees of freedom per node element, which includes membrane, bending and shear deformation effects. The line model has 5 elements and 36 degrees of freedom, the shell model has 400 elements and 2640 degrees of freedom; The agreement of the eigenfrequencies shows, that the line finite element model is good enough for our purpose.

Eigenfreq.	Line model	Shell model	differ.
1. in 2.	4.379 Hz	4.386 Hz	0.15%
3. in 4.	27.364 Hz	27.197 Hz	-0.61%
5. in 6.	76.437 Hz	75.190 Hz	-1.65%
7.	125.514 Hz	125.485 Hz	-0.02%

Table I. Line and shell finite element model

3. Container crane FEM model

Finite element model of the container crane structure has 85 line and 10 point mass elements. The model has 71 nodes and 426 degrees of freedom. The number of elements is the result of several trials and is the least possible for a satisfactory accuracy.

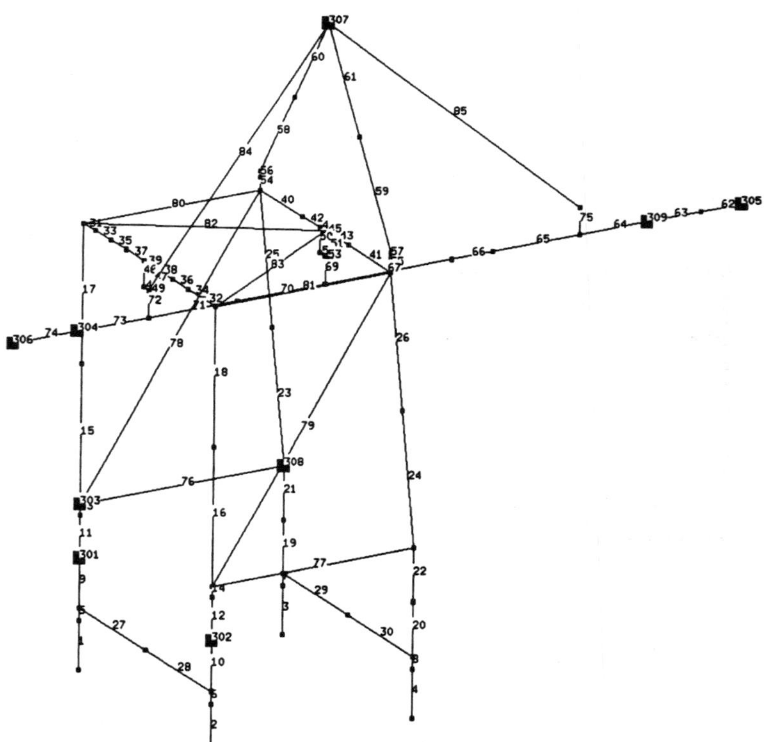

Figure 8. FEM model of the container crane; nodes and elements

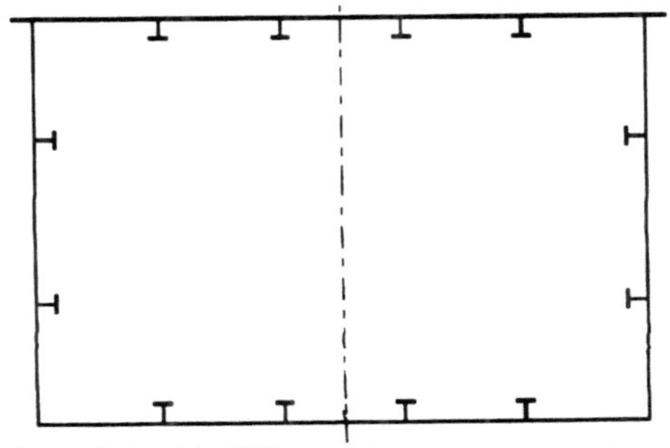

Figure 9. One of the 11 different beam cross-sections (the arm)

3.1. Numerical results

FEM dynamic analysis according to first and second order theory is performed and compared to find out whether the application of second order theory on the analysed model is sensible. According to this theory, the stifness matrix is calculated under the effect of dead weight of the structure (geometric nonlinearity). The eigenfrequencies comparison is in the Table II.

freq.	I. order th.	II. order th.	difference
1.	0.566 Hz	0.556 Hz	-1.7 %
2.	0.634 Hz	0.619 Hz	-2.4 %
3.	0.717 Hz	0.690 Hz	-3.7 %
4.	1.140 Hz	1.129 Hz	-0.9 %

Table.II. The first and the second order theory

Conclusion is that the second order theory is not necessary for our purpose.
The resonance frequencies and mode shapes are on the Fig. 10 and Fig. 11

4. Experimental results

The hoisting appliance, with which the experiment was made, is the container crane with the load carrying capacity 450/500 kN, product of Metalna Maribor, stationed in Koper port, Slovenia.

The purpose of the experiment was the verification of the crane mechanical model, therefore the determination of natural frequencies and modes of vibration and the definition of structural damping. The estimation of dynamic load level (and the shape of load function) encountered in the operating environment was the second goal of the full scale testing.

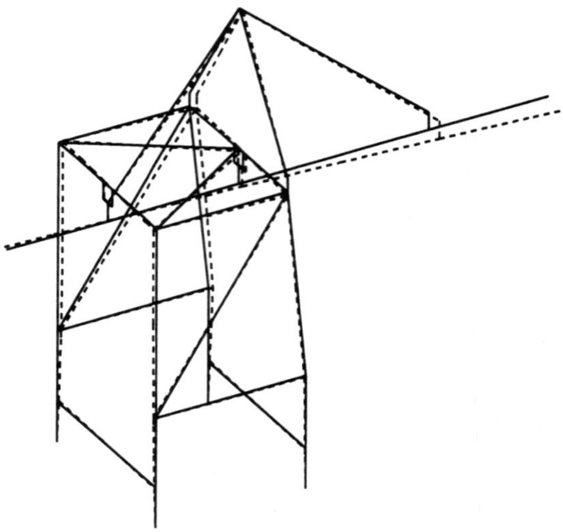

MODE NO. = 1 FREQUENCY = 5.66142E-01 Hz

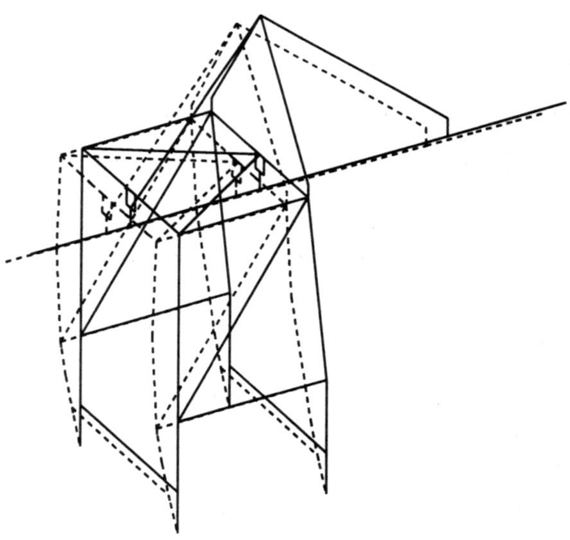

MODE NO. = 2 FREQUENCY = 6.34056E-01 Hz

Figure 10. Eigenmodes and eigenfrequiencies (FEM) of the crane structure

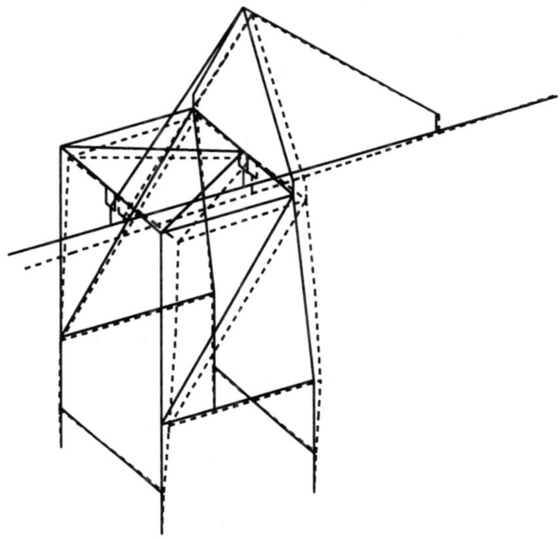

MODE NO. = 3 FREQUENCY = 7.16981E-01 Hz

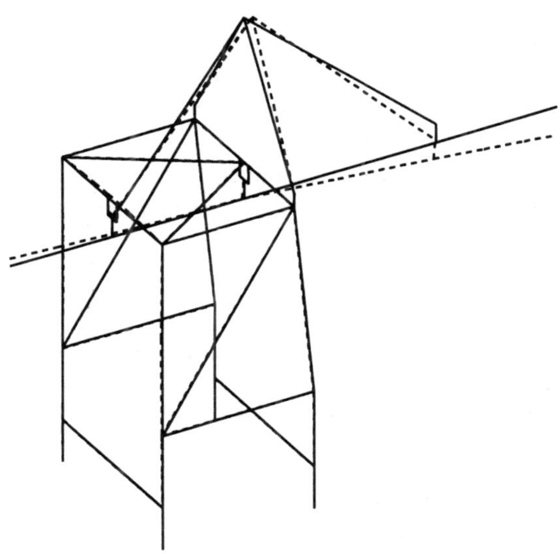

MODE NO. = 4 FREQUENCY = 1.14008E+00 Hz

Figure 11. Eigenmodes and eigenfrequiencies (FEM) of the crane structure

The simplest possible experiment was to be done for achievement of mentioned objectives. Some kind of intersection between forced vibration test with the control of the input (forcing function) and step relaxation test with releasing the structure from a staticaly deformed position was made. Releasing the deformed structure is mechanically difficult to implement. Instead of doing this we tried to force it to the specific vibration mode with limited operational movements using inertial forces of the masses in motion. The effect of the relaxation was in this way more or less achieved. Therefore the input excitation was not directly controlled, but was more or less under control.Transient response (acceleration) of the event was measured. Afterwards the spectral analysis with Fourier transformation was made.

Experimentally measured values of crane natural frequencies are in good agreement with calculated ones (Table III.). Measured damping is ζ=1%.

freq.	FEM result	Experimental	difference
1.	0.517 Hz	0.50 Hz	-3.4 %
2.	0.923 Hz	0.97 Hz	4.8 %
3.	0.783 Hz	0.75 Hz	-4.4 %
4.	0.836 Hz	0.84 Hz	1.5 %

Table III. FEM and experimental Eigenfrequencies of the tested crane

The first numerical results (prior to the experiment) were different only for the second vibration mode where smaller value was computed. The difference was the result of inappropriately modelled bearings and inner reinforcement ribs in the beams above the bearings.
The conclusion is that the mechanical model of the container crane with finite element method is reliable and can be used for other dynamic analyses.

References:

1. Abbassian, F., 'NAFEMS free vibration benchmarks', WS Atkins Engineering Science Report, Issue 2, 1987
2. Dinevski D., 'Mechanical model and vibration analysis of container crane', MSc work, Faculty of mechanical engineering, University of Maribor, Slovenia 1993 (in Slovene language)
3. FEDERATION EUROPEENNE DE LA MANUTENTION standards for heavy lifting appliances, 3rd edition, 1987
4. NISA User's Manual, Version 90.0, Michingan USA , 1991
5. Salawu,O.S., C., Williams, 'Review of full-scale dynamic testing of bridge structures', *Engng Struct. 1995, Volume 17, Number 2, 113-121*
6. White, C.W., 'Dynamic testing of a two-dim. box truss beam' (NASA Contractor Report 4039), NASA, USA, 1987

Experimental and computer simulation of piled embankment behaviour

M.Z. Othman,[a] I.C. Pyrah[b]
[a]*Faculty of Civil Engineering, Universiti Teknologi Malaysia, Jonor, Malaysia*
[b]*Department of Civil and Transportation Engineering, Napier University, Edinburgh, UK*

Abstract

An experimental investigation into the behaviour of piled embankments has been carried out using the pin model apparatus. The development of the experimental model and tests conducted to examine the effects of pile cap area on the model behaviour are discussed. A theoretical simulation of the piled embankment behaviour has also been undertaken using the Finite Difference method.

1 Introduction

The use of piles to help support the fill leading up to road and railway crossings reduces the problems of settlement and stability in several ways. The presence of piles stiffens the soft subsoil and reduces the stresses on the upper subsoil by transferring load through the piles down to a lower firm stratum. As a result settlements are reduced thus allowing the possibility of rapid single stage construction of the embankment without danger of undrained failure of the subsoil. Additionally, the effect of reduced stresses on the subsoil tends to limit the lateral deformation of the soil, thus protecting the bridge support piles from adverse effects. Where the embankment leads up to a structure, differential settlements between the structure and the embankment can be reduced by adopting a construction technique which avoids an abrupt change in soil profile therefore preventing the formation of a step in the road pavement at the junction between embankment and bridge.

In practice embankment piling has been shown to provide a cost-effective method as it obviates the need of side span and pre-loading systems. It also minimises the land intake required for a particular project. However, despite the advantages it offers over the traditional methods of construction, relatively few embankments have been piled in the UK. This is because designers have not been able to predict confidently the proportion of the weight borne by the piles.

Research on this subject, both experimentally and theoretically, is limited. Experimental evidence based on the construction of simple models has been unable to provide complete information on the piled embankment system. Consequently, design methods are based on field experiences, which are not sufficient to take into account some of the parameters considered significant to the performance of the embankment. The presence of embankment piles creates a condition in which the deformation behaviour between the piles and the surrounding subsoil is significantly different. The complexity of the soil-structure interaction problem complicates the analysis, and so far no fully developed theoretical solution exists to quantify the effects of the variables involved.

The paper presents and examines some experimental and theoretical investigations into the behaviour of piled embankments and the effects of the main variables influencing their performance. The simulation of piled embankment behaviour was based on the pin model thus giving the advantage of a photographic recording of the development of arching within the granular fill. Discussion also includes the states of stress in piled embankments produced by the Finite Difference method.

2 Development of experimental model

The experimental investigation was aimed at providing a clear understanding of the principles underlying the behaviour of piled embankments. In order to achieve this, tests were undertaken to examine the principle factors affecting piled embankment behaviour and models were developed to enable a visual study of the development of displacements within the embankment.

The behaviour of piled embankments was simulated using the model shown in Figure 1. A two dimensional mechanical analogue technique was adopted whereby granular soil was represented by steel pins. This technique was used successfully in previous research at Sheffield University to study the development of stresses around anchor piles [1]. The modelling technique offers several advantages :

(i) the boundary condition is automatically and simply satisfied without using any confining structure

(ii) by using small diameter high density stainless steel rods, relatively large forces can be generated in a model, and

(iii) intermediate stresses are eliminated.

The model consists of a steel framework which supports the model components.The base of the framework was securely fastened to the ground to ensure stability of the model structure. An adjustable I-section beam spanned horizontally across the framework and supported the model piles and its assemblies. The length of the embankment to be tested could be varied by

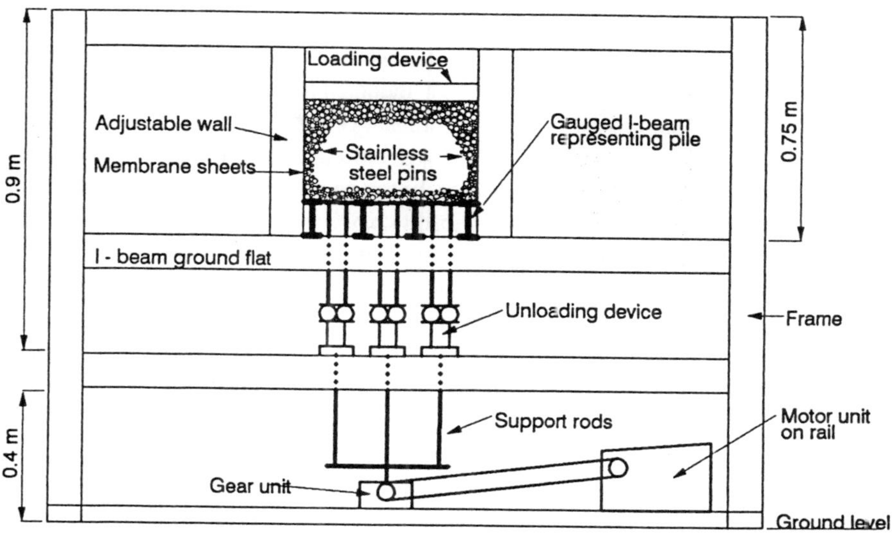

Figure 1: Schematic diagram of pile embankment model

adjusting the position of the vertical walls. The model assumes that no friction occurs between the vertical wall and the pins. Two sheets of rubber membrane were fixed to the vertical surface of each wall and its outer surface greased to form the frictionless boundary.

The pins were fabricated from stainless steel of 3mm and 5mm diameter with length of 75 mm and individually made in the laboratory. These pins were intermixed by hand in proportions of 2 to 1 by weight of large to small pins to form a homogeneous mix as possible. The unit weight of the pin mixture was derived from weighing pins of known volume; this produced a mean value of 6400 kN/m^3. The experimentally determined angle of shearing resistance of the pins was 20°.

The piles were made of extruded aluminium alloy I-sections, with dimensions of 75 mm long, 62 mm deep, 38 mm flange and 2 mm web thickness. The bottom flange of each pile was attached rigidly to the I-section ground steel beam which formed part of the steel frame. The top flange of each pile was attached to an interchangeable pile cap with thickness and depth of 10 mm and 75 mm respectively. In this testing programme a fixed pile cap length of 38 mm was used in all tests and the length of subsoil varied.

The displacement of subsoil between the piles was simulated using an unloading device designed with an interchangeable top plate for varying the subsoil length, Figure 1. The sizes of plates used were 130 x 90 x10 mm, 130 x 110 x 10 mm and 130 x 150 x 10 mm and these represent subsoil length of 90 mm , 110 mm and 150 mm respectively. Two aluminium proving rings, with external and internal diameter of 40 mm and 36 mm respectively,, sandwiched between two horizontal plates, were used to measure the load acting on the subsoil for subsequent displacements. The top horizontal plate on the load cell was connected to the interchangeable plate by four 12 mm diameter vertical rods. The lower plate was bolted to a piston which slides smoothly into a cylinder and in turn was mounted to a horizontal member of the frame. The displacement of the subsoil was measured using a linear displacement transducer fixed under the piston. A reversible and variable speed motor was used for driving the gear system which then converts the circular motion of the gear into vertical displacement of the piston.

The surface of the embankment can be subjected to a uniform surcharge using a loading device which introduces pressure and in turn compresses the pins beneath it. The tests describe in this paper are limited to cases without surcharge.

3 Test methods

The work described in this paper were based on models undertaken using four piles spaced across the length of the embankment. The length of the embankment varied with different test arrangements depending upon the dimensions of the soil components tested. The length of pile caps was fixed at 38 mm and the length of subsoil components varied from 90 mm to 150 mm. The walls at the ends of the model were adjusted into position for testing with the rubber membrane between the plates aligned vertically and greased to provide a frictional boundary. For each test the pins were remixed by hand in a container to form as homogeneous a mixture as possible. The pins were filled between the walls in layers of 100 mm until the entire height. For the purpose of detecting the settlements of pins during the test, the cross sectional tips were coloured forming several horizontal lines across the section of the embankment model.

The test commenced by displacing the unloading device to simulate the displacement of the subsoil between the piles. The displacement of the unloading device was controlled at a constant specified value until the end of the test where no significant change in subsoil load occurred with respect to increments of displacement.

The pattern of movements formed by the pins within the embankment, for a given test condition, was recorded by a lapse photography technique. By this technique the camera photographed the model in multiple exposures on a single frame of film thus enabling the development of pin movements to be recorded.

4 Theoretical studies

The analytical method for assessing the behaviour of a piled embankment is based on the Finite Difference technique in association with the modified Dynamic Relaxation method. A description of the modified Dynamic

Relaxation method for solving soil problems was presented by Pyrah [2]. This approach was adapted for analysing piled embankment problems where the subsoil between the piles is modelled using trap door, comprising of nodes capable of taking prescribed displacements. The soil was modelled with an elastic-plastic stress-strain relationship based on the Mohr- Coulomb yield criterion.

The theoretical analysis considers only an intermediate cell of the embankment and plane strain conditions are assumed. The piles are considered fixed in the horizontal and vertical directions and experiences no rotation. The soil beneath the embankment is assumed to displace vertically and uniformly along and normal to the horizontal plane and regarded as a rigid mass. In the analysis the embankment is constructed instantaneously. The effect of time on soil settlement is ignored and the interaction between the fill and the subsoil at the interface is not taken into account.

In the Dynamic Relaxation method interlaced grids are used for the displacements as such grids gives better finite difference approximation for shear stress. Vertical and horizontal equilibrium are satisfied at the respective displacement points and the stresses are calculated from the strains, defined by the surrounding displacements. The Dynamic Relaxation system of analysis solves the Finite Difference equations by an iterative technique rater than by a relaxation process, where alterations are made so as to always reduce the maximum error. The solution to a static loading problem is obtained when the displacements and stresses satisfy both the stress-strain and equilibrium conditions. In the Dynamic Relaxation technique this is achieved by ensuring that during each iterative cycle the stress-strain equations are always satisfied and the errors in the equilibrium equations are systematically reduced until reaching an acceptable level.

5 Results and discussion

A typical graph of the load borne by the subsoil versus the displacement of the subsoil for the tests undertaken in this research is given in Figure 2. In general, the stresses in the subsoil decreases with increase in settlement, which agrees with the findings of previous researchers [3] and site evidence [4].

A relatively large reduction in stresses occurs immediately after the subsoil is displaced and further settlements lead to a more gradual change in the load-settlement behaviour. The levelling off of the graph indicates that a failure condition in the subsoil has been reached where a continual increase in the settlement causes no significant change in the load. As demonstrated by this test and also by other tests undertaken in this research only a small amount of settlement in the subsoil was required to achieve complete mobilisation of the strength for supporting the load above the subsoil. For example, in this particular test, a settlement of approximately 3 mm can be considered sufficient to produce such a condition. In design practice, the Swedish code [5] assumes complete mobilisation of the soil strength, assuming an immediate transfer of

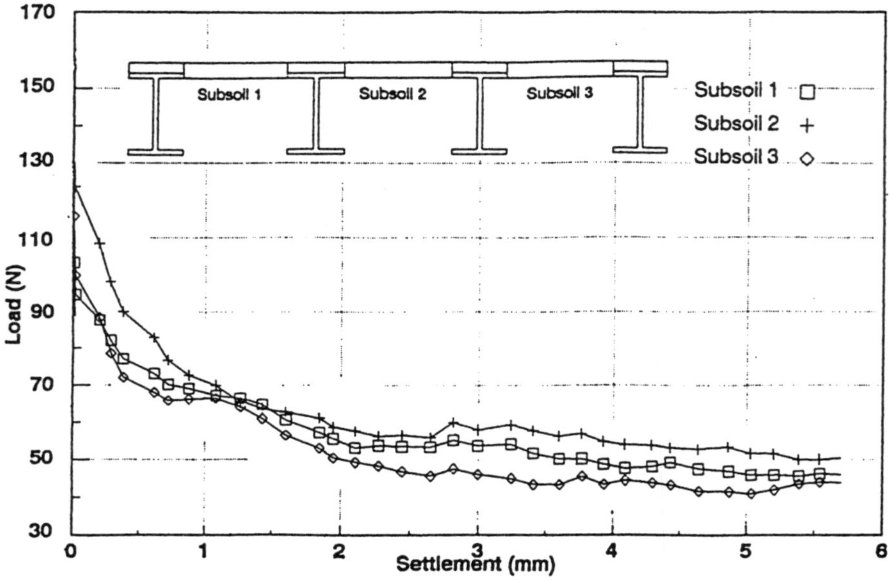

Figure 2: Typical load-displacement relationship of piled embankment

load from the embankment into the piles below it, with the subsoil undergoing negligible displacements.

To some extent, the experimental results shown in Figure 2 are influenced by the displacement of the moveable plates at the initial stage of the test. Although the same initial settlement was prescribed for the three moveable plates, the movements of each plate had to be balanced by its counterparts since the model system was supported about the central axis at the base of the frame. It is also evident from these results that the initial settlement controls the load-displacement path taken by a particular subsoil. However, the load-displacement relationship between the subsoil components appears to be similar.

Figure 3 shows the variation between the percentage of load transferred into the pile and the percentage of pile cap area used in the model. The results were based on 5 repeated tests. In this research the pile cap area was fixed and the subsoil area varied by the changing lengths. The percentages corresponding to subsoil lengths of 90 mm, 110 mm and 150 mm are 29 %, 25% and 20%, respectively. As can be seen in Figure 3, the test results for a particular percentage of the pile cap area vary with differences ranging from 6% to 13%. This is suspected to be due to the effect of the different pin arrangements in each test. Despite these variations, the results illustrate that the percentage of load transferred into the pile increases with percentage increase in the pile cap area used.

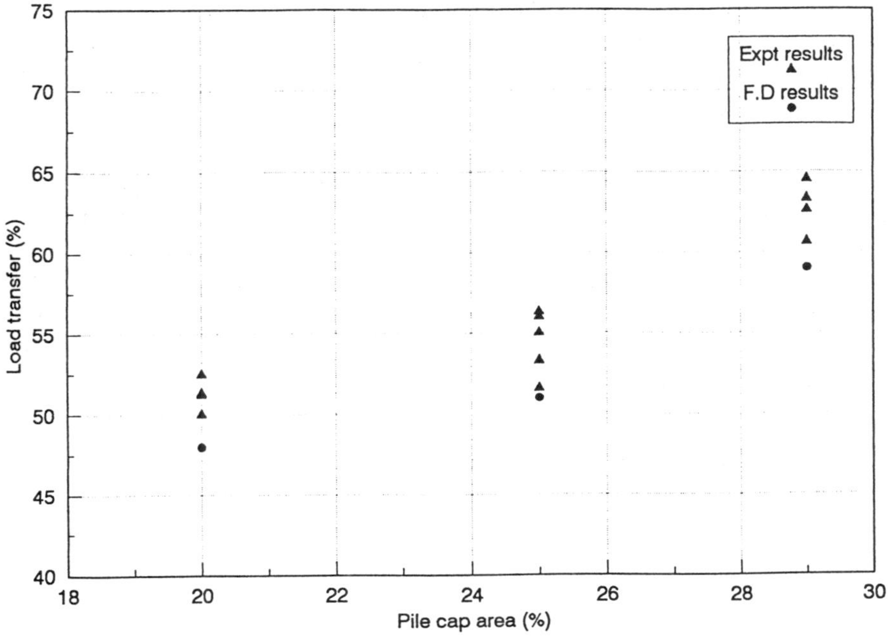

Figure 3: Experimental load transfer of piled embankment model and theoretica predictions of the Finite Difference method.

The transfer of stress from the embankment into the piles is believed to be due to arching taking place within the granular material. As is evident from Figure 3 the relationship between the percentage of load transferred into the piles with the percentage of pile cap area is non-linear. It is also interesting to note that a pile cap area of 30 % was capable of supporting the embankment weight by as much as 63%. An extrapolation of these results indicates that a complete transfer of the stresses from the embankment into the soil may be achieved with further increase of the pile cap area. However, this limiting value is speculated to be less than 100%. When designing pile embankments, previous designers have allowed the piles to be covered by a full slab to ensure a complete transfer of stress from embankment to the piles. These experimental results demonstrate that a relatively small percentage in pile cap area is capable of producing a similar effect.

As illustrated by the results, the differences in the load borne by the piles tend to become less with small percentage in the pile cap area. This is further proven by the theoretical analysis undertaken in this research. The Finite Difference method appears to give lower predictions of the stress transfer compared to the experimental results. Besides the limitations of the experimental model described earlier, other factors could have contributed to the discrepancies between the theoretical and experimental results. The accuracy of the analysis depends on the measured value of the unit weight and angle of shearing resistance of the pins used in the test. In general, comparison of the theoretical solutions for problems in case studies [3,4] indicates the reliability of the Finite Difference method in predicting the behaviour of piled embankment behaviour.

One of the principal aims of the experimental arrangements was to study the development of stress transfer within the pin model. The pin movements indicate that the redistribution of stresses within the embankment system vary across the section, Figure 4. Some movements as great as 8 mm occurred at positions directly above and adjacent to the subsoil. The tests did not indicate any visible movements of pins at the boundaries. From the evidence of the several tests undertaken, the pin movements above the subsoil highlight some form of a semi-circular shape for the arch.

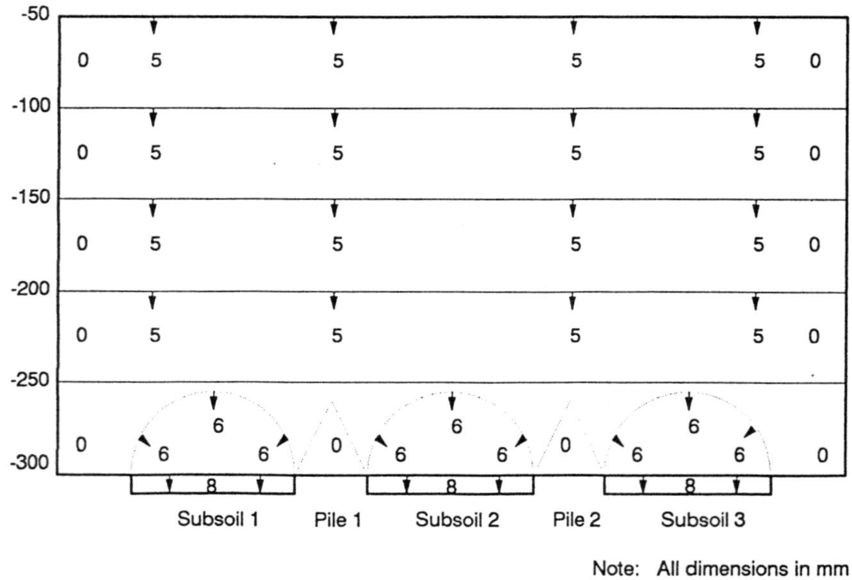

Figure 4: Graphical plotting of pin movements.

6 Conclusions

The pin model is capable of demonstrating the general embankment behaviour reasonably well. In addition there is evidence that arching was formed within the embankment when the subsoil was displaced. The shape of arch or rather the plane of failure in soils has been of much interests to previous researchers because this predetermined shape governs the accuracy of the solution of conventional methods of analysis. As illustrated in this research, the pin model provides an effective method for the visual study of the movement of the granular material which cannot be achieved with a sand model. The Finite Difference method gives a reasonably accurate prediction of the amount of stress transfer occuring in the experiments.

In piled embankment construction, the pile caps need not cover the full ground area in order to transfer load into the piles. Pile caps as little as 30% of the ground area is sufficient to carry a large proportion of the embankment weight.

Acknowledgement

The research reported in this paper was carried out in the Deparment of Civil and Structural Engineering at the University of Sheffield. The support of former colleagues is gratefully acknowledged.

References

1. Abu Taleb, M.G.A, The behaviour of anchors in sand, PhD Thesis, University of Sheffield, 1974.

2. Pyrah I.C., Elasto-plastic analysis in geotechnical engineering using finite difference, pp. 1607-1620, International Conference on Computional Plasticity, Barcelona, Spain, 1987.

3. Hewlett, W.J and Randolph, M.F., Analysis of piled embankment, Ground Engineering, Vol 21, No. 3, 1984, pp 12-18.

4. Reid, W.M. and Buchanan, N.W., Bridge approach support piling, Piling and ground treatment, Proc. Int. Conference on Advances in Piling and Ground Treatment for Foundations, Institution of Civil Engineers, Thomas Telford Limited, London, 1984.

5. Swedish State Road Board, Embankment piles, Report No. TV. 12 ,1974.

Fatigue crack growth analysis of AlZnMg alloy bridge girder

M.H. Aliabadi,[a] Y. Mi[b]

[a]*Wessex Institute of Technology, Ashurst Lodge, Ashurst, Southampton, Hampshire SO40 7AA, UK*

[b]*Department of Aeronautics, Imperial College, London, UK*

Abstract

The boundary element method (BEM) is used to predict the fatigue crack growth in a portable bridge. The predicted results are compared to measurements reported in [1] based on a full scale testing of the bridge. A reasonable agreement is obtained between the BEM and the experimental measurements for the fatigue life prediction. The BEM result for the critical crack size is found to be in good agreement with the test result.

1 Introduction

Aluminium alloys are the most common material used for portable structures mainly because of their light weights and high strength ratio. One example of such structure is a portable bridge used for military purposes. Fatigue crack growth analysis of a welded AlZnMg alloy bridge girder was reported by Webber[1]. In that investigation a full size bridge was constructed and tested. A section of a bridge was subsequently fatigue tested to destruction in the laboratory under conditions simulating those measured in the field. In addition, a fracture mechanics study was also carried out [1] to determine the girder's fatigue tolerance to cracks and to establish regular inspections intervals. However, the fracture mechanics analysis were based on a much simplified model of rectangular bar with a semi-elliptical crack.

In this paper, the incremental fatigue crack growth analysis of a section of T shaped tension chord from AlZnMg alloy is performed using the recently developed dual boundary element method[2-5]. A comparison is made between the boundary element results and the experimental data obtained from ref [1].

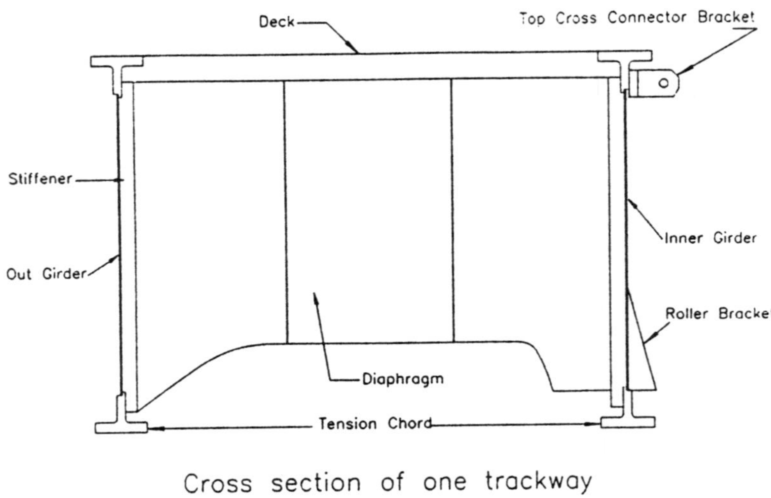

Figure 1: Configuration of one of the two trackways used to construct the portable bridge.

2 Description of the Test

2.1 Description of the Bridge

The configuration for one of the two trackways used to construct the portable bridge is shown in Figure 1. Each of the two trackways of the portable bridge is comprised of two fabricated plate web girders connected by a transverse deep diaphragm. The chords of the girders are made for T section extrusions between which are welded webs as shown in figure 2. Vertical stiffeners are welded on to the webs. For each section of the trackway, two brackets are attached to the inside of the trackway in line with the stiffeners and are used in emplacing the bridge.

2.2 Test Loading

The load used in the laboratory test [1] was established form measurements made on a complete bridge whilst it was trafficked by the heaviest vehicle for which the bridge was intended. A dead load stress of 33MPa (i.e. $\sigma_{\min} = 33MPa$) and a stress range of 133 MPa (i.e. $\sigma_{\max} = 166MPa$) was established at the bottom of the stiffener of the outside girder (see figure 2).

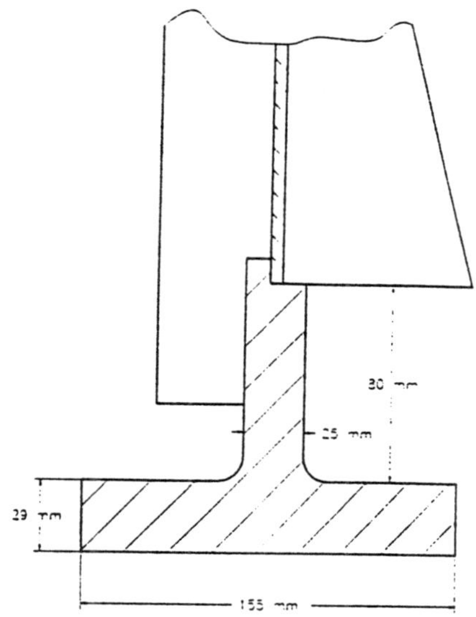

Figure 2: Configuration of the tension chord of the inner girder.

From other measurements a limit load stress of 248 MPa was calculated. In the middle section of the head of the T extrusion, it was assumed that σ_{min} is 37.5 MPa, σ_{max} is 188 MPa and a limit load stress is 281.5 MPa.

2.3 Testing and Results

The test results reported in [1] are briefly described in this section. Before commencing of the fatigue loading, the trackway was given a thorough inspection. A 5*mm* long transverse crack was found at the end of the bottom plate of the bracket in the longitudinal web to flange weld. It was then inspected after the application of every 1,000 cycles and any crack growth monitored. No growth was evident until 17,000 cycles had elapsed, when the crack was found to be 57*mm* long. At the same time, another crack 99*mm* length was found on the other side of the girder extending along the whole length of the stiffener where it was welded to the extrusion and extending to the web. This longer crack eventually resulted in failure. As this length of the crack was greater than the crack length at which failure was predicted, it was decided to apply the limit load (248 MPa) in order to see if the crack was critical at this level of load. Because it was not, testing was continued by applying 1,000 cycles of fatigue loading. At 19,000 cycles and thereafter the limit load was applied until failure occurred at 27,000 cycles on the application of the limit load.

After failure had occurred distinct markings on the fracture surface were observed which coincided with the crack fronts at stages when the limit load

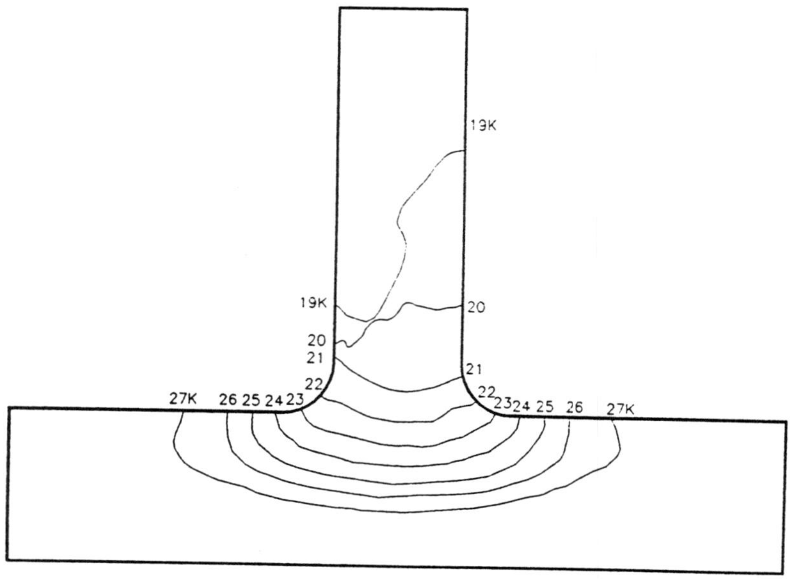

Figure 3: Fatigue crack fronts on fracture surfaces of tension chord T extrusion.

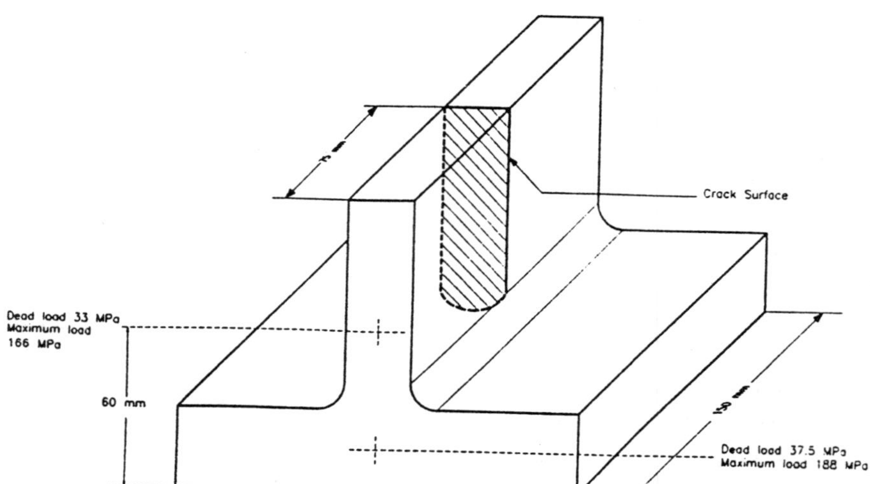

Figure 4: Geometrical model of the T extrusion.

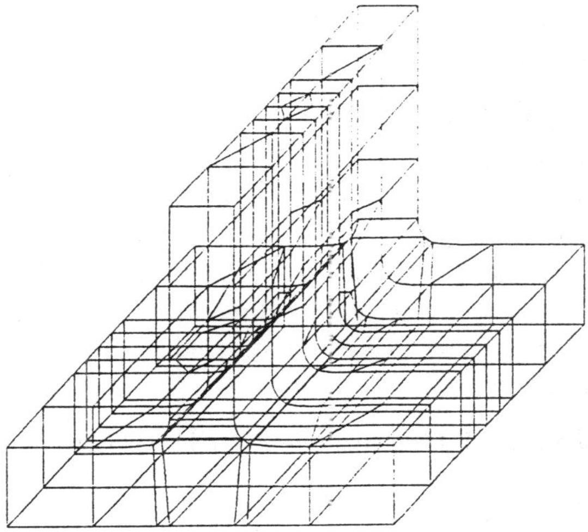

Figure 5: Boundary Element Mesh.

was applied. These fatigue crack fronts are shown in figure 3. In this figure a rapid crack growth can be seen between 19,000 and 21,000 cycles. In fact, from 18,000 cycles onwards, after the stiffener weld toe and bracket weld cracks had joined, the growth became very fast until 21,000 cycles were reached.

3 Fatigue Crack Growth Analysis

As described in the previous section, the cracks initiated in the longitudinal web to flange weld and along the stiffener to the extrusion weld from both sides of the extrusion and eventually joined. The effect of the welding process on the initiation and growth of cracks is complicated and is not covered by the simulation procedure reported here. Moreover, from 17,000 to 19,000 cycles, there was a period of very rapid crack growth with a corresponding growth in stress intensity factors close to the fracture toughness of the material. Only when the crack entered the head of the T extrusion did it behave within the scope of the fatigue crack growth. Therefore, taking these factors into account, it was decided that numerical simulation of the problem should start from a cracked T extrusion corresponding to the stage of 22,000 cycles (see figure 3).

3.1 Geometrical Model and Mesh

A section of a simplified T extrusion with crack is chosen to model the cracked tension chord as shown in figure 4. Shown in figure 5 is the designed mesh for the geometry which is comprised of quadrilateral and triangular quadratic continuous and edge discontinuous elements (see ref[2]). The number of nodes and elements of the mesh are 904 and 143 respectively.

3.2 Applied Load

The applied load is assumed to be linearly varying tensile load along the height of the chord cross section. This is consistent with the data measured and used in the experiments of ref[1]. The linearly varying load (at the middle of the chord is 37.5 MPa and 33 MPa at the bottom of the stiffener) is applied along the longitudinal direction of the chord as shown in figure 4. The corresponding maximum stresses at these levels are 188 MPa and 166MPa respectively.

3.3 Material Properties and Fatigue Growth Law

The Young's modulus and Possion's ratio for this aluminium alloy were chosen as 70,000N/mm^2 and 0.3 respectively. The Paris' law chosen in the fatigue crack growth analysis is due to Maddox and Webber[5], which was also adopted in Webber[1], and can be written as

$$\frac{da}{dN}(mm/cycle) = 3.384 \times 10^{-7} \Delta K^{3.056}$$

where K is in MPa$\sqrt{m}$.

3.4 Results

Starting from a semi-elliptical crack front corresponding to the experimental results at 22,000 cycles, a total of five increments of fatigue crack growth extension were carried out.

The incremental size for each crack increment extension was selected in such a way that new crack fronts would be as close as possible to the markings on the fracture face of the chord, in order to compare with the experimental results.

The stress intensity factors corresponding to the crack fronts are presented in figure 6 for maximum load value (i.e. linearly varying tensile load which is 188 MPa in the middle of the chord head and 166 MPa at the bottom of the stiffener). The crack lengths against number of cycles are presented in figure 7, together with the experimental results. In figure 7, the vertical axis represents the surface half crack length measured from the right hand side of the crack tip at 22,000 cycles. Finally, the predicted crack path and the experimental results are plotted in figure 8.

3.5 Discussion

The BEM results presented in figures 8 and 9 are shown to be relatively conservative life compared to those reported in [1]. Factors which cause the discrepancy between the calculated and experimental results are numerous.

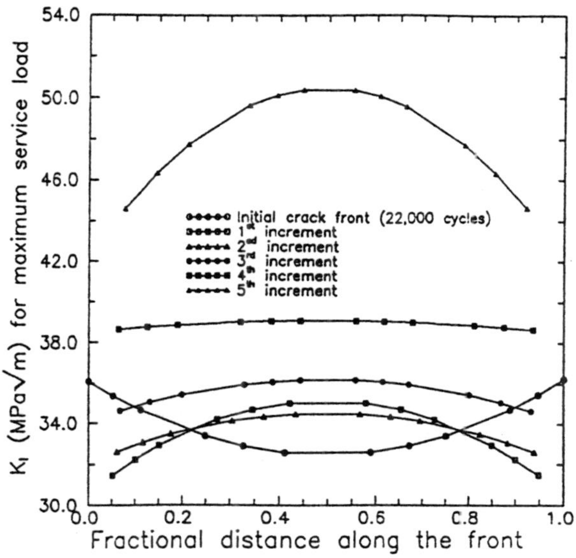

Figure 6: Stress intensity factors for each crack front.

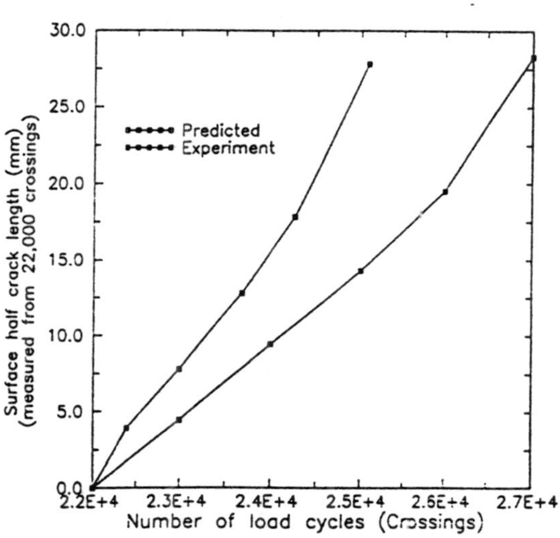

Figure 7: Fatigue life of the chord.

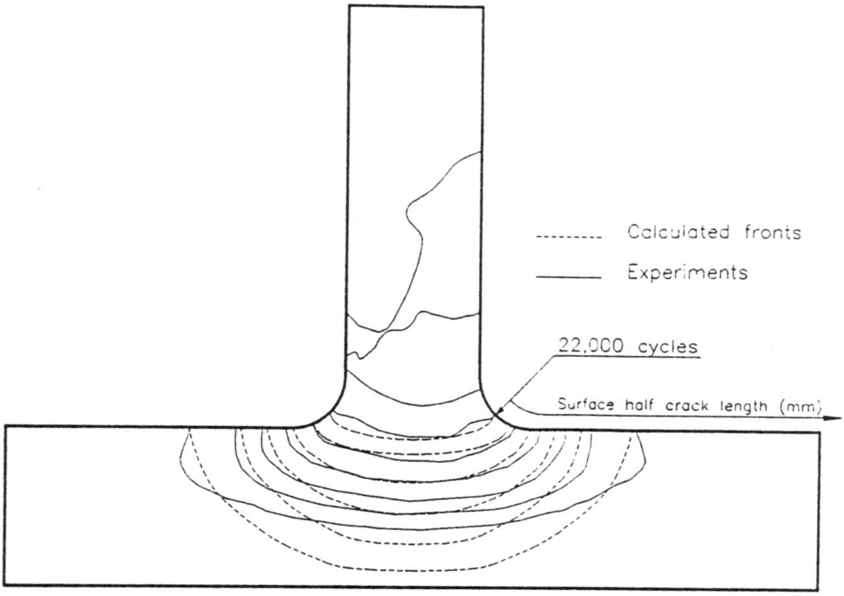

Figure 8: Fatigue crack growth paths.

Some of them are considered here. The most important may be the amplitude of the load. It can be seen from the description of the test described earlier and [1] that the limit load, corresponding to stresses of a magnitude of 248 MPa, is applied after each 1,000 fatigue loadings. The application of this overload in a constant amplitude test can cause the subsequent crack growth comparatively slow due to the introduction of the residual stresses caused by the overload. Hence causing retardation. The retardation was not taken into account due to the lack of definite data.

Another factor may be the assumptions made in modelling of the geometry and the loading. That is the real loading may be more complex than the one assumed in the analysis. Furthermore the welded plate web not modelled may also have some effects on the crack growth rates.

Apart from the above factors, the accuracy of the measurements in the test and the accuracy of the stress intensity factors can affect either test or simulation results. Despite all these uncertainties, the predicted results are of importance.

From the results shown it can be seen that the variation of the stress intensity factors along each incremental crack front tends to be uniform with the increments up to and the fourth increment (see figure 7). Stress intensity factors evaluated along the fifth crack front show a rapid increase and considerable variation. This is mainly because of the lowest part of the front is much closer to the bottom of the head of the chord in comparison with pervious ones. The variation of the stress intensity factors is consistent with the geometry and load. Secondly, the calculated maximum stress intensity factor along the fifth (final) incremental crack front is about 50MPa$\sqrt{m}$ for the maximum value of cycling load and 74.7 MPa$\sqrt{m}$ for the limit load. The fracture toughness for the extrusion under consideration is about 40.7$\approx$ $47MPa\sqrt{m}$. Measured values of the toughness in the test are estimated as 60.8MPa$\sqrt{m}$ and 93.5 MPa$\sqrt{m}$ using a different approach [1]. Therefore, the crack surfaces corresponding to the fifth crack front is close to the critical size for the structure. In fact, failure of the girder happened after 27,000 cycles when the limit load was applied. The crack tips on the chord at this stage are identical with those of the fifth incremental crack front (see figure 8). This coincidence indicates that numerical simulation can be used to estimate the critical crack size for inspection and assessment which is close to reality. As far as the shortfall in the predicted fatigue life is concerned, it may be improved by using alternative fatigue laws.

4 Conclusions

Fatigue crack growth analysis using the three dimensional incremental crack growth procedure developed by Mi and Aliabadi[3] was performed for a cracked T section tension chord of a bridge girder constructed from aluminium alloy. The comparison between the numerical simulation and a full scale test results was made in order to assess and verify the proposed procedure. A reasonable fatigue life prediction was obtained from the calculation and the estimate of the critical crack size was found to be good agreement with the test results.

References

[1] Webber,D. Fatigue design and testing of an AlZnMg alloy bridge girder. *In Proceeding of the third international conference on Aluminium Weldements.* Aluminium-Verlag, Munchen, 1985.

[2] Mi,Y. and Aliabadi,M.H. Dual boundary element method for three-dimensional fracture mechanics analysis. *Engng. Anal. with Bound. Elem.*, **10**(2), 161-171 (1992).

[3] Mi,Y. and Aliabadi,M.H. Three-dimensional crack growth simulation using BEM. *Computers and Structures,* **52**, 871-878 (1994).

[4] Aliabadi,M.H. and Mi,Y. Application of the three-dimensional boundary element method to quasi-static and fatigue crack propagation. *In Fatigue Crack Propagation in Metallic Structures.* Edited by Carpinteri,A. Elsevier Academic Publishers, Oxford (1994).

[5] Mi,Y. and Aliabadi,M.H. An automatic procedure for mixed-mode crack growth analysis. *Communications in Numerical Methods*, **11**, 167-177 (1995).

[6] Madox,S.J. and Webber,D. Fatigue crack propagation in aluminium zinc magnesium alloy filet welded joints. *In Fatigue Testing of Weldements*, ASTM STP 648, editor D.W.Hoeppner, pp159-184, ASTM, Philadelphia, 1978.

Experimental behaviour of reinforced concrete continuous deep beams

A.F. Ashour

Department of Civil and Environmental Engineering, University of Bradford, Bradford BD7 1DP, UK

Abstract

Test results of eight reinforced concrete continuous deep beams are reported. The main parameters considered were the shear span to depth ratio, the amount and type of web reinforcement, and the amount of the main longitudinal reinforcement. The vertical web reinforcement had more influence on the shear capacity than the horizontal web reinforcement. The failure is initiated by a major diagonal crack in the intermediate shear span between the edges of the load and intermediate support plates. Comparisons between test results and the ACI building code (318-89) show little agreement.

Introduction

Reinforced concrete continuous deep beams are fairly common structural elements. They are used as load distribution elements such as transfer girders, pile caps, tanks, folded plates and foundation walls, often receiving many small loads and transferring them to a small number of reaction points. There have been extensive experimental investigations of simply supported deep beams [3,4,5], but very few tests of continuous reinforced concrete deep beams [4,5]. Seventeen two-span reinforced concrete deep beams have been tested by Rogowsky et al. [5], 16 with either vertical or horizontal web reinforcement.

The current codes of practice for shear in reinforced concrete continuous deep beams are based entirely on tests of simply supported deep beams. It was found that the shear strength obtained from the ACI building code (318-83) formula [1] overestimates the measured shear capacity of half of the continuous deep beams tested by Rogowsky et al. [5].

The present paper reports test results of 8 two-span reinforced concrete deep beams. The shear span to depth ratio included two values. Three levels of horizontal and vertical web reinforcement have been chosen. Two main flexural reinforcement ratios were studied. Test results are compared to the predictions of the ACI building code (318-89).

Test specimens

Two series of two-span reinforced concrete deep beams were tested. The overall dimensions of each series are shown in Fig. 1. All tested beams had the same length and width: the length L was *3000 mm* and the width b was *120 mm*. The locations of center lines of loads and supports were the same for all test specimens. Only the beam depth h was varied to obtain two different shear span to depth ratios: for series *I (CDB1,CDB2,CDB3,CDB4,CDB5)*, the depth was *625 mm* to give a clear shear span to depth ratio of *0.8* and for series *II (CDB6,CDB7,CDB8)*, the depth was *425 mm* to give a clear shear span to depth ratio of *1.18*. The details of reinforcement for each beam are given in Table 1.

The ingredients of the concrete were ordinary portland cement, irregular gravel of maximum size *10 mm* and sand. The concrete properties are given in Table 2. The cube compressive strength f_{cu} shown in Table 2 is obtained from the average of compressive strengths of tested cubes *(100x100x100 mm)*: *8* cubes for beams in series *I* and *6* cubes for beams in series *II*. The cylinder compressive strength f_c' given in Table 2 is the average compressive strength of two cylinders *(150x300 mm)*. Table 2 gives the modulus of rupture f_t (obtained from testing one prism of *100x100x500mm* under four point loading system) and fracture energy G_f (obtained from testing two prisms of *100x100x500mm* with notch under three point loading system) of concrete for different beams.

Test results

The test specimens were tested in the Amsler compression machine of total capacity *500 tons*. Special arrangements had been taken to obtain two point loads and three support reactions as shown in Fig. 2. A top steel spreader beam was used to divide the total applied load from the machine head into two equal point loads, one in each span. Another stiffer steel beam was placed underneath the tested specimens to collect the three support reactions to the other head of the machine as shown in Fig. 2. The load was applied in increments of *5 tons* until failure occurred. After each increment, the load was kept constant to allow marking of the new cracks and running the data logger. The test was under load control until the specimen reached its peak strength. After the peak strength of the specimen and during softening, two more readings were recorded by the data logger (ORION 3530) whenever that was possible.

Specimen behavior

The first mid-span crack generally occurred at the same time as the first flexural crack over the intermediate support as shown in Table 3. Then, the first diagonal crack suddenly developed at mid-depth within the intermediate shear span between the applied load and the intermediate support. Significant redistribution of internal stresses clearly occurred after development of the first diagonal crack. As the load increased, more flexural and diagonal cracks were formed and a diagonal crack extended to join the edges of the applied load and intermediate support plates. As the reinforcement was yielding, cracks became wider and the deflection significantly increased. Just before failure, the two spans showed nearly the same crack patterns. At failure, an end block formed, because of the significant diagonal crack connecting the edges of the load and intermediate support plates, and rotated about the end support leaving the rest of the beam fixed over the other two supports. At the separation line between the end block and the rest of the beam, concrete crushing and concrete separation were observed at the top and bottom ends of that separation line respectively as shown in Fig. 3 (Fig. 3(a) for *CDB5* (as an example of beams in series I) and Fig. 3(b) for *CDB7* (as an example of beams in series II))

End support reactions and failure loads

Fig. 4 shows the amount of the load transferred to the end support against the total applied load. On the same figure, the end support reaction obtained from linear elastic finite element analysis (plane stress) is presented. Although the amount of the web reinforcement influences the maximum reaction at the end support, it has no effect on the total load-end support reaction gradient. The gradient is nearly the same as obtained from linear elastic analysis up to the first cracking load, and then the redistribution of stresses increases the end support reaction more than that predicted by the linear elastic finite element. Table 4 gives total failure loads P_t, end support reactions R at failure, maximum shear forces Q within the intermediate shear span, non-dimensional shear strengths $\tau = Q / bhf_c^{'}$ and non-dimensional failure loads in each span $\lambda = P_t / 2bhf_c^{'}$ for tested beams. *CDB3* that had only horizontal web reinforcement, showed the lowest capacity of beams in series *I*.

Mid-Span deflections

The mid-span deflections for different beams against the total applied loads are given in Fig. 5. The deflections given are calculated relative to the intermediate support movement. The mid-span deflections shown in Fig. 5 are those recorded for the failed span. At low load level (up to the first cracking load), the mid-span deflections for each series seem to be independent of the amount and type of web reinforcement. Formation of the first diagonal crack significantly reduced the beam stiffness. All tested specimens exhibited some ductility at failure. The degree of ductility varied depending on the shear span to depth ratio and the amount of reinforcement. *CDB3* that had only horizontal web reinforcement, showed the lowest ductility at failure. Beams having higher shear span to depth ratio (beams in series *II*) produced more ductility.

Steel strains

The variation of strains in different reinforcing bars against the total applied load will be presented for *CDB1* and the steel strain results for the rest of the tested beams can be found in reference [2].

CDB1 steel strains

CDB1 had heavy horizontal and vertical web reinforcement. Fig. 6(a) shows the positions of different ERS gauges attached to the reinforcement. Fig. 6(b) presents the variation of steel strains with the total applied load. Major redistribution of strains occurred at the bottom reinforcement after the first diagonal crack (strains at 9 and 10). After this the bottom longitudinal reinforcement was in tension throughout the length of the beam (strains at points 9, 10 and 11, even over the intermediate support, gauge 8) and yielded at failure (point 9). The top reinforcement was in tension throughout the length of the interior shear span but did not yield and neither did the web reinforcement. The failure was not in the instrumented span.

Test results compared to ACI building code

Comparisons between test results and the predictions of the ACI building code (318-89) [1] are presented in the following. The ACI building code provides special provisions for shear strength of continuous deep beams of clear span to depth ratio less than *5*. Table 5 shows the comparison between the total shear strength $v_t\,(=v_c+v_{sh}+v_{sv})$ of the tested beams, predicted by the ACI building code (318-89) with all safety factors removed, and the test results $\tau = Q / bhf_c'$ given in Table 4. The table shows the contribution from concrete $v_c (= V_c / bhf_c'$, where V_c is the shear force resisted by concrete), vertical web reinforcement $v_{sv} (= V_{sv} / bhf_c'$, where V_{sv} is the shear force resisted by vertical web reinforcement) and horizontal web reinforcement $v_{sh} (= V_{sh} / bhf_c'$, where V_{sh} is the shear force resisted by horizontal web reinforcement). The ratio between the experimental and ACI building code shear strength for different beams ranges from 0.813 to 1.750, with average 1.23, standard deviation 31% and coefficient of variation 25%. The ACI building code predicts that the amount of shear resisted by the horizontal reinforcement is higher than that resisted by the vertical web reinforcement (contrary to test results). The ACI building code predictions are unconservative for two beams (*CDB*1 and *CDB*3). The discrepancy in the results predicted by ACI building code (318-89) may be attributed to the fact that the shear strength equations for continuous deep beams are derived from simple deep beam tests and they are not based on a rational theory. A theoretical analysis of tested beams based on the plasticity theory is given in reference [2].

Conclusions

The parameters studied in this experimental investigation were the degree of vertical and horizontal web reinforcement, shear span to depth ratio and the amount of the main longitudinal reinforcement. The following behavior was observed:

1. All beams exhibited the same type of failure. A major diagonal crack in the intermediate shear span ran between the edges of the load and the intermediate support plates. An end block then formed and rotated about the end support leaving the rest of the beam fixed over the other two supports.

2. The shear capacity is influenced by the type of web reinforcement. For the tested beams presented in this investigation, the vertical web reinforcement had more influence on the shear capacity than the horizontal web reinforcement.

3. The comparison between test results and the ACI building code (318-89) shows poor agreement. The ACI building code (318-89) predictions are unconservative for two beams.

Acknowledgments

The assistance of the staff at Structures Research Laboratory (Cambridge University) and the financial support by Cambridge Overseas Trust are gratefully acknowledged. The author is deeply grateful to Dr. C. T. Morley at Cambridge University for his guidance throughout this project.

References

1. ACI Committee 318. *Building Code Requirements for Reinforced Concrete (ACI 318-89),* American Concrete Institute, Deteroit, 1989.
2. Ashour, A. F. *Behaviour and Strength of Reinforced Concrete Continuous Deep Beams,* PhD thesis, University of Cambridge, 1994.
3. Kong, F. K. *Reinforced Concrete Deep Beams*, Blackie and Son Ltd, 1990.
4. Leonhardt, F., and Walther, R. Deep Beams, CIRIA English Translation, Jan. 1970.
5. Rogowsky, D. M., MacGregor, J. G. & Ong, S. Y. Tests of Reinforced Concrete Deep Beams, ACI Structural Journal, 1986, **4**, 614-623.

Table 1 Details of specimen reinforcement

Beam No.	Main Longitudinal Reinforcement				Web Reinforcement			
	Bottom		Top		Horizontal		Vertical Stirrup	
	Total. No	$A_s f_y^*$ (kN)	Total. No	$A_s f_y^*$ (kN)	Total. No	$A_s f_y^*$ (kN)	Total. No	$A_s f_y^*$ (kN)
CDB1	4ϕ12 mm	56.5	4ϕ12 mm +2ϕ10 mm	56.5	8ϕ8 mm	18.6	29ϕ8 mm	18.6
CDB2	4ϕ12 mm	56.5	4ϕ12 mm +2ϕ10 mm	56.5	4ϕ8 mm	18.6	15ϕ8 mm	18.6
CDB3	4ϕ12 mm	56.5	4ϕ12 mm +2ϕ10 mm	56.5	4ϕ8 mm	18.6	-	-
CDB4	4ϕ12 mm	56.5	4ϕ12 mm +2ϕ10 mm	56.5	-	-	15ϕ8 mm	18.6
CDB5	2ϕ12 mm	56.5	2ϕ12 mm	56.5	4ϕ8 mm	18.6	15ϕ8 mm	18.6
CDB6	2ϕ12 mm +2ϕ10 mm	56.5	2ϕ12 mm +2ϕ10 mm	56.5	4ϕ6 mm	9.80	29ϕ6 mm	9.83
CDB7	2ϕ12 mm +2ϕ10 mm	56.5	2ϕ12 mm +2ϕ10 mm	56.5	2ϕ6 mm	9.80	15ϕ6 mm	9.83
CDB8	2ϕ12 mm	56.5	2ϕ12 mm	56.5	2ϕ6 mm	9.80	15ϕ6 mm	9.83

* $A_s f_y$ is the yield force for a single reinforcing bar.

Table 2 Concrete Properties

Beam No.	f_{cu} (N/mm^2)	f_c' (N/mm^2)	f_t (N/mm^2)	G_f (N/m)
CDB1	36.6	30.0	4.235	148.5
CDB2	39.2	33.1	4.795	177.0
CDB3	25.0	22.0	3.987	215.5
CDB4	31.5	28.0	4.319	223.5
CDB5	32.0	28.7	4.195	254.5
CDB6	26.5	22.5	4.187	254.5
CDB7	30.9	26.7	4.847	173.5
CDB8	29.4	23.6	4.107	219.0

Table 3 First flexural and diagonal cracking load

Beam No.	First Flexural Cracking load at		First Diagonal Cracking Load (tons)
	Mid-Span (tons)	Over the int. Sup. (tons)	
CDB1	40	40	45
CDB2	30	30	35
CDB3	25	25	35
CDB4	30	30	35
CDB5	25	25	35
CDB6	15	15	25
CDB7	15	15	25
CDB8	15	15	25

Table 4 Total failure load P_t, end support reaction R, maximum shear force Q within the intermediate shear span, non-dimensional shear strength τ and non-dimensional failure load λ in each span

Beam No.	P_t (tons)	R (tons)	Q (tons)	τ	λ
CDB1	110.0	19.9	35.1	0.156	0.244
CDB2	95.0	16.9	30.6	0.127	0.191
CDB3	57.0	10.48	18.02	0.109	0.172
CDB4	88.5	15.86	28.39	0.135	0.207
CDB5	82.0	15.2	25.8	0.121	0.190
CDB6	49.5	9.14	15.61	0.136	0.220
CDB7	44.5	8.2	14.05	0.110	0.163
CDB8	38.5	6.88	12.37	0.103	0.160

Table 5 Comparison between Test Results and ACI building code (318-89)

Beam No.	v_c	v_{sh}	v_{sv}	v_t	exp./code
CDB1	0.062	0.087	0.025	0.174	0.897
CDB2	0.059	0.039	0.011	0.109	1.165
CDB3	0.074	0.060	-	0.134	0.813
CDB4	0.064	-	0.013	0.077	1.750
CDB5	0.059	0.046	0.013	0.118	1.025
CDB6	0.040	0.055	0.023	0.118	1.153
CDB7	0.036	0.023	0.010	0.069	1.590
CDB8	0.036	0.026	0.011	0.073	1.410

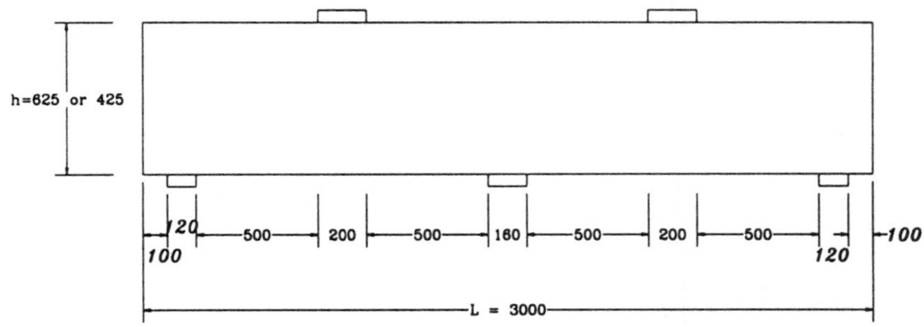

Fig. 1 Geometrical Dimensions of Test Specimens (All Dimensions in mm)

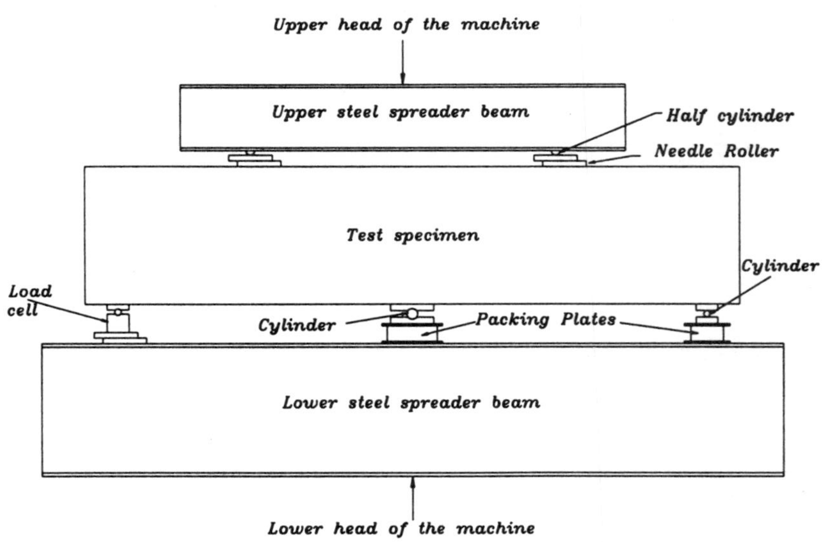

Fig. 2 Test set-up

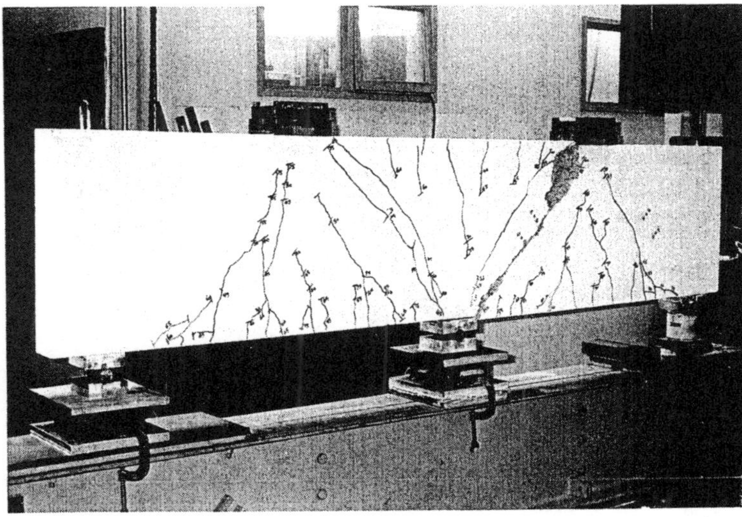

(a) CDB5

(b) CDB7

Fig. 3 Crack Patterns and Failure Zones of Tested Beams

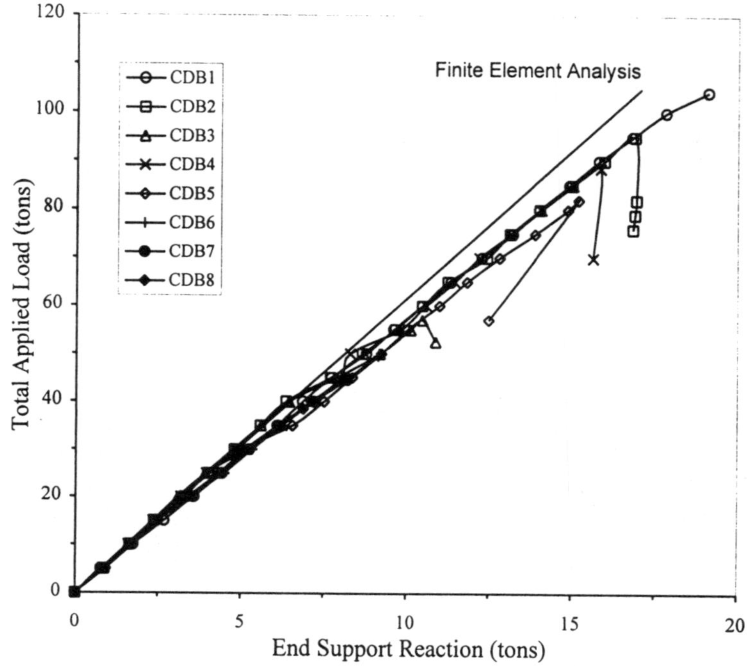

Fig. 4 End-Support Reaction against Total Applied Load

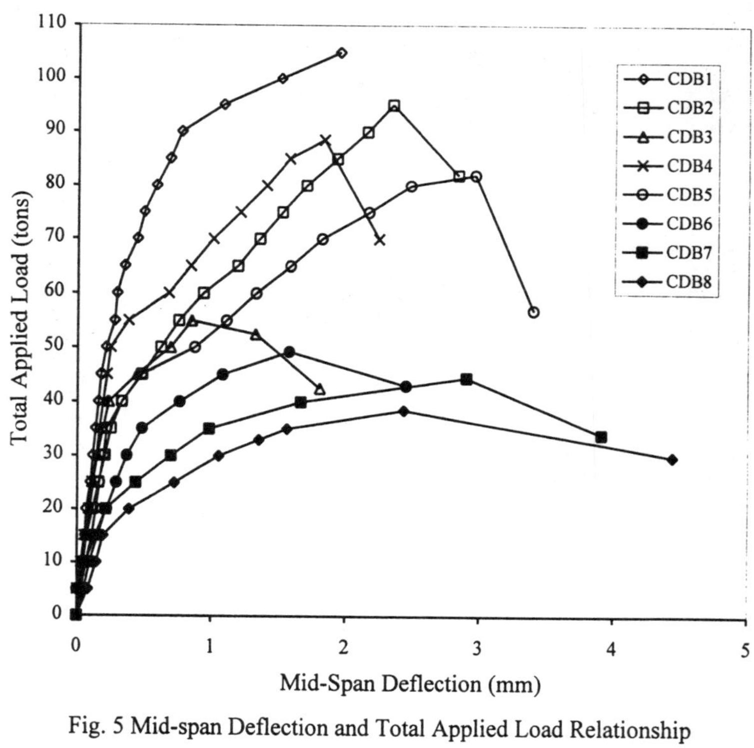

Fig. 5 Mid-span Deflection and Total Applied Load Relationship

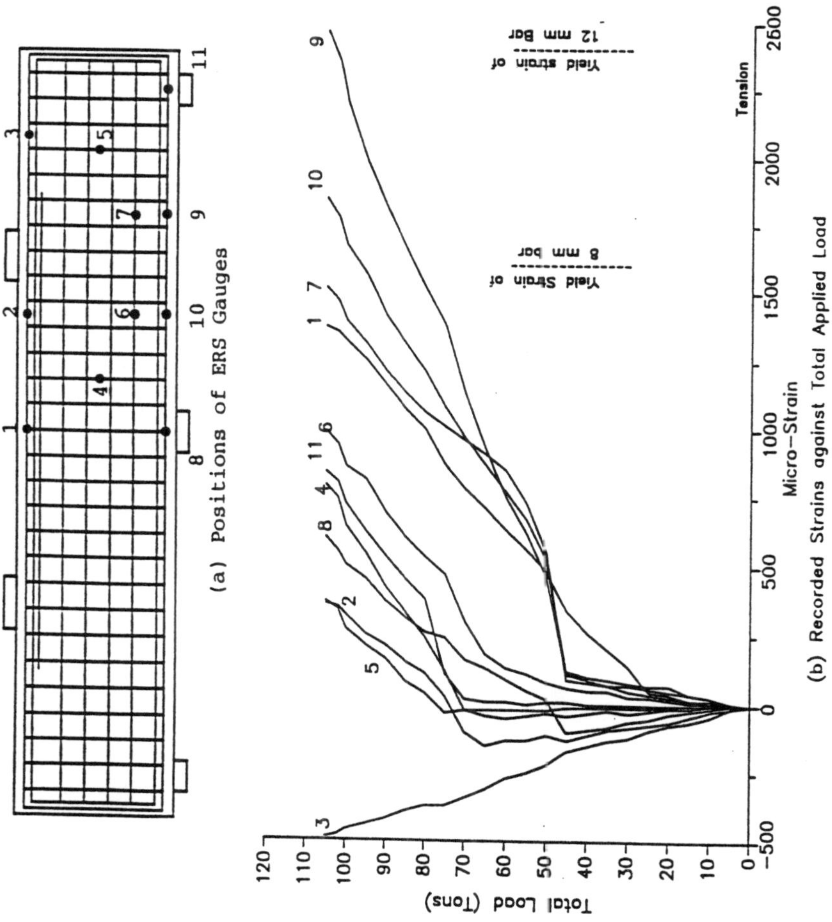

Fig. 6 CDB1 Reinforcement Strains

SECTION 3: ENGINEERING INTEGRITY, DESIGN AND MODEL SIMULATION

Simulation of the wear behaviour of high-speed over-head current collection systems

K. Becker,[a] A. Rukwied,[a] W. Zweig,[b] U. Resch[b]
[a]*Daimler-Benz AG, Research and Technology, 60528 Frankfurt am Main, Germany*
[b]*AEG Bahnfahrwegsysteme, 60326 Frankfurt am Main, Germany*

Abstract

The present work describes a simulation procedure used for design optimisation of high speed train current collecting systems. The optimisation process is focussed on increasing the life time of high speed overhead contact lines. The dynamic behaviour of the catenary is calculated numerically and the results are used as an input parameter for lifetime prediction. The wear behaviour is simulated up to train speeds of 350 km/h by using a specially designed wear machine. The wear performance is measured automatically by a completely computer controlled data acquisition system under the following conditions:

- speeds up to 500 km/h,
- traction currents up to 1000 A,
- contact forces up to 500 N.

First results on the ICE-system (CuAg-contact wire, carbon contact strip) are reported and discussed.

1 Introduction

World record runs of the high speed trains ICE and TGV have shown that the main technical problem with high speed railway systems is the energy transfer from the contact wire to the train. The design of the catenary and the pantograph of the trains is one of most important problems encountered with modern high-speed railway systems.

With respect to the life cycle costs of the catenary evidently the cost efficiency of the system is dominated by the wear behaviour of the highly stressed components of the current collecting system. Considering the fact that overall system costs for replacing the contact wire are much higher than the costs of replacing contact strips on a pantograph, initial priority must in any case, focus on increasing the service life of the contact wire [1].

Presently investigations of wear rate of overhead current collection systems are restricted usually to manual thickness measurements of the contact wire at certain preselected points. It is evident, that this procedure is time consuming and unreliable. Moreover these investigations cannot provide sufficient data on individual influences out of a set of operational parameters as encountered in field service. With respect to the wide range of parameters (materials of the contact strip and the contact wire, contact forces, speed of sliding, traction current and environmental conditions) under which the different railway systems operate, these data are system dependent and not transferable.

The principle design optimisation for high speed current collecting systems procedure used at AEG Bahnfahrwegsysteme is shown schematically in Figure 1. With respect to customer specifications (max. train speed, traction power of the train system, etc.) a draft catenary design is build up by CAD. In a second step the dynamic behaviour is simulated numerically, yielding the contact force between contact wire and contact strip at every point in the catenary during operation. The examination of the wear during service operation of the current collecting system is rather time consuming and in addition unreliable, since important parameters such as contact force and current strength cannot be recorded at the same time.

Thus, a wear testing machine was developed at the Daimler Benz Research Center in Frankfurt. General objectives were the systematic examination of wear behaviour under high-speed conditions and a quantitative description of the wear in dependence of contact force, speed, and current by means of a suitable model. Incorporating these parameters, the major optimising criteria for the current collector/contact wire system can then be determined by simulating wear behaviour and can then be used as draft parameters for the development of an efficient and economic high-speed catenary meeting the requirements of the customer [2,3].

2 Simulation procedure for the dynamic behaviour of the overhead line/pantograph system

Numerical simulations of the dynamics of the catenary and the pantograph are based on a procedure of mathematical modelling [3,4]. The significance of the simulation calculation therefore is linked to the degree of conformance with reality of the model upon which the simulation is based. Mathematical models must be tested and improved using experience gained from operation in the field and from the results of test runs. Once the applicability of a numerical simulation has been assured, it is possible to systematically investigate and optimise the influence of a whole range of parameters affecting the service characteristics of the overhead line/pantograph system. Moreover the simulation can be used to investigate extreme system loads without running the risk of damage and disruptions in test runs in the field.

In modelling up to four tensioning sections are required in order to simulate realistically catenary arrangements, such as catenary segments above switches. In addition, the simulation model and procedure must be sufficient flexible in order to optimise special designs such as lowered contact wires under structures. Particularly important are simulations for uninsulated overlaps, catenaries above switches with crossing bars and special designs. The contact wire is subjected to very rapid wear if high contact forces or even arcing occurs in these areas. Within a brief period of service, it would be necessary to replace the contact wire for the entire tensioning section.

If we look at the catenary dynamics from the point of view of the trailing power unit, it is clearly apparent that the droppers bend outwards. Consequently, the droppers no longer perform adequately their carrying function. This and other important non-linear characteristics of the overhead line and the pantograph must also be incorporated in any meaningful simulation procedure.

By taking into consideration up to four pantographs of different spacings and designs, it is possible to investigate various pantograph arrangements, such as double traction by two locomotives or motor train sets.

2.1 Model of wave propagation in the catenary

As the pantograph passes along the overhead line, it generates transverse mechanical waves in the overhead line. These waves propagate forward and backward with respect to the direction of travel. With currently used high speed overhead lines, the simple model of the oscillating string can be used to describe wave propagation in the overhead line.

In the oscillating string model, the contact wire, catenary wire and stitch wires are described as piano wires under tension. The analytical treatment results in differential equations that can be solved by various means, for example using the d'Alembert method. Using this model it is possible to describe several tensioning sections.

2.2 Pantograph model

A description of the dynamics of the pantograph is of equal importance compared to the dynamics of the catenary itself. The pantograph is described using an equivalence model consisting of discrete masses, springs and dampers. Figure 2 shows the equivalence model for the DSA350SEK pantograph. The springs and dampers interconnect the discrete masses. The flexible upper frame of the pantograph and the individual spring of the contact strips result in a differential equation system with six degrees of freedom. Three degrees of freedom are required in order to describe the translation motion of the lower frame, the elastic upper frame and the pan collector. The other three degrees of freedom are needed for the rotary motions of the pan collector and the elastic upper frame. The non-linear characteristics of the pantograph dampers and the contact strip suspension must not be neglected when establishing the model. Aerodynamic loading of the pantograph can be taken into consideration by adding forces to the discrete masses.

3 Simulation of the wear behaviour of the overhead line/pantograph system

Wear tests at railway current collecting systems are much more critical in comparison to other technical test procedures when it comes to the correct transfer of system parameters to laboratory tests [5-9]. The wear test machine used here was especially developed for this purpose at the Daimler-Benz Research Institute in co-operation with AEG-Bahnfahrwegsysteme. Figure 3 shows the general outline of the wear machine. The machine is based on a revolving disc of 1.8 m diameter. The contact wire is fixed laterally at the outer edge of the disc allowing wear tests to be performed on original contact wire geometries according to international standards. The disc is driven by a variable-speed 55 kW DC motor mounted on the 10 t steel frame. The maximum speed of 1525 rpm results in a sliding speed of 500 km/h on the circumference. In order to ensure service-near-test conditions, original ICE-contact strips are used. The contact force is applied pneumatically and can be varied in the range between 0 and 500 N. The stagger between contact wire and the contact strip is simulated by means of an eccentric drive operating in conjunction with a cross table. An AC power source (0-1000 A) is used to supply the traction current through the sliding contact.

The wear rate is measured in situ by two laser sensors employing the principle of triangulation. The high resolution system in the μm range of the wear test puts high demands on the mechanics of the test stand. In order to ensure that no critical vibrations occur in the working range of the test stand the individual components of the machine as well as the complete unit have been optimised already at an early state of design with respect to the resonance behaviour employing the Finite Element Methods. The first characteristic frequency was calculated to occur at 47 Hz which ensures that during regular operation at rotating frequencies of maximum 25 Hz no resonant vibrations take place.

Figure 4 shows schematically the data acquisition. The machine is completely computer controlled, allowing fully automised wear tests. The wear machine has been designed to operate in the following ranges:

Sliding speed:	0...500 km/h
Traction current:	0...1000 A
Contact force:	0...500 N

4 Results

4.1 Contact Force Simulation

The results of the simulation must be compared with on-the-spot measurements in order to confirm the mathematical model of the pantograph and overhead line. Test runs involve the use of strain gauges to measure the cutting force between the contact strips and the contact strip spring [10,11]. Figure 5 shows a comparison between the forces that were measured and those calculated from the string model of the contact wire. The forces were applied over a length of 500 m. The measured data was taken during test runs with the Re250 and DSA350S pantograph on the Cordoba-Cantillana line. The test speed was 240 km/h. The vertical lines mark the supports. The lower curve shows the difference between the righthand and lefthand forces of the strain gauges, thereby allowing the contact wire stagger to be recognised. The stagger shows a bend of the line at km 383.

The middle curve plots the sum of the calculated forces. The measured forces are shown in the top curve. The values are shifted by 300 N in order to improve the representation. It can be seen that the forces compare favourably. The rotational motions of the contact strips are described correctly.

The measured curve and the calculated curve are not identical because the motion of the locomotive roof and aerodynamic effects have not been taken into account.

4.2 Wear Rate

The corresponding wear measurements have initially been performed on the ICE system in the speed range up to 350 km/h. Figure 6 shows the mean wear rate for a sliding speed of 150 km/h as a function of the traction current and the contact force. The results have been standardised with the reference test (v=100 km/h, I=300 A and F=250 N) to show more clearly the effects of the parameter variation.

The results show, that the mechanical wear is dominating the wear of the contact wire. The wear rate increases monotonically with the contact force when no current is applied. With additional moderate traction current of 100 A the wear rate reaches a minimum at the present speed of 150 km/h. This effect can be explained as a smearing effect of the contact strip graphite in combination with the applied current density. Inspections of the contact surface after the wear tests confirmed a thin graphite layer on the copper sliding surface. With a further increase of the current density, in addition to the mechanical wear, the electrical wear leads to increase of the wear rate. No arcing was included in this investigation. Further tests at speeds up to 350 km/h show a decrease of the wear rate (Figure 7).

No increase of the wear rate was observed even in the tests with a high current of 300 A. It might be concluded, that the smearing effect (Figure 7, tests at 100 A) would be shifted to higher currents, since the contact time decreases with increasing speed. The contact strip wear was monitored by measuring its weight loss after each test cycle. Figure 8 shows the wear rate at a speed of 350 km/h. The results show that the contact strip wear is dominated by electrical wear. The wear rate also increases with increasing contact forces, but maximum values are reached at high current densities and no minimum was found at moderate current densities.

All the results have been summarised in a wear model to estimate, in combination with the contact force simulations, the life time of existing contact line systems. The wear rate is described as a function of speed, current density and the specific contact force. The application of the model shows, that the life time of the contact strip under high speed conditions will exceed 100.000 km and the contact wire meets the requirement of 2 million pantograph passes. These results of the contact strips are close to the observations of the Deutsche Bahn AG in practice. Since results for the contact wire from service will be available in a few years only, no comparison can be done at this time.

5 Conclusion

The simulation procedure will be used by AEG Bahnfahrwegsysteme for determining the significant optimisation criteria of the overhead line/pantograph system and for application as a design parameter for future commercially viable high speed contact lines.

The measurements have been extended to the high strength contact wire materials CuMg as used with the newly developed Re330 for the Deutsche Bahn AG /12/.

Future aspects of the program are investigations of metallised pantograph slipper materials and the systematic variation of atmospheric conditions.

References

1 Schneider, F. & Lerner, F. - Economically oriented quality criterion for the overhead contact line/current collector system, *ZEV Glas. Ann.*, 105, 1981, **9**, 265-270.

2 Becker, K.; Resch & U., Zweig, B. W. - Optimising high-speed overhead contact lines, *Elektrische Bahnen,* 92, 1994, **9**, 243-248.

3 Becker, K.; König, A., Resch & U., Zweig, B. W. - Systematic development of a high-speed overhead contact line, *Railway Technical Review*, 40, 1995, **3/4**, 47-55.

4 Resch, U.: Simulation of the dynamic behaviour of overhead lines and pantographs at high speeds, *Elektrische Bahnen*, 89, 1991, **11**, 139-145

5 Hinkelbein, A.: Wear of the contact line. *Elektrische Bahnen,* 40,1969, **9**, 210-213

6 Kasperowski, O.: Contact materials for current collectors of electric motive power units. Monthly Bulletin IRCA, XV,1964, **2**, 47-69.

7 Klapas, D.; Benson, F.A.; Hackam, R. & Evison, P.R.: Wear in simulated railway overhead current collection systems, *Wear*, 1988, **126**, 167-190

8 Klapas, D.; Benson, F.A. & Hackam, R.: Simulation of wear in overhead current collection systems, Rev. Sci. Instrum., 56, 1985, **9**, 1820-1828.

9 Oda, O.: Wear of current collecting system in Shinkansen, JNR Quarterly Reports, 25, 1984, **2**, 72-75.

10 Kluzowski, B.: Device for measuring the contact force between contact wire and pantograph, *Elektrische Bahnen,* 47, 1976, **5**, 112-114.

11 Ostermeyer, M. & Seifert, R.: Monitoring systems for assessing the relationship between pantograph and overhead line, *Elektrische Bahnen,* 88, 1990, **11**, 122-126.

12 Kießling, F.; Semrau, M.; Tessun, H. & Zweig, B.-W.: New high-performance overhead line type Re 330 used by the German Railways, *Elektrische Bahnen,* 92, 1994, **8**, 234-240.

Figures

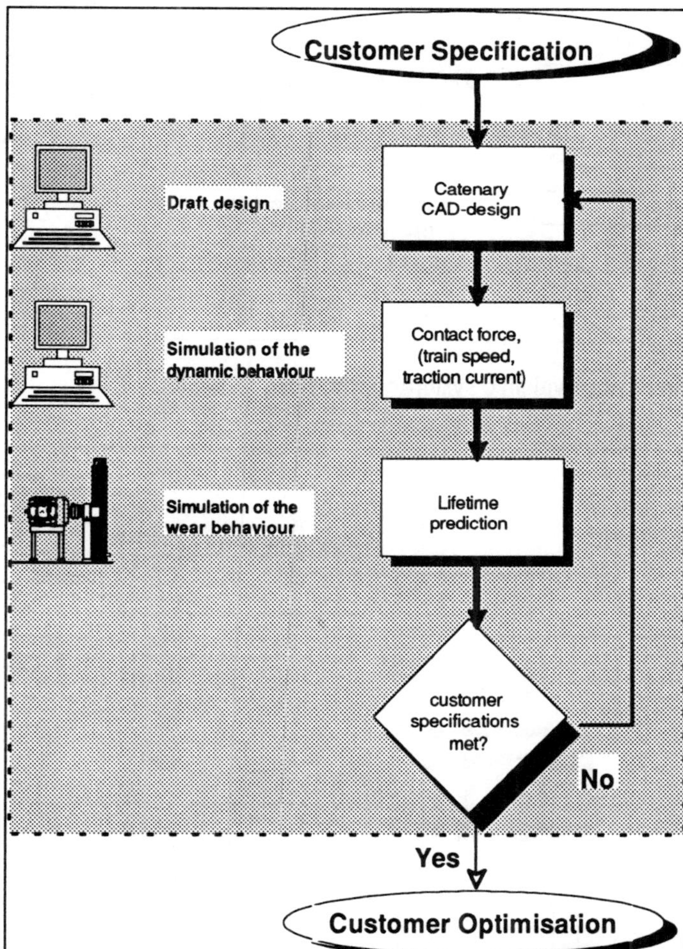

Figure 1: High-speed catenary design optimisation

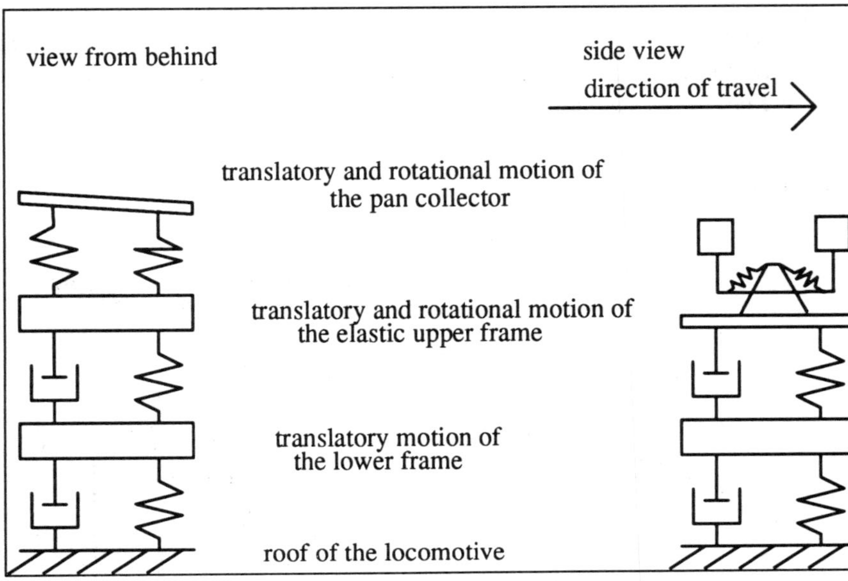

Figure 2: Equivalence system for the DSA350SEK

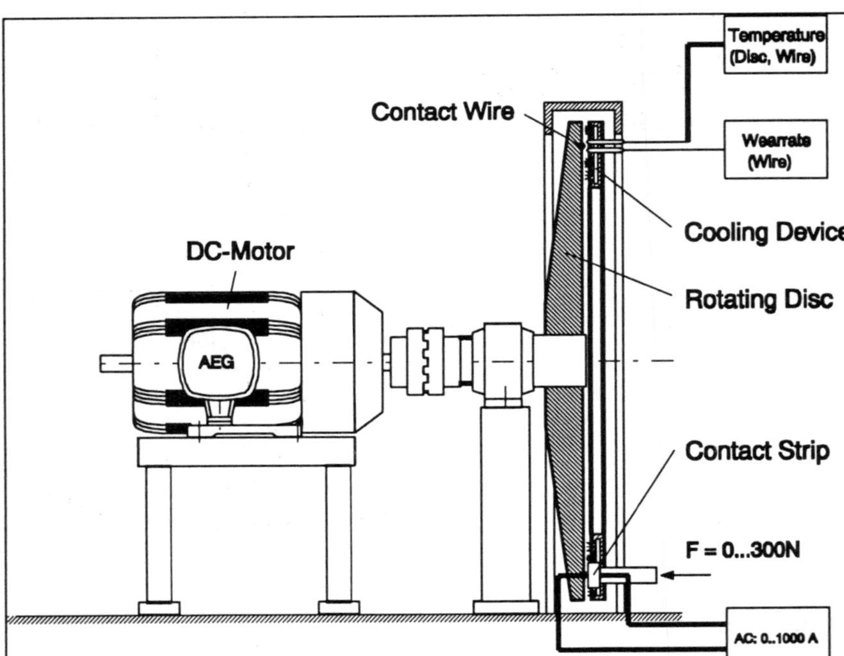

Figure 3: Wear machine

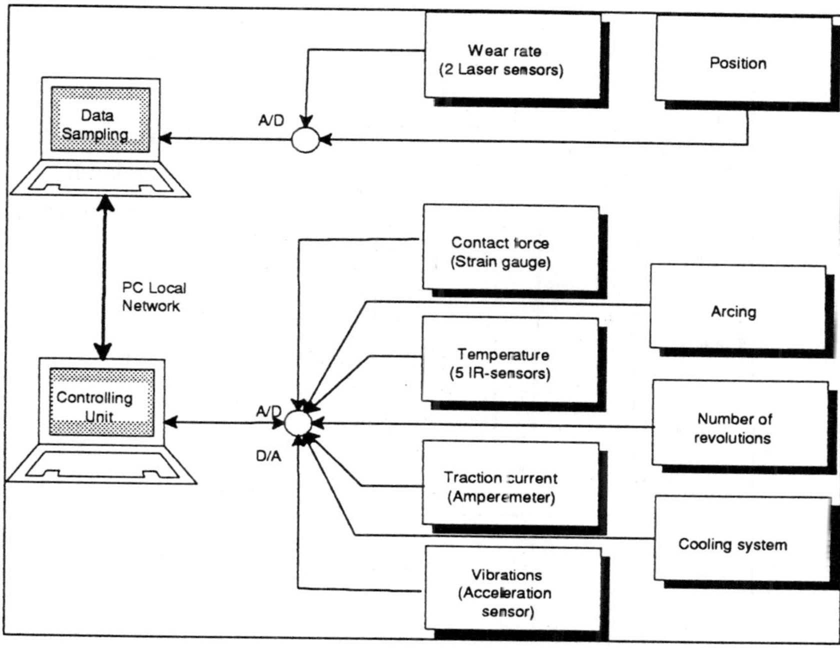

Figure 4: Data aquisition

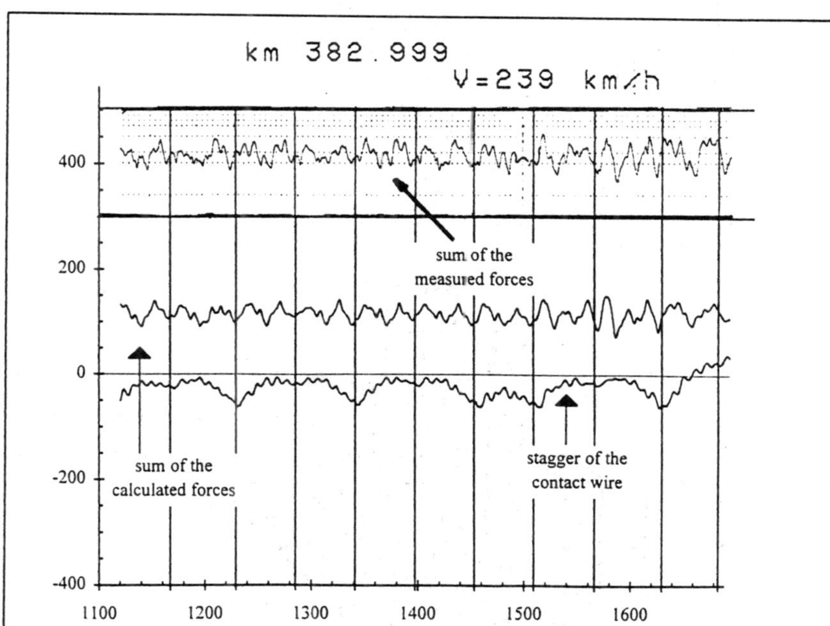

Figure 5: Comparision between simulation and measurement (contact forces)

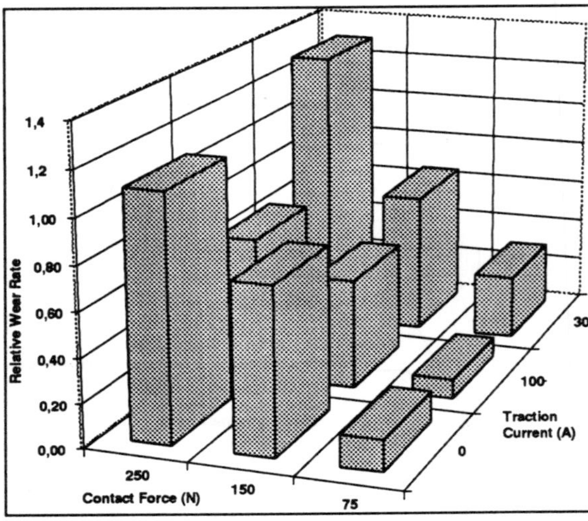

Figure 6:

Wear rate at 100 km/h (contact wire)

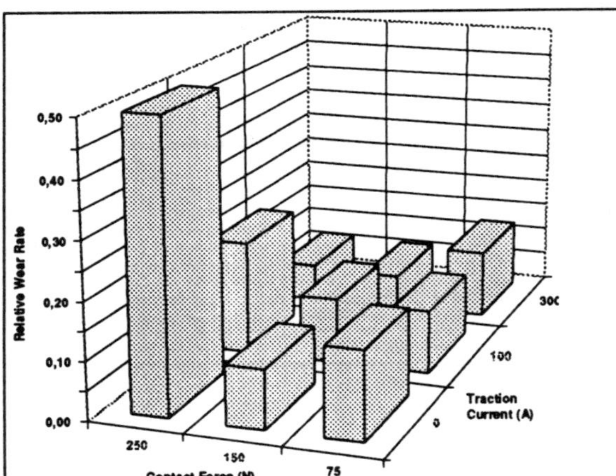

Figure 7:

Wear rate at 350 km/h (contact wire)

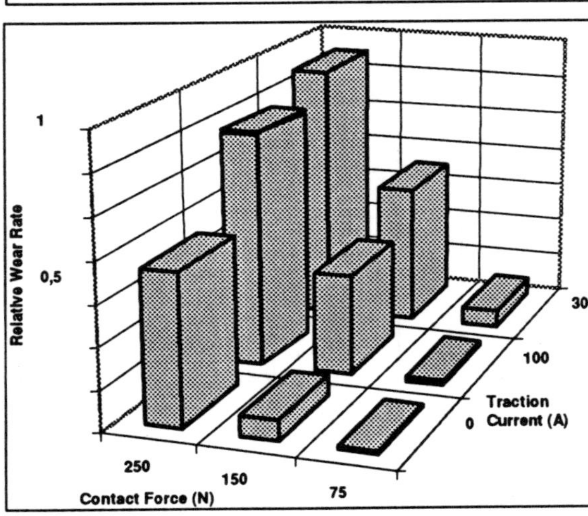

Figure 8:

Wear rate at 150 km/h (contact strip)

Reliability based integrity assessment of offshore platforms

S. Dharmavasan
NDE Centre, University College London, Torrington Place, London WC1E 7JE, UK

Abstract

Fatigue crack growth is one of the main deterioration processes in offshore structures. The fatigue life of welded connections in offshore structures is therefore an important design criterion as well as a governing factor for the planning of inspection and maintenance actions. Due to the considerable costs associated with inspection and maintenance of offshore structures it is necessary to optimise the inspection and maintenance plans such that the costs are minimised and at the same time the risk kept within acceptable limits. This implies that the estimation of fatigue crack growth as well as the effect of inspection and maintenance actions must be modelled accurately. This paper will describe the development of a reliability based structural integrity assessment tool based on fracture mechanics models.

1 Introduction

The offshore industry currently requires that the structural integrity of fixed offshore platforms is ensured by inspecting periodically. In the past, decisions on inspection, repair and maintenance (IRM) have been made by experienced engineers applying their judgement in the form of rules-of-thumb together with appropriate deterministic analyses. However, it is now expected that, by employing recently developed techniques based on structural reliability methods considering the effects of uncertainties, inspection and maintenance scheduling can be made more rational.

In order to be able to apply reliability theory, it is vital to be able to model the failure criteria as well as the uncertainties involved in the process. In the case of fixed offshore structures in areas such as the North Sea, the major problem is that of the deterioration of the welded joints due to fatigue. Furthermore, in-service inspection data and its interpretation play a major role in the whole process. Thus for fixed offshore structures it is necessary to have

- accurate fatigue fracture mechanics modelling
- validated databases for all the modelling parameters, such as material properties
- information on NDT reliability such as accuracy and sensitivity

This paper describes the development of a fracture mechanics based limit state function which can be used to provide reliability based structural integrity assessment for fixed offshore platforms.

2 Reliability Techniques

2.1 Overview

Within the last decade, reliability based methods have become recognised tools in offshore engineering. This is most significantly reflected through the increasing role played by reliability methods in the formulation of codes, classification notes and also recently in the formulation of "safety cases".

For offshore structures in the North Sea, the dominant deterioration process is fatigue crack growth. The fatigue life of welded connections in offshore structures is therefore an important design criterion and a governing factor for the planning of inspection repair and maintenance actions. Due to the considerable costs associated with IRM of offshore structures, it is desirable to optimise maintenance planning such that costs are minimized whilst ensuring that risks are kept within acceptable limits. This implies that the estimation of fatigue crack growth, as well as the effect of IRM actions, must be accurate.

Unfortunately the fatigue life of a welded connection such as a tubular joint is influenced by a large number of uncertainties, such as wave loading, stress concentration factors, material properties and the size and number of initial defects. Furthermore, inspections and repair actions of joints are subject to significant uncertainties mainly due to the difficult conditions under which they are performed.

Deterministic analysis methods in combination with, e.g., the partial safety factor code formats have proven their usefulness in assuring safe structural designs. However one area where probabilistic analysis methods provide better information is when the task is to identify which parameters are dominant. For this it is necessary to take into account all the involved uncertainties and to correctly model the mutual dependencies between the uncertainties. Modern reliability methods serve as a tool for the consistent handling of uncertainties and, when combined with state-of-the-art models of fatigue crack growth, can provide a powerful tool for reliability-based IRM planning of offshore structures.

In order to ensure and maintain the safe operation of offshore structures inspection and maintenance actions are performed with adequate intervals. The costs associated with inspection and maintenance of offshore structures are very considerable mostly due to the difficult conditions under which inspection and maintenance must be performed. Inspection and maintenance actions should therefore be optimised with respect to inspection intervals, inspection techniques and potential repair actions.

2.2 Description of limit state

It is usually possible to describe whether a structural component is in a safe or failed condition with the use of a deterministic function (the so called *state function*). Component reliability analysis is used in the calculation of the probability of this function being in the failure state, under the influence of uncertainties in the input values. This probability is usually known as the probability of failure (POF).

The idea of structural reliability can be demonstrated by considering a simple situation where loading, *L*, interacts with strength, *S* (Figure 1). Here the state function (*g*), also known as the *safety margin*, could be formulated as,

$$g = S - L \qquad (1)$$

When $g > 0$, the structural component is said to be in the *safe state*. Failure occurs when $g < 0$. In the limiting case where $g = 0$, g is usually known as the *limit state function*.

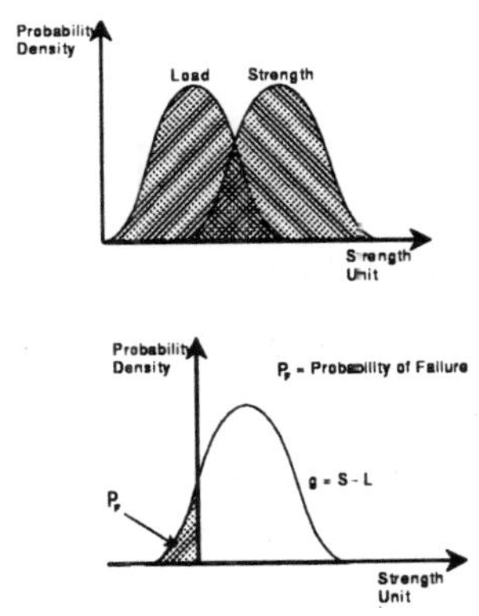

Figure.1: Component reliability analysis - Interaction between Strength and Loading

In order to estimate the reliabilities and the expected costs several probabilities must be estimated. To this end modern reliability methods such as first order reliability method (FORM) and second order reliability method (SORM) (Faber et al, 1992) are adequate. These methods are based on the concept of a limit state function.

When fatigue crack growth events are considered the limit state functions are formulated as

$$g(x) = a_{CRIT} - a(t,x) \qquad (2)$$

where a_{CRIT} is the critical crack size and $a(t,x)$ is the actual crack size at time t. Alternatively and often more appropriately, the limit state function can be formulated as

$$g(x) = \frac{N}{N(a_{CRIT},X)} - 1.0 \qquad (3)$$

where N is the anticipated lifetime and $N(a_{CRIT},X)$ is the time before a crack of size a_{CRIT} is obtained.

The theory used to predict the crack growth in tubular welded joints is fracture mechanics.

2.3 Reliability analysis results

Several proposals for IRM planning can be found in the literature, see for example a review in (Faber et al, 1992). Some strategies are purely reliability-based (Pedersen et al, 1992), whereas others combine costs with reliability (Sorenson et al, 1992).

The necessary input for IRM planning is:

- joint and weld geometry
- stress history
- material characteristics
- initial defects
- previous inspection results
- inspection reliability
- costs of failure, inspection and repair

The output for each joint and over a period of time, will be:

- expected remaining life
- probability of failure and reliability measures
- expected costs of for maintenance plans
- reliability sensitivity measures

An example of a reliability-based IRM plan is illustrated in Figure 2. Inspections are planned whenever the reliability index decreases to a certain code specified value. At the planned inspections it is assumed that no damage is detected and the reliability is updated using this assimption. This approach assures that a minimum reliability is maintained throughout the lifetime of the structure.

Using a costs based approach only the next inspection time is planned but the costs associated with failure, repair and inspection are taken into account

(Figure 3). After an inspection is performed the inspection results are taken into account when planning the next inspection. This scheme is often referred to as the adaptive scheme and is described in detail in Faber et al (1994).

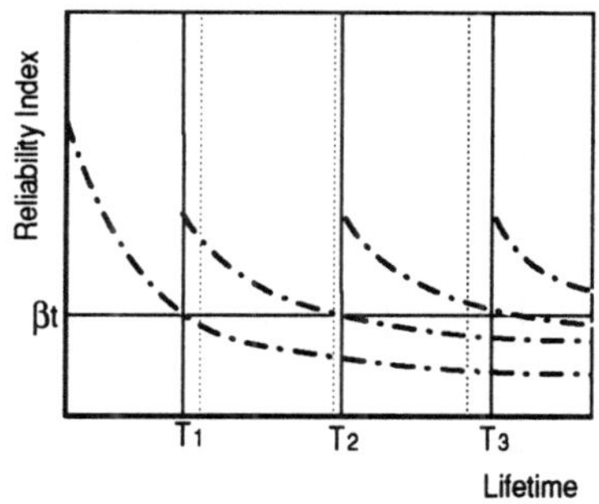

Figure 2: Reliability-based IRM planning

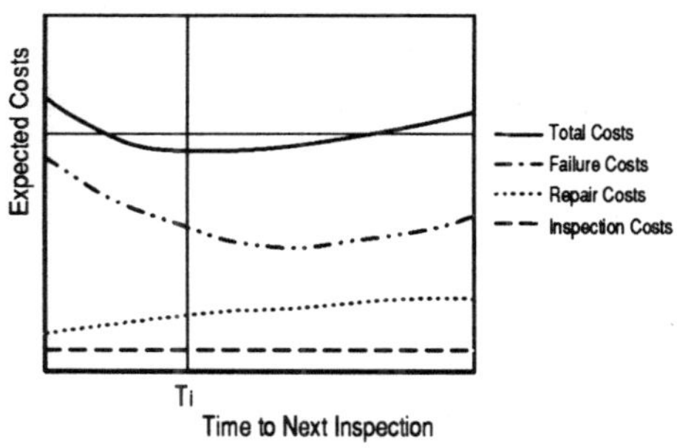

Figure 3: Costs-based IRM planning

3 Fracture mechanics modelling

3.1 General

Fatigue crack growth models for welded structures, based on linear elastic fracture mechanics have been described recently by several authors (Straalen and Dijkstra, 1991, Dharmavasan et al, 1991). In general, the crack growth model gives the relationship between the crack growth rate (da/dN) and the fatigue loading parameter (ΔK) (stress intensity factor range).

$$\frac{da}{dN} = f(\Delta K) \qquad (4)$$

This relation has in general a sigmoidal shape as given in figure 4, with a threshold value ΔK_{th} at the lower ΔK values and an unstable crack growth part at high ΔK values.

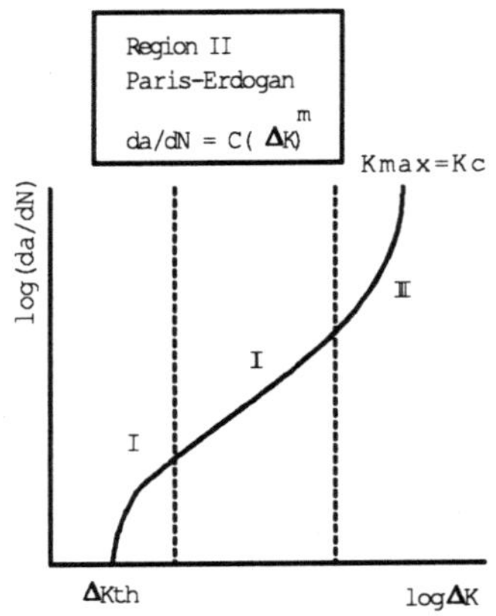

Figure 4: Crack growth relationship

The stress intensity factor range is the difference between the maximum SIF and the minimum SIFduring a load cycle. The SIF is a measure for the magnitude of the stresses near the crack tip and depending on the geometry, the load level and the crack dimension.

$$K = Y\sigma\sqrt{\pi a} \qquad (5)$$

where:

σ = remotely applied stress
Y = correction factor depending on geometry and loading conditions
a = crack depth

Integration of the crack growth relation (equation 5) from an intial defect a_i to the final size a_r gives the lifetime N of the detail.

$$N = \int_{a_i}^{a_r} \frac{da}{f(\Delta K)} \qquad (6)$$

A fracture mechanics module called the Component Fatigue Analyses (CFA) was developed by combining the FACTS program developed at University College London (Dharmavasan et al, 1991) and the FAFRAM program of TNO (Dijkstra and Straalen, 1991). A brief description of the basis of the CFA is given below. A more detailed description can be found in a paper on the CFA (Dijkstra et al, 1994).

3.2 Application to tubular joints

The general purpose fracture mechanics model has to be modified to suit offshore tubular structures (Straalen and Dijkstra, 1991).

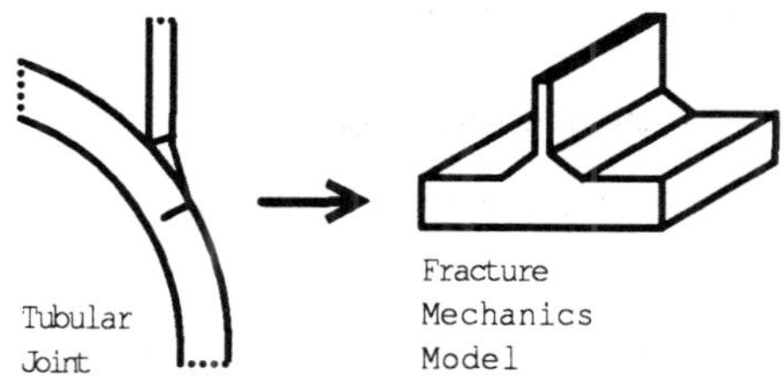

Figure 5: Fracture mechanics model for tubular joints

Fatigue cracks in tubular joints generally start at the weld toe and grow in the depth and width direction into the tube wall. Therefore, the CFA assumes a semi-elliptical crack at a weld toe and the real geometry is translated to a simplified model as shown in figure 5. With the crack growing into both depth and width direction, crack extension rules for both directions are needed. In the CFA this can be done in two different ways, namely:

1. Crack growth model for both crack depth and width direction. In this model equation 2 is used in both directions with the appropriate SIF range (ΔK_a for the depth direction and ΔK_c for the width direction).

2. Forcing function for crack width direction. In this model the following equation (7) is used only in the depth direction and the crack width is defined as a function of the crack depth.

$$c=f(a) \tag{7}$$

Both methods have their own advantages and disadvantages. The first approach is more correct from a theoretical point of view. However, the problem here is how to deal with multi initiation and crack coalescence. The latter approach needs a forcing function of the type shown in equation 7. However all the available forcing functions are based on experimental observations. One of the questions requiring further investigation is whether the forcing functions are applicable for situations different from the situations on which the formulae are based.

The crack growth model mostly used is the Paris-Erdogan relation (equation 8), which represents the linear part of the crack growth curve of figure 4.

$$\frac{da}{dN}=C(\Delta K)^m \tag{8}$$

where:

C = crack growth coefficient
m = crack growth exponent

The material constants C and m are dependent on the material and the environmental conditions (seawater).

The CFA has a number of crack growth relations available and the most appropriate method used in the analysis.

3.3 Stress intensity factor solutions

The stress intensity factors implemented within the CFA system fall into two main categories:

- empirical equations derived from fatigue crack growth data obtained from tubular joints.
- analytical and empirical solutions dervied for flat plates with semi-elliptical cracks modified by other factors to model crack growth in tubular joints.

3.4 Modelling of multi-segment crack growth curve

The behaviour of steels in a corrosive environment leads to a multi-segment Paris curve as shown in Figure 6.

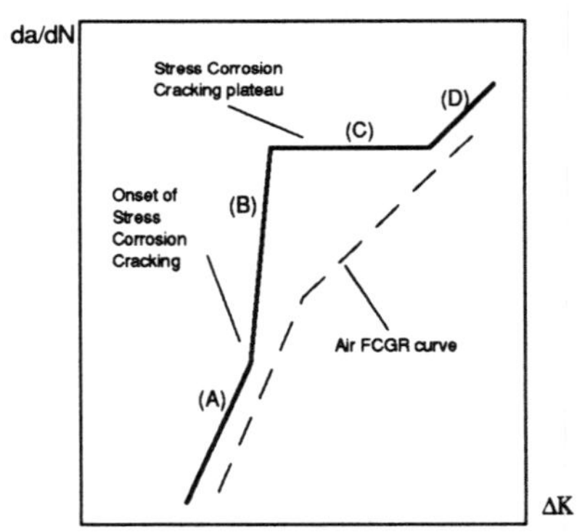

Figure 6: A multi-segment crack growth curve

The corrosion crack growth modelling is complicated by random loading experienced by offshore structures in service. Assuming that there is no load interaction, a calculation procedure was developed by Kam and Dover (1989) which adequately predicted the mean values of the experimental data. The procedure involves calculating the mean crack growth rate by summing all the crack growth rates (using the relevant C and m material properties for the segment), each caused by a particular stress range, and weighted by the probability of that stress range occurring in the loading history. This procedure is described by the equation below:

$$\left(\frac{da}{dN}\right)_{av} = \sum_{j=1}^{ksegments} \left\{ C_j \left(Y\sqrt{\pi a}\right)^{m_j} \int_{\Delta S'_{j-1}}^{\Delta S'_j} (\Delta S)^{m_j} p(\Delta S) d(\Delta S) \right\} \qquad (9)$$

Once the average growth rate is determined, the crack growth curve is determined by numerical integration.

4 Implementation of software

The reliability analysis was implemented as a set of FORTRAN modules. The overall framework is shown in Figure 7. In order to enable the analysis module to be embedded within other analysis modules including a knowledge based system, the communication with the analysis modules is by text files. The results are also output to text files in a specific format.

The program takes as input probabilistic models for the initial crack geometry, joint geometry, material characteristics and inspection reliability in addition to the long term statistics of the stress processes which are specific to the hot-spot. The latter is stored in a hot-spot specific file and includes analysis options such as stress intensity factor solution, inspection method etc.

The output consists of updated reliabilities, expected cost estimates and crack growth estimates. More detailed results such as partial derivatives, algorithm monitoring, are also possible. A graphical user interface has also been developed to aid with creating input files and to display results.

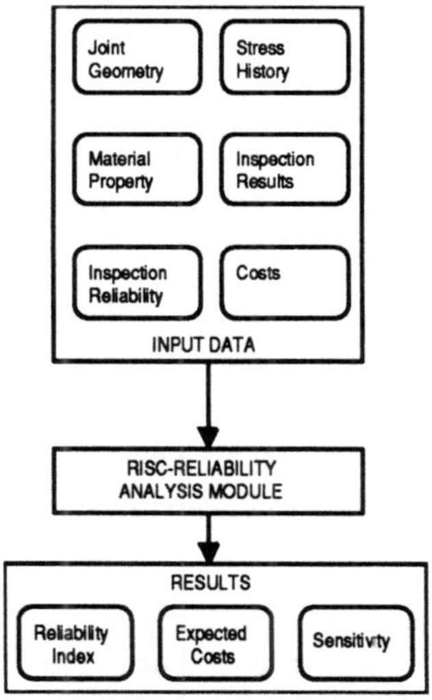

Figure 7: Structure of Reliability Analysis Module

5 Conclusions

This paper has described how a fracture mechanics based model developed for crack growth predictions in tubular welded joints has been modified so that it can be used for reliability based structural integrity assessment. One of the main problems in applying reliability based methods is in the availability of good statistical databases especially for material properties, geometric variations, inspection reliability etc. The methods developed here underwent substantial validation as part of a European Community funded research project.

References

Dharmavasan, S., Reynolds, A.G. and Topp, D.A., 1991. *The Development of a PC-Based Integrated Fatigue Analysis Software Package for Offshore Structures,* Proc. Offshore Mechanics and Arctic Engineering, ASME, Stavanger.

Dijkstra, O.D., Foeken, R. Van., Dharmavasan, S., 1994. *Fracture Mechanics Limit States for Reassessment and Maintenance Planning of Offshore Structures*, Proc Symp. Offshore Mechanics and Arctic Eng., ASME, Houston.

Dijkstra, O.D., Straalen IJ.J. Van, 1991. *Fatigue Crack Growth Program FAFRAM (FAtigue FRActure Mechanics),* TNO Building and Construction Research, Report BI-91-051.

Faber, M.H., Dharmavasan, S., Dijkstra, O., 1994. *Integrated Analysis Methodology for Reassessment and Maintenance of Offshore Structures*, Proc Symp. Offshore Mechanics and Arctic Eng., ASME, Houston.

Faber, M.H., Sorensen, J.D. and Kroon, I.B., 1992. *Optimal Inspection Strategies for Offshore Structural Systems*, Proc Symp. Offshore Mechanics and Arctic Eng., ASME, Calgary.

Kam, J.C.P. and Dover, W.D.: Corrosion Fatigue of Welded Tubular Joints: Fracture Mechanics Modelling and Data Interpretation, Proc. 8th Int Offshore Mechanics and Arctic Eng. Symp., ASME, The Hague, 1989.

Pederson, C., Nielsen J.A., Riber, J.P., Madsen, H.O. and Krenk, S. 1992. *Reliability based Inspection Planning for the TYRA* Field, Proc Symp. Offshore Mechanics and Arctic Eng., ASME, Calgary.

Sorenson, J.D., Faber, M.H., Thoft-Christensen, P. and Rackwitz, R. 1991. *Modelling in Optimal Inspection and Repair*, Proc Symp. Offshore Mechanics and Arctic Eng., ASME, Stavanger.

Straalen IJ.J. Van., Dijkstra, O.D., 1991. *Application of the Fracture Mechanics Approach to the Fatigue Behaviour of Welded Tubular Steel Structures*. International Symposium on Tubular Structures.

Proposal of a computer method to solve the Zener-Hollomon equation

I. Rieiro, M. Carsí, F. Peñalba
Centro Nacional de Investigaciones Metalurgicas, CSIC Avda. Gregorio del Amo, 8 28040 Madrid, España

Abstract

We have developed a integrated computational method in order to solve the Zener-Hollomon equation. This method solves the fitting problem of the equation of Zener-Hollomon for wide ranges of applied stress, strain rate and work temperature by means of the progressive application of three algoritms kinds.In first place a program, by means of a reiterated linear regression by planes, it allow to get some values that will be later the initials values of a second block of algoritms and it permit also purify the experimental data set. The second block of algoritms carries out an fitting by nonlinear least squares using a variation of the of modified Gauss-Newton method. In third place is chained a group of programs that they permit iterate starting of the solution of the previous program approaching (if is necessary) for you line in RxR until reach the good. All this with a support of statistical analysis for they determine the goodness of each step, of the data the residuals and of the last fitting gotten.

1 Introduction

The behavior of different kinds of steels subjected to torsión test under variable conditions of applied stress , temperature of work and strain rate, it is of great interest for that it allow to determine the mechanic characteristics of the same, like mechanic resistance or ductility. Also for that one could get micro-structural characteristics of the materials, like size of grain and of subgrain ,volumetryc fraction of recrystallized material and recrystallization time and of everything it one could carry out predictions. The conditions of stress and temperature of work, they are generally of the fluency for work in hot with polycristallyne material in situation of equilibrium, between processes of recovery and recrystallization, for homologous temperatures higher to 0,5. For the zone of constant strain rate, several empirical laws have been used that relates stress, temperatures and strain rate , [p.e].: Power-Law, Brek-Down and another. Different according to the range of stresses and of the material of that treat each other. But these laws don't have a general validity in all the ranges of work. The equation of Garofalo or of Zener-Hollomon it seem to give a correct answer to wide ranges of σ(stress), ϵ (strain rate), T (tempera.tura). Also this equation believes that this supported by physical theoretical models (the who they justify it), that they are based on the structure of the material to level of atomic lattice. The theoretical models are diverse, well slip for boundary of grain is been about, diffusion for vacant interstityall, slip of dislocations, climbed of dislocations, etc. This physical support for the equation of the Zener-Hollomon:

$$\dot{\epsilon}=A.e^{\frac{-Q}{R.T}}.\sinh[\alpha.\sigma]^{n} \qquad (1)$$

$\dot{\epsilon}$=*strain rate*(sec^{-1})
σ=*aplied stress (MPa)*
T=temperature (ºKelvin) **(2)**
A,Q,α,n=constants independents of T
R=cte. perfect gases

it allow you to give of physical meaning to the constants that appear in the same and that makes more interesting the obtaining of the values of the equation that better they fit the same to experimental data sets, and it makes necessary that the obtaining of these constants is the more precise possible. It believe that the A constant include exponentials factors of entropy variation by the kind of process that develops the slip, as well as parameters of physical interest like the vector of Burgers and according to the models, frequencies of proper vibration of the atomical lattice , or you of them line of dislocatión. The constant Q, it is an energy of activation ([entalpy]) proper of the predominant process and it believe that it could include values of the field of internal stress generating an activation not thermal. It believe that the constant $\{\alpha\}$ represents a certain volume of activation pondered by the average number of element that participates in the basic process of slip and the constant "n" it is possible that it am related with the average number of constituent element that they participate in the basic slip. But the equation of Garofalo is complex equation from the point of view of the quality of the values gotten starting from the same for all possible range of the values of interest in stress and temperatures. It is strongly nonlinear for two of the parameters (α and n) and conditionally linear for another two (Q and A). Therefore the mathematical problem is reduced to fit the better function of the family of those that they have the form (1) to a group of data produced by torsión or traction$\{\epsilon(i),\sigma(i),T(i)\}$

with i= 1. .n, getting like result{ A, Q, n, α}the good values of the fitting. In the analysis of nonlinear regression some topics exist to the who it should fit the mathematical treatment of the problem, the who they are:1) The expectatión functión is correct. 2) The response is expectatión functión plus disturbance. 3) The disturbance has a normal distributión. 4) Each disturbance has a normal distributión. 5) Each disturbance has a zero mean. 6)The disturbance have equal variance. 7) The disturbance are independently distributed. 8) The disturbance are experimentals errors. The usual methods utilize the linearization of the problem using a method of Gauss-Newton, with or without factor of step, with convergence controls like a relative offset convergence criterion. Also is usual the direct method for sum of square errors , the method of Newton- R., and the modifications of Levenberg-Marquardt to the method of descending steps. Nevertheless they in the literature are been accustomed to utilize "iterative methods" diverse, not very specified. It is known that to the criterions exposed one must add in the nonlinear problems two very intense conditions. On one hand the great dependence for the convergence and the speed of the same of the initial values and for another the possibility of obtaining of stationary values that they is not minimums in the optimization process of the function of error, well by being points of seat, or outputs spurious owed to the characteristics of the data or of the fitting surface . In the case that is interested in us the surface of fitting is configurate for abrupt zones with very intense values of gradient. The function to optimal obtain is:

$$SSE(\vec{\theta}) = \sum_{i=1}^{i=n} (\dot{\varepsilon}^*_i - \theta_0 . e^{\frac{-\theta_1}{x1^*_i}} . \sinh(\theta_3 . x2^*_i)^{\theta_2})^2 \quad (3)$$

the star superindex indicates experimentals values and the vector{θ_i},i=0. .4 substitutes to the explained constants. The form of the expression for a dates sets of several ten it makes obvious the great difficulty of the form of the fitting surface .

2 Construction of the codes

In first place has constructed a code that linearized]the equation of Zener-Hollomon for all the constants, except for one of them,{θ_3} .These is traveled in an interval defined by the values physicaly allowables of the problem, with a step pass of adequate discretization, so that[$\Delta\theta_3/\theta_3$]<τ, being{τ} determinant of the precision of the good(optimal). The code carries out multiple fittings for the successive values of the mesh of the selected constant, of all they fittinga we select the better by means of the coefficient of global determination (corrected) Gotten R^2. the good passes another routine of the code the who it determine the board of the ANOVA and of her by means of the obtaining of the function of distribution experimental F of Snedecor controlled the goodness of the adjustment by means of a hypothesis test , and with the corresponding hypothesis test the goodness of each one of the gotten constants is determined, making use of the experimental distribution of the [t-] of Student. Starting from the anterior gets intervals of confidence for each one of the of the parameters of the equation of Zener-Hollomon gotten by means of an adjustment linearized by plans. This code to their time delivers some residuals e(i)= ϵ^{*}(i)- ϵ(i), for i= 1. .n, to those that it undergo to a study of normality, of independence and of presence of "outliers." . Is gotten besides of other parameters a function of frequencies of the same and a function of autocorrelation of the residuals. This first code allows to characterize the problem reference of the quality of the set of experimental data, of the function of distribution to that they respond and of a possible selection and elimination of points that could introduce disturbances of experimental base.

The second develloped code bases on some routines that apply a method of nonlinear fitting of least squares Newton-modificado kind, with a step factor of and convergence control that takes how initials values the good delivered by the previous code . This second code is also supplied of routines for the exhaustive analysis of the residuals that is compared with the anterior. Terminated this second code has been gotten a set of solutions with enough safety. These outputs undergo to the treatment of a third code that it consist in essence in the work tied of two routines, each one of the who it allow to carry out fittings on the R^4 space, moving us in cuts on two dimension, R^2, it in each one of the cuts have been selected one of the constants of the adjustment of the equation of Zener-Hollomon, well Q ó well{α}, taking the last gotten value for her, varying only two parameters and we for this move for you line, it in each line remain free a different parameter, entering so in iteration, in a sorrounding of the good searched for. By being quite close the delivered value in the second code the work of the third code is not but a fine adjustment of the convergence and only for some parameters to fit. On the other hand sometimes the use of the third code results superfluous then with the two first has been arrived to an exact solution and enough precise.

3 Employed Algorítms

The basic algoritm for the first code for the search of initials values fit the model:

$$Z_i = a_0 + a_1 . X_{1.i} + a_2 . X_{2.i}(a_{3.k}) + e_i \tag{4}$$

where $Z(i) = Ln[\epsilon]$, $a_0 = Ln[A]$, $a_1 = -Q$, $X_{1.i} = 1/R.T_i$, $a_2 = n$,, $X_{2.i} = Ln[Sinh[a_{3.k}.\sigma_i]]$, with $a_{3.k} = \alpha_k$.Being R=constant of perfect gases .The objetive function on fitting by mean a linear least square regression to the experimental data set $\{\epsilon_i, \sigma_i, T_i\}$, i=1..n,travelling $a_{3.k}$ a interval $[a_{3.min}, a_{3.max}]$, with a step size Δa_3

For each value K=1..p of the constant a_3 a vector of estimate is gotten $\{\alpha\}_{i=0,,2}$,by means $\alpha=(X'.X)^{-1}.X'.Z$, where X' indicates transposition of matrix and the notation is the traditional in these cases ($X=\{x_{i,j}\}$,i=0,1,2 j=1...n matrix of data , Z=vector of data). The algoritm construction , the statistical controls and the analysis of the residuals, like the ANOVA board, and the application of the pertinent hypothesis test (F-Snedecor, [t-Student]) could consult in the paper of the authors[5]. The determination of the good becomes with the determination coefficient, R^2_k , that it vary with{ k} to the same as $a_{3,k}$, being $R^2_k= (\Sigma(Z_{i,est}- E(Z))^2)/(\Sigma(Z_i-E(Z))^2)$, with $Z_{i,est}=X.\alpha$. Selected the maximal value of the determination coefficient, after analyzing and purify the experimental data set , and checked the goodness of the fitting , it make a file with the outputs $\{A,Q,n,\alpha\}$ that it will be utilized for the following code like initial values set . The following algoritmo tries the problem of nonlinear form . Given $Z(i)=f(X(i)_k,\alpha^0)+e(i)$, expounded in form of matrix for i= 1.2. .n (data) and $\alpha^0 \in \mathbb{R}^4$, we applied a function of minimization

$$SSE(\alpha)=\|Z-f(\alpha)\|^2 \qquad (5)$$

Being the statatistics of estimation that guarantees the good in theapproach of the method of modified Gauss-Newton the following:

$$\hat{\alpha}=\alpha_0+(F'.F)^{-1}.F'.e+O_p\left(\frac{1}{\sqrt{n}}\right) \qquad (6)$$
$$e=Z-\hat{Z}$$

being in these notation $f(\alpha)=\alpha_0.e^{(-\alpha 1/X(1,i))}.[\text{Sinh}(\alpha_3.X(2,i))]^{\alpha 2.}$

The Z variables and X are the noted previously, F' indicates trasposition and the matrix F is the gradient reference of the variable$\{\alpha\}$. Starting from the initial values contributed by the previous code is advanced in the address of the gradient with a search of the size of step carried out by means of a optimization linear univariate for each step utilizing the method of the golden ratio and the of the numbers of Fibbonacci faintly, could use a falling sequence of power of { 1/ 2}, with good result. They work hard precise methods of the control of the speed and of the convergence. Also the problem has dealings with statistical adequate controls, analysis of the residuals and study of the goodness of the fitting. The details could be consulted in the paper of the authors[6]. In occasions the combined work of the two first codes previously detailed it is enough precise and one could stop the process and take like good the gotten result. Nevertheless it in other occasions is not so, and in spite of having gotten a value, this is not the good, then is conditioned by the consecutive development, we for this have constructed a iterative method that works to starting from the last result, carrying out a "fine tunning" of the previous result. This code this formed fundamentally for two routines chained of form iterative, it in each one of them are carried out a linear regression of the form ZA(i)=AP+ N.XA(i), being in each one of them different the meaning of the variables. In the first ZA(i)=Z(i)+Q/X(2,i), with Z(i)=Ln[ϵ(i)], X(2,i)=R.T(i), AP=Ln[A],N has the given meaning in the eqn. (2), XA(i)=Ln[Sinh[α(k).X(1,i)]], with X(1,i)=σ(i), all the notations not specified they have the given meaning in the eqn. (2) .En this routine α(k) travels all them value of k, resultants of dividing the interval of values physicaly acceptable with an adequate step so that the result of the good guarantees a precision of 0.001. It are determined the good starting from the determination coeffcient and it write values them of the good in a file that it take the following routine in which ZA(i)=Ln[A]+Q(k)/X(2,i), XA(i)=Ln[Sinh[α.X(1,i)]], and all the rest values are the same. In this routine the constant that travels an interval divided in steps and it on the one which they are carried out the progressive regressions is the Q(k) parameter , which becomes with the same criterion of suitable precision. In the first Q= [cte]., and in the secondary α= [cte]. The details of control of iterations, variation of ranges of traveling and visual inspection of the evolution of the determination coefficients, as well as the adjustment of precisions could be consulted in the work of the authors[5],[6].

4. Applications.Material and experimental procedure.

In order to control the operation of the group of codes they have been utilized two types of materials. A steel microalloyed with Vanadium and a small percentage of Titanium. The second material is the very well-known alloy Ti-6Al-4V that it is one of the alloys more thoroughly studied by their interest like structural material in aerospatial applications . The microalloyed steel has been characterized in diverse work,[7],[8],[9]. The alloy of Titanio has been previous studied[10],[11]. The experimental data have been gotten by means of torsion test in a torsion bank described in[[8]. The experimental gotten data are show in Tables 2,3 and the characteristics of the steel in the Table 1

Tabla 1.Chemical compositión of used steel

C	Mn	Si	S	P	Cr	Ni	Mo	V	Ti	Cu	Sn	Al	N ppm	O ppm
0.29	1.28	0.34	0.03	0.02	0.13	0.08	0.01	0.09	0.02	0.13	0.02	0.04	106	45

The sample of Titanio this formed by 33 data the one which is statisticaly significant for this kind of work. The of steel this formed by 16 data, for the one which if it introduce any kind of problem in the same, this would would that sample was scarce Nevertheless all work in this field owes adapt to samples relatively scarce.With the Titanio are worked with low strain rates , while it in the case of the steel are worked with tall sevalues. The rounds for temperatures and tensions are similar

Table 2. Experimental datas for the Titanium

ϵ(seg(-1))	T(°K)	σ(MPa)
0.00007	1073	23.91
0.0002	1073	43.92
0.0005	1073	73.75
0.001	1073	104.12
0.003	1073	147.69
0.0002	1123	12.82
0.0005	1123	223.08
0.001	1123	35.76
0.003	1123	56.37
0.007	1123	80.23
0.0002	1173	8.38
0.0005	1173	15.33
0.001	1173	21.41
0.003	1173	32.48
0.003	1173	52.59
0.0005	1179	9.59
0.001	1179	15.78
0.003	1179	25.34
0.007	1179	35.44
0.07	1200	72.3
0.2	1200	91
2	1200	169
0.07	1220	63.9
0.2	1230	67.4
0.5	1230	95.4
2	1230	131.3
1	1230	112.4
0.5	1250	68.6
2	1250	104
5	1250	130
0.07	1243	39
0.2	1270	51.5
0.5	1270	67.6

Table 3. Experimental datas for themicroalloyed steel

ϵ(seg(-1))	T(°K)	σ(MPa)
2	1353	75.76
2	1398	64.75
7	1353	96.21
7	1398	80.18
14	1353	107.2
14	1398	91.87
21	1353	113.71
21	1398	99.59
1	1423	53.55
14	1423	86.07
21	1423	91.69
7	1100	260.74
2	1373	67.63
7	1373	84.83
14	1373	97.12
21	1373	107.2

5 Results

For the microalloyed steel and for the Titanium the application of the sequence of the three codes it show the next result

Table 4. Results by codes for the microalloyed steel

MATERIAL	A	Q(J/MOL)	N	α	R(COEF.DET.)
CODE 1	3.16X10^(19)	277028 ±13900	5.48±0.24	3.65X10^(-4)	0.9957
CODE 2	3.161X10^(19)	276900	5.47	3.65X10^(-4)	
CODE 3	4.46X10(26)	275931	5.46	1.8X10^(-5)	0.9989

Table 5. Results by codes for the Titanium alloy

MATERIAL	A	Q(J/MOL)	N	α	R(COE.DET.)
CODE 1	7.08X10^(16)	422871 ±22200	1.96±0.12	0.016	0.993
CODE 2	7X10^(16)	422867	1.94	0.016	
CODE 3	4.22X10^(15) 5X10^(16)	372671 419000	2.01 1.96	0.005 0.016	0.981 (1) 0.982 (2)

Table 6. Comparison of the results

MATERIAL-REF.	N	Q(J/MOL)	ΔT(K)	$\Delta\epsilon$(SEG^(-1))	TEST MODE
Ti (this work)	1.97	419000	1073-1270	7x10^(-5) - 5	torsion
Ti[12]	2.03	451000	1173-1243	7x10^(-2)-8.7	torsion
Ti[13]	2.0	400000	1073-1173	10^(-3) - 10^(-1)	compression
Ti[14]	1.56	522000	1073-1223	5x10^(-2) - 50	torsion
Steel(this work)	5.46	275931	1100-1423	1 - 21	torsion
Steel[15]	4.76	339000	1323-1473	1- 14	torsion

The hipotesys test for the goodness of the fitting in the case of the steel and for the code of nonlinear fit they introduce a value of the F function of experimental Snedecor and with the parameters of the case of 751.72, which it makes significant the fitting. The study of the residuals allows to affirm that although in any zone a seccession of the normality is detected in general they behave correctly (the data). They are not detected "[outliers]." It seem that the residuals introduce a certain correlation [serial] (consequence of the method of experimental work). The function of density (in frequencies) it of the residuals could be seen in the graphs[3] and[9] .The serial representation of the residuals pattern in the graphs[2] and[8]. The outputs of the normal graphs of the Chi-cuadrado test for the residuals are shown in the graphs[4] and[10]. In the graph[1] one could observe the behavior if it fit leatory of the values of the determination coefficient for the running of the third code for the iterative method . This behavior indicates that the iterative methods with answer surfaces of as complex as it is the of the

present work they could take line of advance that very close to the better line and they carry to solutions not good, like it so happens in many occasions. For the contrary if one is next to the good and it make itaration for you line adequate the result will be correct. However the two first codes working on the whole guaranteeing, besides more and better information, a global result and generally more correct.
In the graphs[6] and[7], one could observe the representations of the equation of Garofalo in three dimension for the solutions contributed by our codes.

6. Conclusions

The comparison with the bibliographical data , and they with the physical conditions of the problem allow to affirm that the gotten outputs not are only good but rather they are more complete and they are contributed with precision, could determine confidence intervals of individual or groups for the gotten constants. The statistical analysis of the code allows to debug the bases of data to posteriori eliminating with certitude that data that could perturb the result, of form but not making that about arbitrary form with the one which it would be lost information. It on the whole the package of programs work without necessity of initial data for the one which contributes a big autoconsistence to the resolution of the problem. Although the characteristics of the groups of special data are reached an enough precision, to the less in the frame of the current necessities in the investigation in this field. The exactness of the solutions is questionable anyway and this related with the more o less grade of verisimilitude of the equation of Garofalo. Any theoretical deepening at the present time on the models that they support this equation puts of evident that is necessary elect more complete equations, more complex and maybe with modifications in their structure. Diverse models introduce relationships hyperbolic sine differents and it would become‘ precise keep in mind that they in some ranges of work can to act several process that contributes to the fluency. In spite of everything the anterior seems us that the direct methods not lineal they are a guarantee of being working with more security that with the iterative methods

REFERENCES

1. M.J.D. Powell . Department of Applied Mathematics on Theoretical Physics University of Cambridge . Siulver Street, Cambridge CB39EW, England, pp 122 to 141, Proceedings of Numerical Analisys of the 10th Biennal Conference held a Dundee, Seatland, June 28 July 1, 1993
2. Fletcher and C.M. Reeves "Function minimization by conjugate gradients", The Computer Journal, Vol 7 pp 149-154 1980
3. Douglas M. Botes and Donald G. Watts "Nonlinear regression Analisys and its Applications " John Wiley & Sons . NewYork 1993
4. A. Donald Gallant . Nonlinear Statistical Models. John Wiley & Sons. NewYork 1986
5. I.Rieiro, M.Carsí. Paper Submited for printer.
6. I.Rieiro, M.Carsí. Paper Submited for printer
7. C.Garcia, M.Carsi, F. Peñalba and M.P. Andres.Journal of Materials Science, 27 nº 17 Sep92, pp 4567-4576
8. F. Peñalba , M.Carsí ,F. Zapirain, Mem et Sci.Rev. Mat. 2 (1991) 99 79-90
9. F. Peñalba , M.Carsí ,C.Garcia ,M.P.Andres and F. Zapirain,ISIJ Interactions , Vol 32 (1992) nº 2 pp 232- 240
10. F. Peñalba, M.Carsí . "Influence of Titanium on the static recrystallization of a medium carbon microalloyed", Journal of materials Science, 30 (1995) pp1750-1753
11. M.Carsí, F. Peñalba, J.Ibañez,G.G.D. " Prediction of superplastic behavior high strain rate test", Material Letters, 20 (1994) pp 119-123
12. .Varga, Central Research Institute for Physics of the Hungarian Academy of Science, Budapest , Hungary pp 161- 166 , Colloquia Mathematics of Societatys Janos Bolyai . Colloquiem on Numerical Methods in Titany (Hungary) from 2 to 6 September 1968

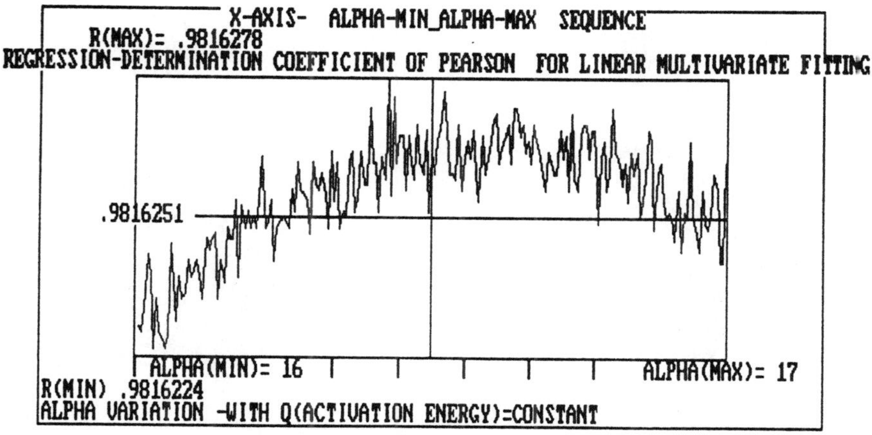

GRAF[1] - STRONG ALEATORY BEHAVIOR OF THE VALUES OF THE DETERMINATION COEFFICIENT FOR THE RUNNING THE THIRD CODE - IN THE CASE OF TITANIUM ALLOY - THIS BEHAVIOR MAKE STRONGLY INSTABLE THE PARAMETERS ATTAINMENT

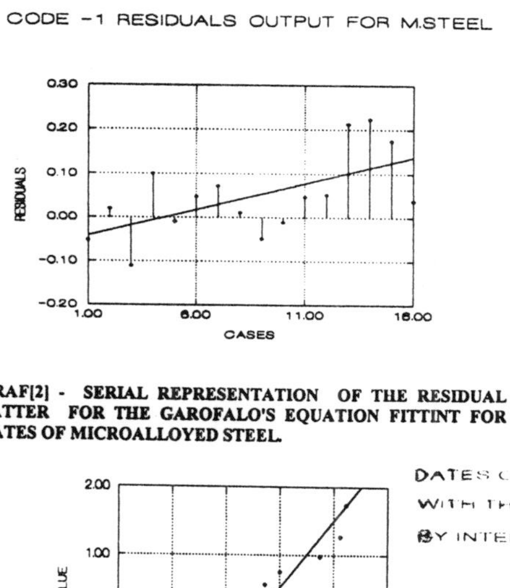

GRAF[2] - SERIAL REPRESENTATION OF THE RESIDUAL PATTER FOR THE GAROFALO'S EQUATION FITTINT FOR DATES OF MICROALLOYED STEEL.

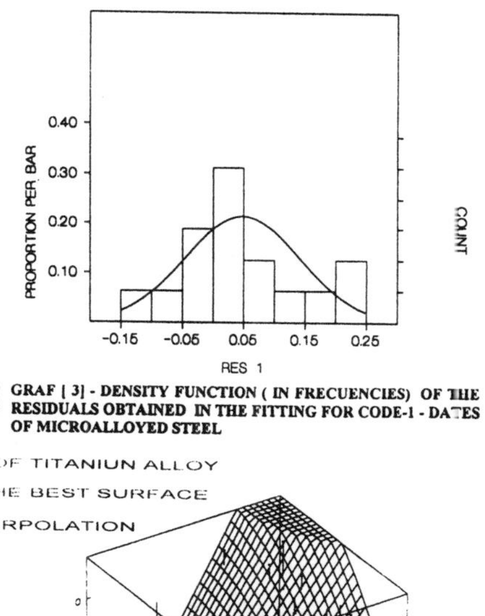

GRAF [3] - DENSITY FUNCTION (IN FRECUENCIES) OF THE RESIDUALS OBTAINED IN THE FITTING FOR CODE-1 - DATES OF MICROALLOYED STEEL

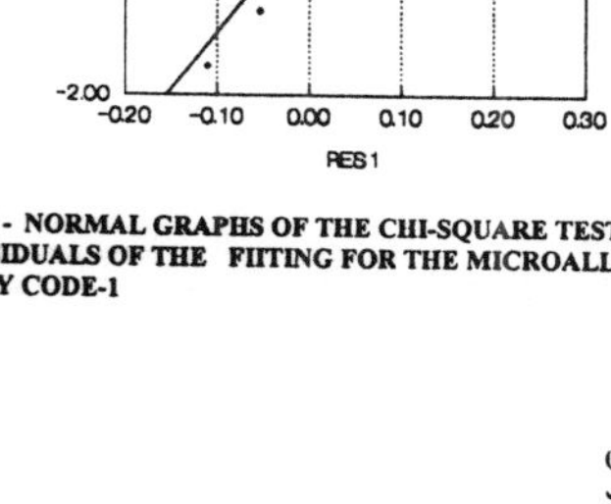

GRAF [4] - NORMAL GRAPHS OF THE CHI-SQUARE TEST FOR THE RESIDUALS OF THE FIITING FOR THE MICROALLOYED STEEL BY CODE-1

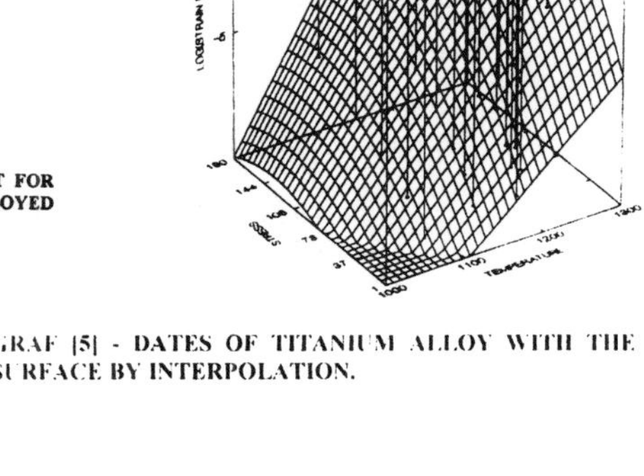

GRAF [5] - DATES OF TITANIUM ALLOY WITH THE BEST SURFACE BY INTERPOLATION.

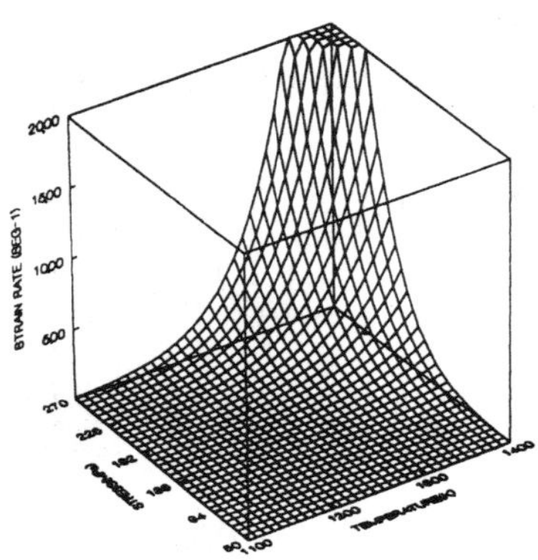

GRAF[6] - THREE DIMENSIONAL REPRESENTATION OF THE GAROFALO'S EQUATION FOR THE SOLUTION CONTRIBUTED BY CODE - 1 - MICROALLOYED STEEL.

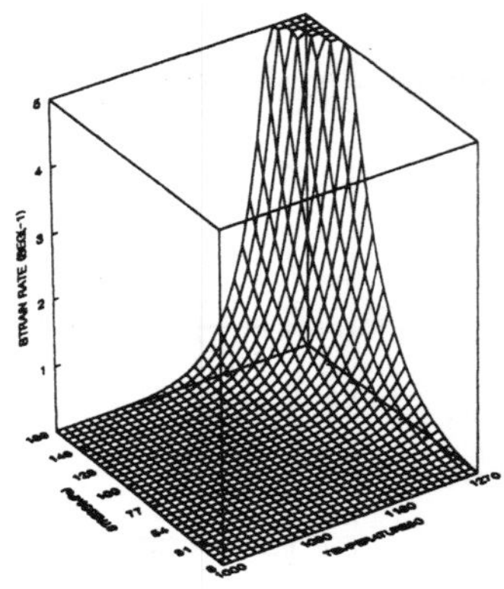

GRAF[7] - THREE DIMENSIONAL REPRESENTATION OF THE GAROFALO'S EQUATION FOR THE SOLUTION CONTRIBUTED BY CODE - 3 - TITANIUM ALLOY

CODE -3 RESIDUALS OUTPUTS FOR TITANIUM ALLOY

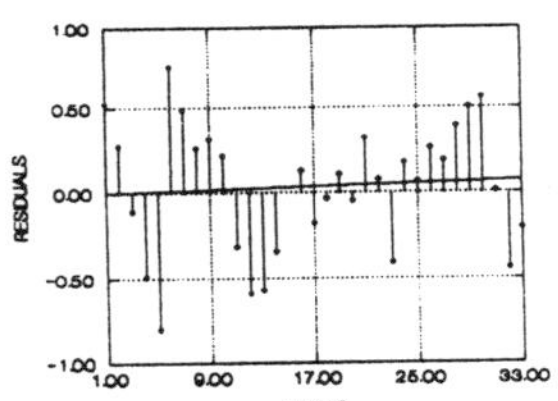

GRAF [8] - SERIAL REPRESENTATION OF THE RESIDUAL PATTER FOR THE GAROFALO'S EQUATION FITTING FOR DATES OF TITANIUM ALLOY

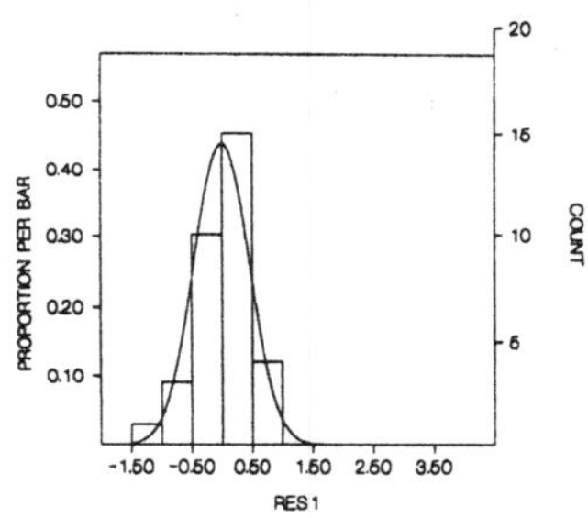

GRAF[9] - DENSITY FUNCTION (IN FRECUENCIES) OF THE RESIDUALS OBTAINED IN THE FITTING FOR CODE-3 - DATES OF TITANIUM ALLOY

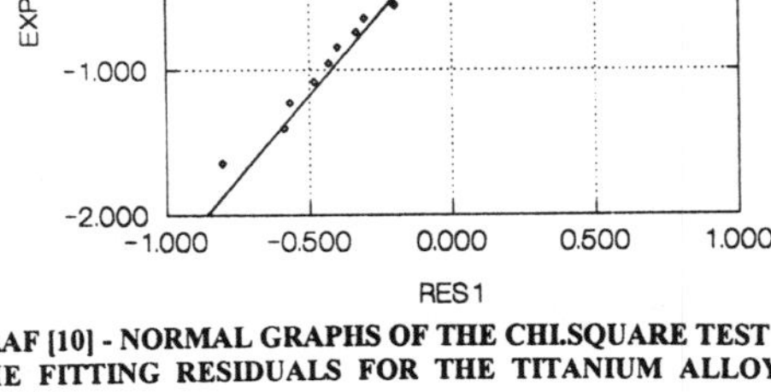

GRAF [10] - NORMAL GRAPHS OF THE CHI.SQUARE TEST FOR THE FITTING RESIDUALS FOR THE TITANIUM ALLOY BY CODE-3

On elastoplastic analysis of 1-D structural member subject to combined bending and torsion

M. Moestopo
Structural Mechanics Laboratory, Inter-University Center for Engineering Sciences, Bandung Institute of Technology, Jl. Ganesha 15, Bandung 40132, Indonesia

Abstract

A new model of an elastoplastic rectangular beam element is presented. The model is used in an elastoplastic analysis of the beam due to combined bending and torsional moment by introducing a plastic-depth factor. Using the plastic node method, the yielding process of a section can be included in the analysis without modeling the beam as a plane or a solid structural member. Therefore, the computational cost on an elastoplastic analysis is reduced. Examples shows that the present analysis gives a more realistic and conservative collapse load.

1. Introduction

In an elastic-plastic analysis, the yielding in a beam element is usually assumed to be controlled only by bending. However, for member that also carries torsional moment such as in grillage structures, the interaction between bending and torsion at yielding element should be considered.

The plastification in a beam member is often assumed to lump at a plastic hinge section which has a zero length while the material is assumed to be elastic perfectly-plastic. The yielding initiates on edges of a section when the bending and the torsion satisfy the fully-plastic condition, i.e. the analysis involves only the elastic and the fully-plastic section. A collapse load obtained by these assumptions is expected to be higher than the load obtained by considering the plastification process of the section.

An elastoplastic analysis of a beam element is presented by the author which includes the interaction between bending and torsion during the plastification at the plastic hinge. To accomodate the plastification of layers at the yielding section, an elastoplastic beam model is introduced so that the beam does not need any mesh or layer refinement.

2. Plasticity and Plastic Node Method

Yielding is assumed at any point of a beam element due to combined normal and shear stresses when the von Mises yield condition is satisfied :

$$\sigma^2 + 3\tau^2 = \sigma_y^2 \tag{1}$$

where σ_y is the normal yield stress of the material due to axial tension. This condition at a fully-plastic beam section is satisfied when the bending moment, M, and torsional moment, T, at that section satisfy the yield interaction [1] :

$$\left(\frac{M}{M_p}\right)^2 + \left(\frac{T}{T_p}\right)^2 = 1 \tag{2}$$

where M_p is the fully-plastic bending moment of the section in pure bending and T_p is the fully-plastic torsional moment of the section in pure torsion. The yield function, f, can be written as

$$f = \left(\frac{M}{M_p}\right)^2 + \left(\frac{T}{T_p}\right)^2 - 1 \tag{3}$$

where $f = 0$ refers to a yield condition.

The Plastic Node Method was introduced by Ueda [2] for an elastic-plastic analysis of structure by assuming the plasticity is lumped as a fully-plastic section at beam-ends (i.e. plastic node).

The elastic force-displacement relation can be written as :

$$\mathbf{k}\ \mathbf{u} = \mathbf{r} \tag{4}$$

where $\mathbf{u}$ is the displacement vector of the beam element, $\mathbf{k}$ is the beam stiffness matrix, and $\mathbf{r}$ is the force vector. The displacement increment, $\delta\mathbf{u}$, can be written as

$$\{\delta\,\mathbf{u}\} = \{\delta\mathbf{u}^e\} + \{\delta\mathbf{u}^p\} = \begin{Bmatrix}\delta\mathbf{u}_i^e \\ \delta\mathbf{u}_j^e\end{Bmatrix} + \begin{Bmatrix}\delta\mathbf{u}_i^p \\ \delta\mathbf{u}_j^p\end{Bmatrix} \tag{5}$$

where subsricpts i and j refer to two beam-ends, superscript e and p refer to the elastic and the plastic part of the displacement. For the associated flow rule the plastic displacement increment is expressed as

$$\{\delta \mathbf{u}^{p}\} = [\lambda]\left\{\frac{\partial f}{\partial \mathbf{r}}\right\} \tag{6}$$

$$\left\{\frac{\partial f}{\partial \mathbf{r}}\right\} = \begin{Bmatrix}\Psi_i\\ \Psi_j\end{Bmatrix} = \begin{Bmatrix}\left(\dfrac{\partial f}{\partial \mathbf{r}}\right)_i\\ \left(\dfrac{\partial f}{\partial \mathbf{r}}\right)_j\end{Bmatrix} \tag{7}$$

and $[\lambda\,]$ is the scalar factor of proportionality matrix or plastic multiplier matrix. The increment of forces can be expressed as

$$\{\delta \mathbf{r}\ \} = \begin{Bmatrix}\delta \mathbf{r}_i\\ \delta \mathbf{r}_j\end{Bmatrix} = \begin{bmatrix}\mathbf{k}_{ii} & \mathbf{k}_{ij}\\ \mathbf{k}_{ji} & \mathbf{k}_{jj}\end{bmatrix}\begin{Bmatrix}\delta \mathbf{u}_i^{e}\\ \delta \mathbf{u}_j^{e}\end{Bmatrix} = \begin{bmatrix}\mathbf{k}_{ii} & \mathbf{k}_{ij}\\ \mathbf{k}_{ji} & \mathbf{k}_{jj}\end{bmatrix}\begin{Bmatrix}\delta \mathbf{u}_i - \delta \mathbf{u}_i^{p}\\ \delta \mathbf{u}_j - \delta \mathbf{u}_j^{p}\end{Bmatrix}$$

$$= \begin{bmatrix}\mathbf{k}_{ii} & \mathbf{k}_{ij}\\ \mathbf{k}_{ji} & \mathbf{k}_{jj}\end{bmatrix}\begin{Bmatrix}\delta \mathbf{u}_i - \lambda_i \psi_i\\ \delta \mathbf{u}_j - \lambda_j \psi_j\end{Bmatrix} \tag{8}$$

and

$$\{\delta f\} = \begin{Bmatrix}\delta f_i\\ \delta f_j\end{Bmatrix} = \begin{bmatrix}\Psi_i^{T} & 0\\ 0 & \Psi_j^{T}\end{bmatrix}\begin{Bmatrix}\delta \mathbf{r}_i\\ \delta \mathbf{r}_j\end{Bmatrix} \tag{9}$$

where the zero value of δf_i or δf_j is assumed if node i or j is plastic.

For a plastic node of a standard beam element where $\mathbf{r} = [Q_k\ \ M_k\ \ T_k\ ]^T$, Ψ_i^T and Ψ_j^T can be written in the form of

$$\Psi_k^T = \left\{\frac{\partial f}{\partial \mathbf{r}_k}\right\}^T = \left[0\quad 2\frac{M_k}{M_p^{\,2}}\quad 2\frac{T_k}{T_p^{\,2}}\right] \tag{10}$$

where subscript k refers to the plastic node i or j.

The two equations, Equation 8 and Equation 9, can be used to construct the force-displacement relation for the 1-D beam element,

$$\{\delta \mathbf{r}\ \} = \begin{Bmatrix}\delta \mathbf{r}_i\\ \delta \mathbf{r}_j\end{Bmatrix} = \begin{bmatrix}\mathbf{k}_{ii}^{ep} & \mathbf{k}_{ij}^{ep}\\ \mathbf{k}_{ji}^{ep} & \mathbf{k}_{jj}^{ep}\end{bmatrix}\begin{Bmatrix}\delta \mathbf{u}_i\\ \delta \mathbf{u}_j\end{Bmatrix} \tag{11}$$

where superscript ep indicates the elastic-plastic stiffness matrix.

As illustrations, let consider two cases of plasticity of a beam element.
Case 1 : Node i is fully-plastic, node j is still elastic.
In this case, $\delta f_j = 0$ and $\lambda_j = 0$, and the elastic-plastic stiffness matrix becomes

$$\mathbf{k}^{ep} = \begin{bmatrix} \mathbf{k}_{ii} - \mu_1 \mathbf{k}_{ii}\Psi_i\Psi_i^T\mathbf{k}_{ii} & \mathbf{k}_{ij} - \mu_1 \mathbf{k}_{ii}\Psi_i\Psi_i^T\mathbf{k}_{ij} \\ \mathbf{k}_{ji} - \mu_1 \mathbf{k}_{ji}\Psi_i\Psi_i^T\mathbf{k}_{ii} & \mathbf{k}_{ij} - \mu_1 \mathbf{k}_{ji}\Psi_i\Psi_i^T\mathbf{k}_{ij} \end{bmatrix} \tag{12}$$

where :

$$\mu_1 = \frac{1}{\Psi_i^T \mathbf{k}_{ii} \Psi_i} \tag{13}$$

or

$$\mathbf{k}^{ep} = \mathbf{k} - \Omega_i \tag{14}$$

where :

$$\Omega_i = \begin{bmatrix} \mu_1 \mathbf{k}_{ii}\Psi_i\Psi_i^T\mathbf{k}_{ii} & \mu_1 \mathbf{k}_{ii}\Psi_i\Psi_i^T\mathbf{k}_{ij} \\ \mu_1 \mathbf{k}_{ji}\Psi_i\Psi_i^T\mathbf{k}_{ii} & \mu_1 \mathbf{k}_{ji}\Psi_i\Psi_i^T\mathbf{k}_{ij} \end{bmatrix} \tag{15}$$

The matrix Ω_i describes the stiffness matrix decrease due to the fully-plastic section (plastic node) which is assumed at end-node i.

Case 2 : Node i is still elastic, node j is fully-plastic.
In this case, $\delta f_i = 0$ and $\lambda_i = 0$. The elastic-plastic stiffness matrix becomes

$$\mathbf{k}^{ep} = \mathbf{k} - \Omega_j \tag{16}$$

where :

$$\Omega_j = \begin{bmatrix} \mu_2 \mathbf{k}_{ij}\Psi_j\Psi_j^T\mathbf{k}_{ji} & \mu_2 \mathbf{k}_{ij}\Psi_j\Psi_j^T\mathbf{k}_{jj} \\ \mu_2 \mathbf{k}_{jj}\Psi_j\Psi_j^T\mathbf{k}_{ji} & \mu_2 \mathbf{k}_{jj}\Psi_j\Psi_j^T\mathbf{k}_{jj} \end{bmatrix} \tag{17}$$

and

$$\mu_2 = \frac{1}{\Psi_j^T \mathbf{k}_{jj} \Psi_j} \tag{18}$$

3. Elastoplastic Analysis

The plastification of a section due to combined bending and torsion is different from the plastification in pure bending or in pure torsion individually. The complexity in the analysis may arise if the analysis must include the yielding of the material along the length, the width, and the thickness of the beam element. Also, the ratio of bending to torsional moment will vary during the plastification.

An elastoplastic section that models the plastification of a rectangular beam section is introduced as shown in Figure 1. This model accomodates the yielding of the section via a plastic depth factor, α, which describes how far the plastification has occured from both edges of the section toward the center of the section, due to combined bending and torsion. At an elastoplastic section, the normal stress, σ* and the shear stress, τ* acting on the yielding region must satisfy the yield condition in Equation 1.

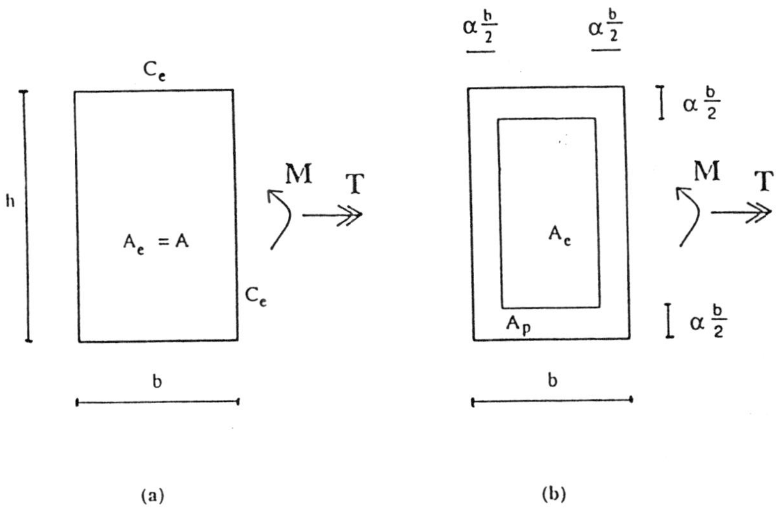

Figure 1 : A Rectangular beam section model.
a. Elastic Section b. Elastoplastic section

The bending moment, M, acting on the section can be written as a summation of M_1 , M_2 and M_3 , as shown in Figure 2 :

$$M_1 = (h-\alpha\, b)\left(\alpha\, \frac{b}{2}\right) b\, \sigma^* \tag{19}$$

$$M_2 = 2\left\{\tfrac{1}{4}\left(\alpha\, \frac{b}{2}\right)(h-\alpha\, b)^2 \sigma^*\right\} \tag{20}$$

$$M_3 = \tfrac{1}{6}\, b\, (h-\alpha\, b)^2 (1-\alpha)\, \sigma^* \tag{21}$$

Although the model does not satisfy the compatibility on the yielding section, nevertheless the bending moment contribution from normal stress near the neutral axis is relatively small compared with the one from normal stress acting near the bottom and top layer of the section.

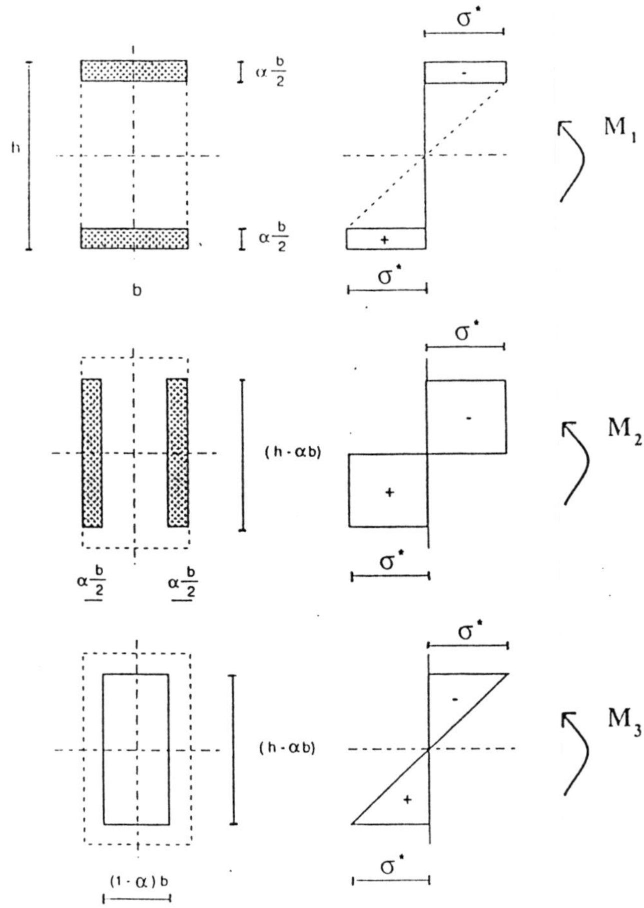

Figure 2 : Bending moment, M, acting on the elastoplastic section model.
$M = M_1 + M_2 + M_3$

The torsional moment, T, can be written as a summation of T_1, T_2, and T_3 :

$$
\begin{aligned}
T_1 = & \; 4\left\{\tfrac{2}{3}\left(\frac{\alpha b}{2}\right)^2 . \frac{\alpha h}{4} . \tau^*\right\} \\
& + 2\left\{(b - \alpha b) . \frac{\alpha b}{2} . \frac{\alpha b}{4} . \tau^*\right\} \\
& + 2\left\{(h - \alpha b) . \frac{\alpha b}{2} . \frac{\alpha b}{4} . \tau^*\right\} \\
= & \; \tfrac{1}{2}\alpha^2 b^2 \left(b + h - \tfrac{4}{3}\alpha b\right) \tau^*
\end{aligned}
\tag{22}
$$

$$T_2 = 2\left(h - \alpha b\right)\left(b - \alpha b\right)\left(\frac{\alpha b}{2}\right)\tau^* \tag{23}$$

$$T_3 = c_s\left(h - \alpha b\right)\left(b - \alpha b\right)^2 \tau^* \tag{24}$$

where T_1 and T_2 are equal to two times the volume of the solid bodies shown in Figure 3, and c_s can be approximate by the Roarks's formula [3] :

$$c_s \cong \frac{1}{3}\left[1 - 0.6095\,\frac{b}{h} + 0.0065\left(\frac{b}{h}\right)^2 - 1.8063\left(\frac{b}{h}\right)^3 + 0.9100\left(\frac{b}{h}\right)^4\right] \tag{25}$$

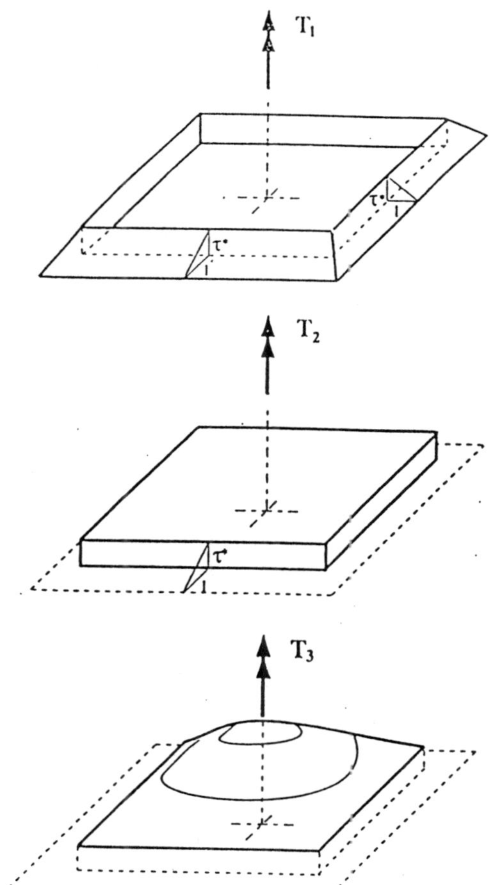

Figure 3 : Torsional moment, T, acting on the elastoplastic section model $T = T_1 + T_2 + T_3$

The bending and the torsional moment at a yielding section can be written as

$$\begin{aligned} M &= M_1 + M_2 + M_3 \\ &= \tfrac{1}{6}\, b\, h^2\, \sigma^* \cdot \vartheta_m \end{aligned} \tag{26}$$

$$\begin{aligned} T &= T_1 + T_2 + T_3 \\ &= b^2\, h\, \tau^* \cdot \vartheta_t \end{aligned} \tag{27}$$

where :

$$\vartheta_m = 1 + \tfrac{1}{2}\alpha\left[1 + (1-\alpha)\left\{\tfrac{b}{h} - \alpha\left(\tfrac{b}{h}\right)^2\right\}\right] \tag{28}$$

$$\vartheta_t = \left(1 - \alpha\frac{b}{h}\right)(1 - \alpha)\left\{\alpha + c_s(1 - \alpha)\right\} + \tfrac{1}{2}\alpha^2\left\{1 + \left(1 - \tfrac{4}{3}\alpha\right)\frac{b}{h}\right\} \tag{29}$$

When the edges begin to yield, $\alpha = 0$, $\vartheta_m = 1$ and $\vartheta_t = c_s$, while for a fully-plastic section, $\alpha = 1$, $\vartheta_m = 1.5$ and $\vartheta_t = (3 - b/h)/6$. These values agree with the analytical values for elastic and fully plastic section.

Substituting the values of σ^* and τ^* in Equation 26 and Equation 27 into Equation 1 will give a new yield criteria for an elastoplastic section :

$$\left(\frac{1.5}{\vartheta_m}\right)^2\left(\frac{M}{M_p}\right)^2 + \left(\frac{3 - \frac{b}{h}}{6\,\vartheta_t}\right)^2\left(\frac{T}{T_p}\right)^2 = 1 \tag{30}$$

The plastic-depth factor, α, can be computed numerically for any applied bending and torsional moment on an elastoplastic rectangular section, by using Equation 38 and Equation 39.

The stiffness matrix of a yielding member decreases. To obtain a conservative collapse load, the values of moment of inertia of yielding members are approximated by $I_p = \omega\, I$ and $J_p = \omega\, J$, where $\omega = (1\text{-}\alpha)^4$. Thus, the elastic-plastic stiffness matrix of the beam element, $\mathbf{k}^{ep}$, is then computed as

$$\mathbf{k}^{ep} = \mathbf{k} - (1 - \omega)\, \Omega \tag{31}$$

Once a section becomes fully-plastic, $\omega = 0$, and Equation 31 becomes Equation 14 or Equation 16.

4. Some Examples

a. A right angle bent structure due to a concentrated vertical load shown in Figure 4 was analysed by Ueda [2], Heyman[4], and Hodge[5] to obtain the collapse load by assuming the elastic and the fully-plastic section. The beam properties are $E = 3\ (10)^7$ psi, $G = 1.2(10)^7$ psi, $\sigma_y = 20{,}000$ psi, $b = h = 0.75$ in., $L = 20$ in. Previous analysis gives a collapse load P = 499 lbs when the vertical deflection at node B is about 1.0 in. The present elastoplastic analysis gives P = 480 lbs with the load-displacement curve shown in Figure 5. The first yielding occurs at node D when P = 220 lbs.
In present analysis the variation of bending and torsional moment is non-linear with a smooth transition from elastic to elastoplastic state (Figure 6). After the edges of the section yields at P = 220 lbs, the rate of change of the bending moment decreases while the torsional moment increases faster than in elastic state. Figure 7 shows the change of $(M/M_p)^2 + (T/T_p)^2$ and the plastic-depth factor, α, of the section in node D.

b. An inelastic lateral buckling of intersecting connected beams shows in Figure 8 was analysed using the present analysis. The beam properties are : $E = 3\ (10)^7$ psi, $G = 1.16(10)^7$ psi, $\sigma_y = 20{,}000$ psi, $b = 0.125$ in, $h = 5$ in., $L_1 = 20$ in, $L_2 = 10$ in. All supports are fixed.
The stability analysis by Morchi [6] gives an elastic buckling load P = 5,020 lbs, while the virtual work method gives an upper bound collapse load P = 4,688 lbs. By including the stability matrix in the analysis, the present elastoplastic analysis gives an inelastic buckling load P = 3,925 lbs as shown in Figure 9.

5. Conclusion

A model of plastification of a rectangular beam element has been presented. The model can be used to perform an elastoplastic analysis of structure with beam member where the yielding of the beam section is governed not only by the bending but also by the torsion at the beam-ends. The inelastic buckling can also be detected using the analysis.

The present analysis gives a more realistic and conservative collapse load. Since the beam element is analysed as a 1-D element and not by a plane element or a solid element to include the yielding proccess at a beam section, the analysis could save considerably large computational cost.

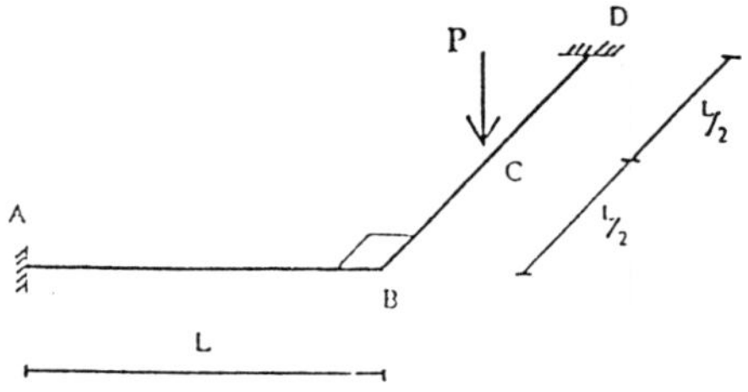

Figure 4 : A Right angle bent structure under a vertical load

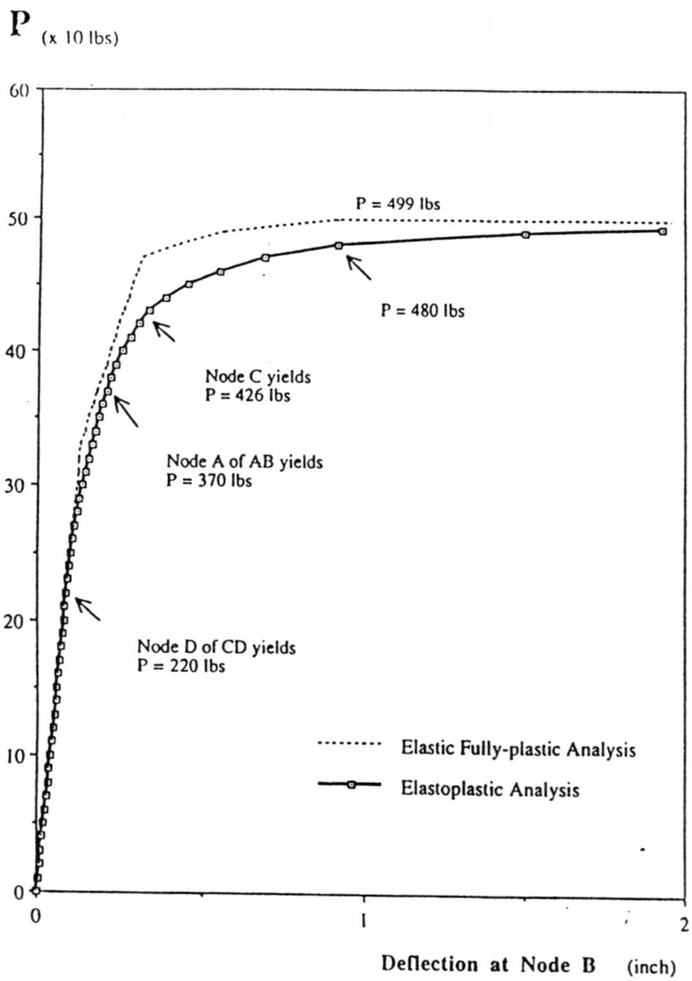

Figure 5 : Load-displacement relation of structure in Figure 4

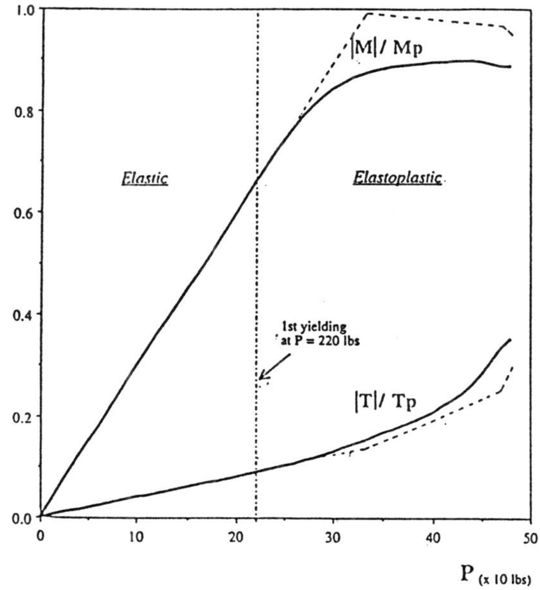

Figure 6 : Bending and torsional moments at Node D in Figure 4. Dash lines corresponds to previous analysis.

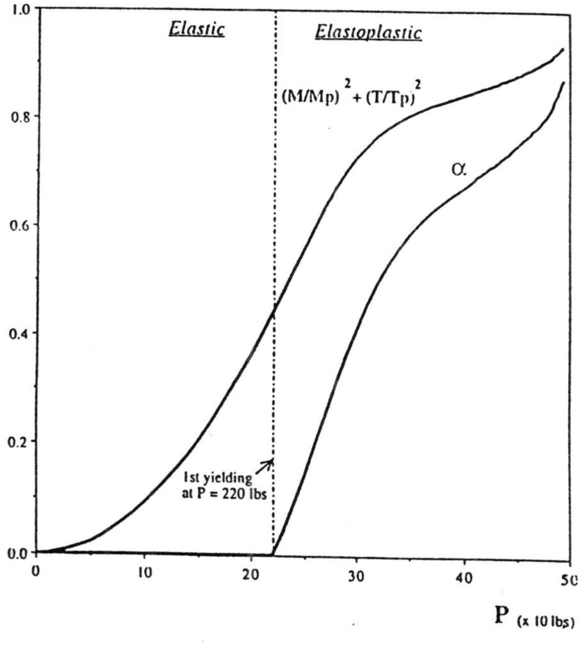

Figure 7 : The variation of plastic depth factor, α, on yielding section at node D

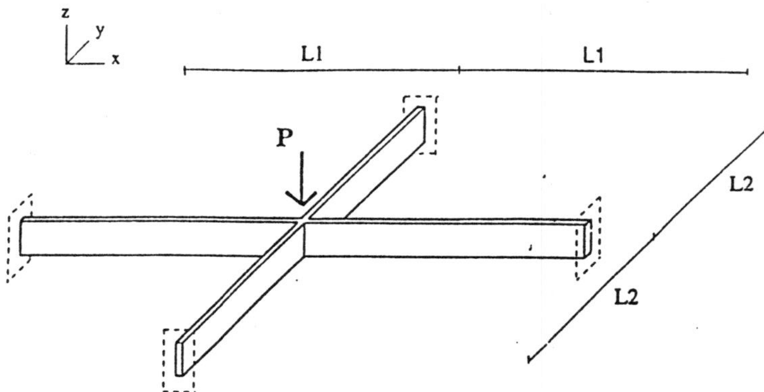

Figure 8 : A Rigidly connected cross-beams.

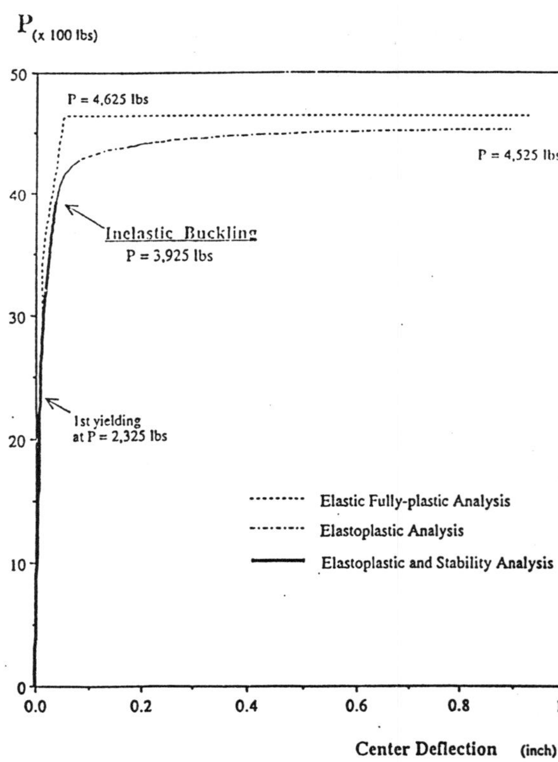

Figure 9 : Load-displacement relation of structure in Figure 8

References

1. Chakrabarty, J., *Theory of Plasticity*, McGraw-Hill Inc., New York, 1987.
2. Ueda, Y. and Yao, T., The Plastic Node Method : A new method of plastic analysis, *Computer Methods in Applied Mechanics and Engineering*, 1982, **34**, 1089-1104.
3. Young,W.C., *Roark's Formulas for Stress and Strain*, McGraw-Hill Inc., New York, 6ed., 1989.
4. Heyman, J., The limit design of space frames, *Journal of Applied Mechanics*, 1951, **18,** 157-162.
5 Hodge, P.G., *Plastic Analysis of Structures*, McGraw-Hill Inc, New York, 1959.
6. Morchi, S.P. & Lovell, E.G. Lateral Buckling of intersecting connected beams, *Journal of the Engineering Mechanics Division*, 1978, **104**, EM2, 367-381.
7. Moestopo, M., On Ultimate Strength Analysis of Grillages Structures, *PhD. Thesis*, University of Wisconsin-Madison, 1994.

The dynamic analysis of light aluminium/steel structures subjected to wind excitation

G.T. Taylor, P. Wu

Department of Energy and Environmental Technology, Glasgow Caledonian University, Cowcaddens Road, Glasgow G4 0BA, UK

Abstract

In recent years an expanding market for light skeletal structures such as walkways, shelters and large bus stations has required constructors and customers to seek information on their behaviour when subject to wind excitation.

The design of structures, taking into consideration the criteria associated with the dynamic response to wind excitation, is a specialised and highly complex task. The most significant guide for designers available at present is that produced by ESDU International plc. This is however apparently limited to particular types of structures and requires considerable interpretation in its use.

The paper reports on an examination of the dynamic response of light aluminium/steel structures, where the required dynamic analysis was carried out using the finite element method and long walkway shelters, upwards of 300m, are a particular focus. Significant idealisation of the physical problem is established and the paper presents the 'best' model for the analysis. The results of a parametric study are presented and combined with the wind excitation criteria, prescribed by ESDU, design curves are proposed. The finite element results are underpinned at this stage by more traditional methods of dynamic analysis, however, future experimental work is described to further validate results.

The paper concludes on the validity of the finite element modelling of the light structures and the implications of the work on design methodology.

1.0 Introduction

The last decade has seen an increasing market for light, skeletal structures such as long, covered walkways, simple bus shelters and larger bus stations constructed from extruded aluminium sections with steel section inserts riveted together. Southside Engineering Ltd. in Glasgow, Scotland, have achieved considerable success with their products.

The primary strength of most of these structures is derived from inserting a thin, steel, folded plate channel into an extruded thin, walled aluminium box section each then riveted together. A combined section gallows is shown in Fig 1.

Structures that are particularly long 200 - 300m, usually a walkway, require to be checked in regard of wind excitation. This poses a difficult problem for the design engineer who may wish to specify particular combinations of dimensions. Very long structures have to be regarded as 'line structures' for which natural frequencies have to be calculated and then checked with regard to the criteria presented by ESDU wind engineering code Ref 1.

Dynamic analysis of structures of this scale and nature can only be effectively carried out by finite element analysis and only beam/plate elements can be used. The steel and aluminium sections are represented by beam elements and aluminium and polycarbonate or toughened glass panels by plate elements in finite element models. This in itself becomes a problem, since for very long structures a large number of elements are required to replicate the actual structure which becomes uneconomical.

The objectives of the present research work are firstly to idealise the skeletal structures with a view to reducing the number of finite elements required to carry out a dynamics analysis. Secondly the creation of design guides for long walkways as an aid to the designer is a priority and thirdly to investigate the appropriateness of the ESDU criterion to long walkways.

This paper presents the results of the research relating to the first two objectives.

2.0 Dynamic Analysis and the Engineering Problem

In this study a modal analysis is required to be performed to predict natural frequencies and mode shapes of typical structures. The criteria invoked from the ESDU design guide requires the fundamental natural frequency to be identified.

The study focuses on cantilevered gallows type walkways which can be constructed for any length without a break in the length. They also may have roof widths varying from 1m to 2m. The height of the gallows column is normally constant at 2m with a gradual rake on the roof for drainage purposes. Over its total length, a walkway will be constructed in sections (bays) - one bay is the distance between consecutive gallows; in this study the length of each bay of the walkway is 4.9 m. The roof sections are also constructed per bay length and is a fabrication of aluminium box or T-sections covered either by aluminium plate or polycarbonate sheet. Side walls are constructed from T-sections with polycarbonate or toughened glass panel inserts.

3.0 Beam/Plate Finite Element Models

The finite element software package used was the ANSYS software with pre and past processors. Typical beam/plate element models are shown in Fig 3 and Fig 4. The beam element representation of the combined aluminium box-section and the steel channel is shown in Fig 2.

The 'coupling' of nodes at rivet locations is achieved by using a generalised constraint which allows specification of relationships between degrees of freedom at node points and the selection of appropriate freedoms reflecting the restraining effect of the rivet.

This part of the study focused on finding an idealised structure which reduced the number of elements significantly but was accurate in the prediction of natural frequencies. Three issues were considered:

i) the need to include the side wall in the model;
ii) the reduction or idealisation of the roof section to an equivalent plate;
iii) the idealisation of the side wall to an equivalent plate.

3.1 Model of Complete Structure without Side Wall

Finite element models of complete structures were created one with and one without the side wall. The results presented in Table 1 show that the difference between the fundamental natural frequencies of the two structure configurations is 11.3%. This indicates that the side wall cannot be neglected in a 'best' idealised finite element representation.

Table 1: Side Wall Effect

	Shelter with side panel		Shelter with no side panel	
MODE	1 Bay	2 Bay	1 Bay	2 Bay
1	2.518	2.240	2.263	2.098
2	4.747	3.359	2.543	2.308
3	6.514	5.868	5.299	3.629
4	11.289	6.650	6.028	5.917

3.2 Determination of an Equivalent Plate for Roof Sections

The roof of the walkway shelter is of hollow-box sections and T-sections in a rectangular criss-cross configuration. These are covered by 12.5mm sheet aluminium. This section is idealised as one single equivalent plate reducing the required number of finite elements from 30 to 8. Over a full-length structure upwards of 300m this leads to a significant economy in a dynamics analysis.

i) Calculation of Equivalent Roof Plate Thickness

For a rectangular plate, calculation of the thickness of the equivalent roof plate can be made by Roark's Formula, Ref 2.

$$\delta_{max} = \frac{-\alpha q a^4}{D} = \frac{-\alpha q a^4}{\dfrac{Et^3}{12(1-\mu^2)}} = 12 \text{ x } \frac{-\alpha q a^4 (1-\mu^2)}{Et^3} \qquad (1)$$

Where:

q	=	Uniformed pressure applied to the roof plate, here $q = 100N/m^2$ (Pa).
a	=	The length of the plate.
t	=	The thickness of the plate, here is unknown.
E	=	Modules of Elasticity, For aluminium E = 7.1E10.
μ	=	Poisson's ratio, for aluminium $\mu = 0.33$.
α	=	Factor refer to the ratio of b/a.
δ_{max}	=	Maximum deflection of the plate. It can be obtained from a finite element analysis of the roof section only with q applied.

From equation (1)

$$t = 3\sqrt{\frac{12 \text{ x } \alpha q a^4 (1-\mu^2)}{E\delta_{max}}} = 3\sqrt{1506 \text{ x } 10^{-8} \text{ x } \frac{\alpha \text{ x } a^4}{\delta_{max}}}$$

$\delta_{max} = 0.00298$m at the centre of the free edge of plate from the original roof finite element analysis.

$\alpha = 0.00515$ of **a** (the length of roof) = 4.9m, **b** (the width of roof) = 1.9m, in the condition that the roof is treated as a plate of three edges simple supported and one edge free.

$$t_1 = 3\sqrt{1.506x10^{-8} x \frac{\alpha x a^4}{\delta_{max}}}$$

$$= 3\sqrt{\frac{1.506x10^{-8} x 0.00515 x 4.9^4}{0.00298}}$$

$$= 0.02466\text{m} = 24.7\text{mm}$$

ii) Calculation of Equivalent Roof Plate Density

The density of the equivalent plate is:- $D = \frac{W}{V}$

Where: W = Mass of the original roof V = Volume of the equivalent plate

Mass of the original roof is:-
$V_1 = a \text{ x } b \text{ x } h = 4.9 \text{ x } 1.9 \text{ x } 0.0125 = 0.1164(m^3)$ (aluminium sheet)

$V_2 = 5 \text{ x } b \text{ x } A + 3 \text{ x } a \text{ x } A$
$= 5 \text{ x } 1.9 \text{ x } 0.000834 + 3 \text{ x } 4.9 \text{ x } 0.000834 = 0.02(m^3)$ (aluminium sheet)

The total volume of the original roof is:-
$V = V_1 + V_2 = 0.1164 + 0.02 = 0.1366(m^3)$

Hence:-
Mass = density x volume = $d \text{ x } V = 2700 \text{ x } 0.1366 = 368.8(kg)$

The volume of the equivalent plate is:-
$V = a \text{ x } b \text{ x } t$

Where: t = thickness of the equivalent plate.
a = length of the roof.
b = width of the roof.

Hence:-

$$D = \frac{W}{V} = \frac{368.8}{a \text{ x } b \text{ x } t} = \frac{368.8}{4.9 \text{ x } 1.9 \text{ x } 0.0247} = 1603.8(kg/m^3)$$

Comparison of the natural frequencies between the model with original actual roof and that with the equivalent roof plate is shown in Table 2.

Table 2: Equivalent Roof Effect

	1 Bay		2 Bays		4 Bays	
MODE	Original model	With equivalent roof	Original model	With equivalent roof	Original model	With equivalent roof
1	2.518	2.515	2.240	2.224	2.085	2.078
2	4.747	4.920	3.359	3.446	2.582	2.615
3	6.514	6.410	5.868	6.064	4.979	5.291
4	11.289	11.230	6.650	6.876	5.634	5.831

In a similar way, the actual side wall was idealised and an equivalent plate with appropriate thickness and density determined thus making further reductions in finite element numbers.

Comparison of the natural frequencies between the actual structure model (original roof and side wall configurations) and the 'best' model (equivalent roof and side plate) is given in Table 3. It can be seen that extremely good results are obtained.

Table 3: Actual Structure Versus Idealised Structure

	1 Bay		2 Bays		4 Bays	
MODE	Original model	With equivalent roof & side plate	Original model	With equivalent roof & side plate	Original model	With equivalent roof & side plate
1	2.518	2.515	2.240	2.238	2.085	2.078
2	4.747	4.944	3.359	3.454	2.582	2.615
3	6.514	6.351	5.868	6.022	4.979	5.291
4	11.289	11.329	6.650	6.827	5.634	5.831

4.0 Proposed design Guides

Having established a convenient idealised 'best' finite element model predicting accurate value of natural frequencies a parameter study was then undertaken to establish an aid to the design of cantilevered walkways.
Varying structure length and roof width, natural frequencies were obtained from many computer runs. Fig.5 presents the results of this study as a design guide where linear interpolation for roof widths not shown is valid, Fig 6.

5.0 Use of ESDU Wind Engineering Guide

The nature of the frequency of wind loading is that it mainly affects structures with low natural frequencies. The possible worst case obviously happens in situations of extreme weather conditions. Structures susceptible to wind excitation are usually extremely flexible with very low natural frequencies.
ESDU (Engineering Sciences Data Unit) Wind Engineering Ref 2 code of practice is used in this study to determine the stability of the structures under wind excitation.
From the ESDU guide, the criterion for the structures susceptible to wind excitation is:-

$$\frac{f_n}{\sqrt{\zeta_s}} < 30$$

where ξ_s is the damping ratio; the damping ratio for the structures considered here selected as 0.005 in the guide and f_n is the fundamental frequency.

Thus, $f_n \leq 2.12$ is the critical value derived from the criterion:- $\frac{f_n}{\sqrt{\zeta_s}} < 30$.

The criteria noted here is shown on Fig 6 as a line at frequency 2.12. This allows selection of structure lengths to avoid wind excitation problems. The relevance of this criterion to the type of structure under consideration requires to be investigated.

The criterion was selected on the basis of matching the walkway with the class of structure for which the criterion was established. However, the class of structure may include a broad range of structure configurations and there is thus doubt as to the appropriateness of the criterion to the walkway.

6.0 Proposed Experimentation

While dynamic analysis by finite elements has been well established, experimental verification of the finite element results requires to be undertaken. This, however, is envisaged as being difficult to carry out. It is possible to construct a 1 bay cantilever structure in the laboratories but the excitation and the confirmation of acceleration and deformation profiles would prove difficult and time consuming.

In the first instance a cantilever gallows will be set up and tests carried out. This will assist in confirming the validity of the corresponding finite element analysis.

7.0 Conclusions

Economic finite element models for the dynamic analysis of long, cantilever walkway structures constructed of aluminium/steel members have been established and shown to give excellent results.

Design curves as an aid to the designer of long walkway structures have been produced based on a study of the main parameters. They include the ESDU criterion for structures susceptible to wind excitation. Design curves based on finite element analysis can be produced for other similar type structures and is the preferred way forward.

The ESDU criterion involved herein requires to be examined as to its appropriateness for the walkway type structures focused on in this study.

Experimentation is difficult and requires a clear rationalisation.

References

1. ESDU Engineering Data, Wind Engineering Sub-series Volume 3a, ESDU International plc. December 1989.

2. Roark's Formulas for Stress and Strain, Warren C. Young, 6th Edition, McGraw Hill, 1989.

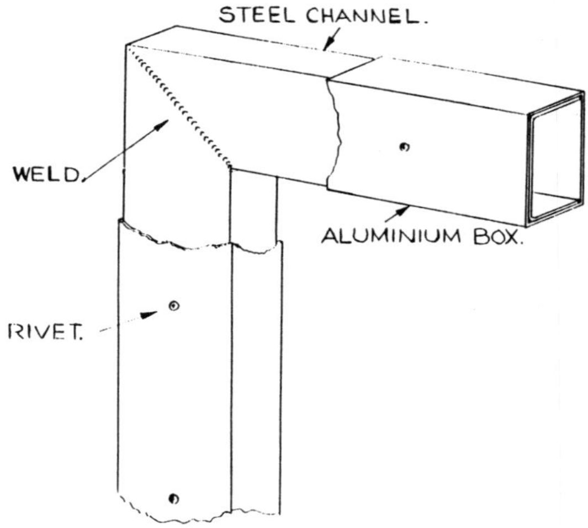

Fig 1 Arrangement of Riveted Al/Steel Combined Section Gallows.

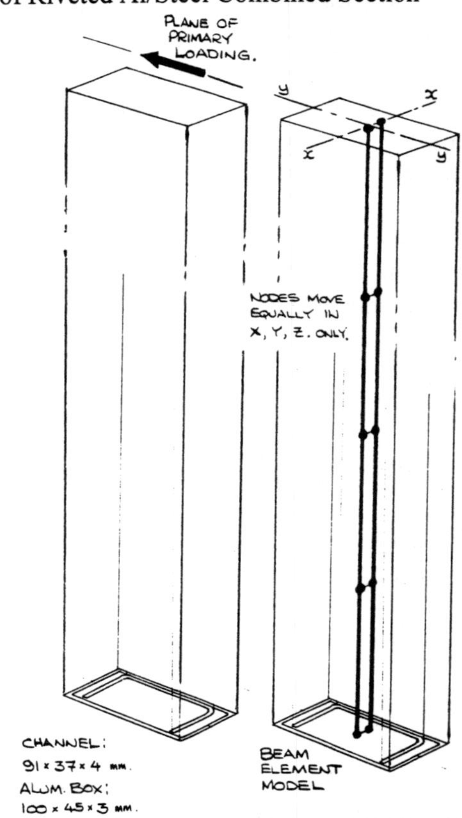

Fig 2 Beam Finite Element Representation of Combined Section Member.

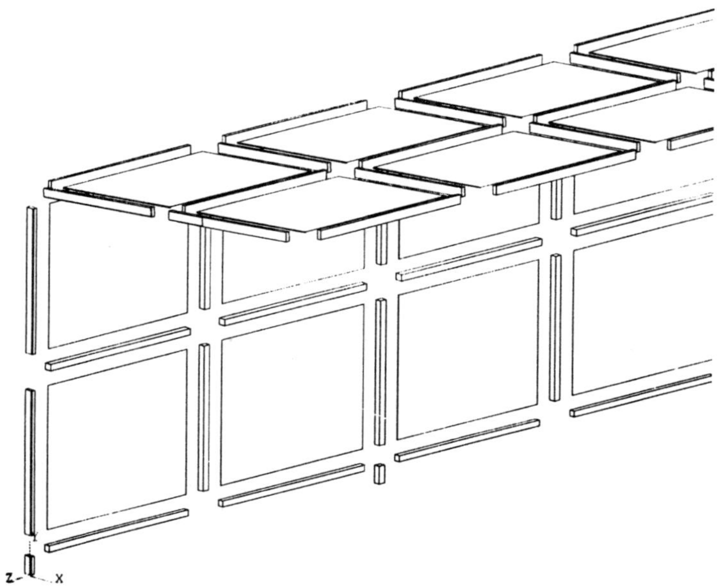

Fig 3 Finite Element Model of Complete Structure.

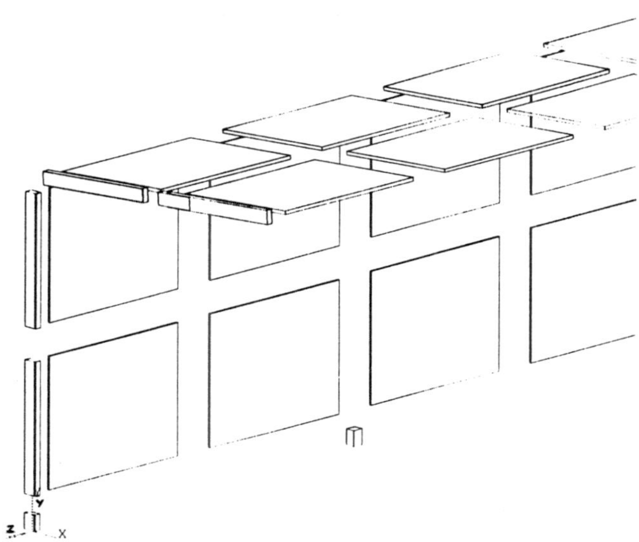

Fig 4 Finite Element Model of Idealised Structure.

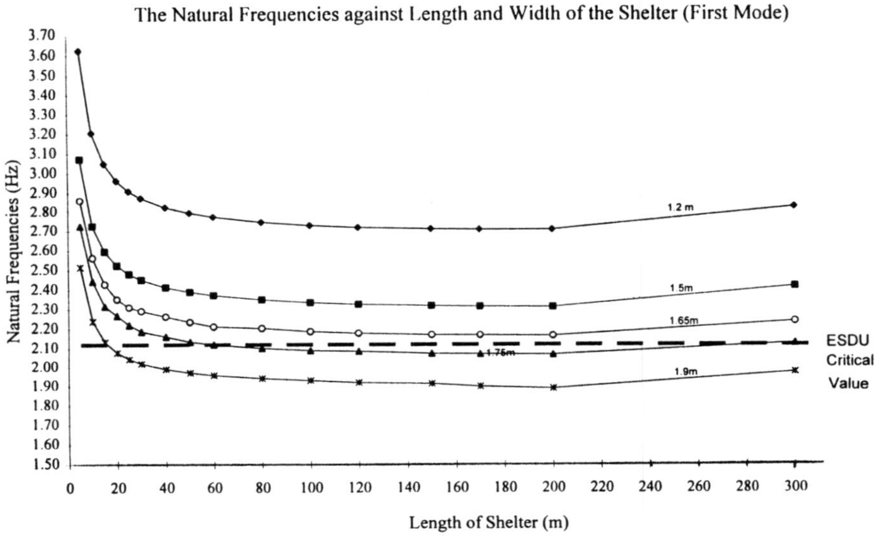

Fig 5 Design Curves for Cantilevered Structures (ESDU).

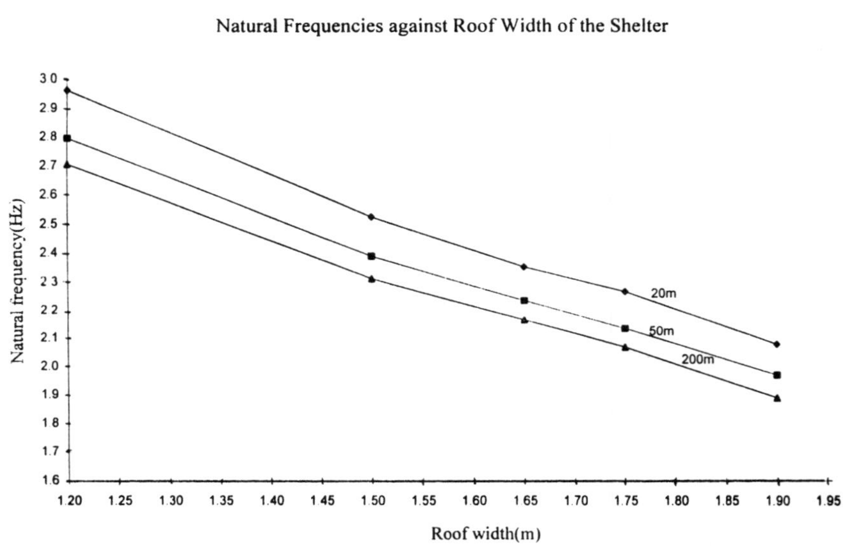

Fig 6 Variation of Natural Frequency and Roof Width.

Engineering integrity: a controversial issue in Germany

R. Adolfs, T. Auf dem Brinke, K. Detert
Faculty of Mechanical Engineering, Institute of Materials Technologie, University of Siegen, D-57068 Siegen, Germany

Abstract

Public acceptance has become a decisive factor for promoting technical progress. This is especially true in Germany, due primarily to political pressure to ensure checks on ecological consequences and on the integrity of technical products and processes. As a consequence, sophisticated testing procedures have been developed to prove engineering integrity. Stress analysis and materials testing procedures are typical examples of such developments. However, modern testing procedures require the employment of the computer, not only for the control of the testing process, but also for data acquisition and analysis. These will be outlined for the modern versions of well known classical materials testing procedures such as the tensile test and the impact bending test. The difficulties of obtaining results acceptable by Germnan authorities for approval to build and employ new technical apparatus and materials are discussed. The concept of developing simple test procedures which are transparent and easy to check is emphasised.

Introduction

In the past, it was almost unanimously accepted, that technical progress would improve the welfare of mankind, hence representing an effective instrument to advance civilisation. In the meantime the attitude towards technique has changed, primarily in the industrialised countries. The hazards of the technical progress to the ecology and to survival of mankind are well understood. Therefore any technical invention, which might represent a risk to the public will be seriously considered in regard to its consequences to public concern, before its production will be decided. This issue has obtained an important attention of the industrial management especially in Germany, where the Green-movement has an important influence on the media. Any new technical construction or process must be thoroughly assessed and checked, whether any harmful consequences are possible and what measures have to be taken, to secure a safe service. Likewise great effort is required to prove it, before a licence for service is given by the authorities. Special technical institutions are installed, with the mere objective to survey and check the technical devices and appliances in regard to the design, to its intended function and performance in service. The TÜV is in Germany the best known institution of this kind.

Modern testing procedures

In order to test the integrity of a technical device or appliance more and more sophisticated procedures have been invented in the different fields of engineering applications. A few will be mentioned here.

Stress analysis

One calls a procedure to verify in mechanical loaded structures, that the local stresses in the whole construction will not exceed the limits of the materials selected for this construction during service as *stress analysis*. Today this procedure requires the employment of very complicated computer-programs for finite element calculations. In order to compare the results of the theoretical calculation of stresses with the reality, a very subtle experimental method of stress analysis by applying strain gauges to the structure has been developed. In order to convert the measured data from strain gages into data of local stress values, again a very complicated computer program is required.

Materials testing

In order to define the limits of mechanical properties under various loading conditions, mechanical testing had been developed into more and more sophisticated procedures. - Even the simple tensile test requires today a computer, for data logging during the test, for converting it into the required test data of materials properties and for documenting it [1]. Figure 1 shows an example of a document of a tensile test, recorded by employment of a computer. In many cases the computer is today also used to control the prescribed testing program performed by the testing machine. - Materials testing according to the methods of fracture mechanics requires the

employment of a computer for data-logging as well as for the control of the testing machine, as demonstrated in other papers given in this conference [2]. Analysing the test data and calculating the materials behaviour from the test data is another task performed by a computer program.
- Another example, where computer are necessary for the testing procedure, are the different methods of fatigue tests. Fatigue is even still today the most common origin of failure of industrial components of machinery. Complicated test procedures are invented, which require the employment of a computer for control, analysing and calculating the test results [3].
- A further example of development of a well known classical test into a very sophisticated testing procedure by the employment of a computer represents the so-called Instrumented Charpy-V Impact Bending test [4]. In the past it was sufficient to register the value of absorbed energy during the impact, which eventually led to fracture of the test piece. Today data logging during the impact is preferred by applying a special strain gauge to the fin of the hammer. So one can record the actual load applied to the specimen during the very short time of the impact. From this record one can calculate a stress-strain diagram during the impact of the hammer [5]. Integrating of this diagram should result in the total energy consumed by the test. [As demonstrated in the figure 2].

Computers are essential in testing for the integrity of structures.

All those examples mentioned, demonstrate, how much the methods of calculation in design and the testing procedures of the material employed have been developed, in order to secure safe performance of technical structures. Furthermore those examples demonstrate likewise, that employment of a computer is inevitable for modern testing procedures. Of course one could demonstrate other technical fields as well, which can prove in the same manner, that employment of a computer plays a vital role for calculating and testing. Special methods to assess the safe performance of machinery had only become possible by the utilisation of a computer as e.g. the recently developed methods of simulation of the mechanical behaviour of a real structure in service on a computer-screen by an animation program.

Since the introduction of the New Natural Science by Galilei and Sir Francis Bacon [6], it was realised that mathematical description is the most appropriate procedure to understand and to exploit the processes known in nature and technology. Therefore the invention of the electronic calculator and the methods of programming the calculation procedure in the middle of this century have become a strong impact to the technical progress. Several new technologies like space- and nuclear energy-technology had depended on the progress of computer technology for the necessary calculations. One can predict, that other new technologies are still to come with the employment of computer technology. On the other hand, Galilei already knew that theoretical calculations in mechanics require the verification by experimental techniques, since reality deviates from theoretical predictions [7]. Hence, experimental engineering attracted much attention since the last century. Technical progress is therefore accompanied by developing testing procedures, which again require a mathematical description of the testing process. As a consequence the application of computer technology had been introduced into the testing procedures of the integrity of man-made structures. The following reasons made the employment of computers inevitable for testing procedures of today:

- The employment of a computer is required for complex modern testing procedures in engineering.
- Only the application of a computer with its appropriate software program has made complicated testing methods economically feasible.
- Furthermore the application of modern computer techniques reduces the sensitivity to human errors, one had to face, when all the testing had been conducted by hand.
- In many cases the application of a computerised program reduces the limits of scatter and errors due to individual bias introduced by man handled procedures.

The employment of a computer for testing has also disadvantages.

- Computer-programs become more and more sophisticated and loose transparency. The whole procedure of data logging, transforming test data into the required results by calculation may loose trustworthiness to non-experts, who do not understand the program.
- From the experience in the commercial section of the banking system, where computer are in service for some time, one has realised the

sensitivity of results, obtained by the use of a computer to all sorts of criminal assaults to falsify data.

- The legal issue has not yet been resolved, who can be made liable for damage caused by malfunction of a computer-program utilised for a technical procedure [8].

As a consequence of those disadvantages one will run into difficulties, when testing results obtained by employment of a computer are required to be approved as demonstration of a safe performance of the technical item at stake.

1: It is difficult to prove that the mathematical description of the testing procedure is in agreement with the reality.

2: It is likewise difficult to prove, that the hardware and the software operate without error.

3: Furthermore one runs into the difficulty to assure, that the measures are sufficient to protect the operation of a computer against criminal assaults to falsify its operation.

As a consequence of those disadvantages it has been almost impossible to obtain approval of German authorities responsible of surveillance, when the results of proving integrity are obtained by assistance of a computer. One of the reasons is the present lack of acquaintance with computer technology of the people involved in the surveillance business of testing integrity. Another reason are the legal regulations of testing integrity, which hardly take notice of the progress of information technology. As an example, the experience should be mentioned, when it was required to perform certified tensile tests in our university-laboratory. There had been no difficulties, to get approval to make those tests with the old mechanically driven testing machines after the surveillance-institution had verified on their own tensile specimens, that the recorded values had been correct. After the replacement of the mechanically driven tensile test facilities by electronic control, no approval was firstly obtained to use any of these machines. Finally the people from the surveillance-institution put a seal to all the amplifiers of the measuring gages in the electronic control, then we could use those machines for certified tests. However, even up to now a tensile test performed by a machine, which is controlled together with the calculation of test results by a computer program is not accepted as certified test.

Therefore it is still not resolved, how one can get approval for service, when testing the integrity of a structure or the safe operation of a technical unit requires the employment of a computer with a sophisticated program. The solution to this problem as mostly verified today in Germany is the appointing of a well known and highly estimated special institution for the surveillance, which may attain the authorisation by the government, to perform those tests on their own machines with their own computer and software programs in their own responsibility. This solution however transfers only the problem of trustworthiness to another institution, it will add extra costs to the testing procedure but by no means increase the integrity of the structure.

Conclusion

Proving the integrity of the structure is likewise important as all the effort to provide integrity. For providing the integrity of structures or of the performance of devices and appliances the utilisation of modern information technology has become inevitable. The same goes for complicated testing procedures. The difficulties however arise, when the results of computerised testing must be approved by surveillance-institutions. One must admit that none of the computerised procedures are sufficiently protected against different methods of falsification. Sophisticated computer-programs are characterised by a lack of transparency. It is therefore very difficult to discover falsification of the results obtained by such a program. It is therefore highly recommended to put much effort in developing simplified testing procedures, when approval of the results of testing is required. Such simplified testing procedures should demonstrate transparency to such an extent, that even non-experts can easily discover sources of falsification of results, to avoid uncertainties of the trustworthiness of the results obtained by such testing procedures. The above mentioned Charpy-V Impact Bending test, to prove ductility of steels or other materials, is such a simple and transparent test, when the absorbed energy is measured by the classical procedure [9]. The instrumented Charpy-V Impact Bending test requires the employment of complicated data-logging and calculation, so that transparency is missing. It will not be appropriate for getting approved results. It offers however a very useful information, which supports the reliability of the result in measuring the absorbed energy as indication of ductility of the material.

Public acceptance of a new technical solution of a certain process or a specific appliance or construction represents today an important issue to put those new items into realisation. Today one can not count anymore on the former general confidence of the industrialised society, that technical progress will bring a better welfare to the people, so that a certain risk is justified for its realisation. At present a pronounced distrust prevails, that technical progress may lead to more harm than benefit to the welfare of environment and nature and eventually to the survival of the human society. Each step of technical progress requires therefore today a simple and transparent test of integrity in order to overcome the distrust of the public against the technical progress.

List of References

[1] **EN 10 002-**1 [1990], EN 10 002-2 [1991], EN 10 002-3 [1994], EN 10 002-4[1994], EN 10 002-5 [1991] Tensile Test- Zugversuch, CEN, Brüssel

[2] **ANSI/ASTM E 468** - 82 Standard Practice for Presentation of Constant Amplitude Fatigue Test Results For Metallic Materials, and **ASTM E 606** - 80 Standard Recommended Practice for Constant-Amplitude Low-Cycle Fatigue Testing, Philadelphia,Pa

[3] **ASTM E 813-**88 Standard Test for J_{Ic}, A Measure of Fracture Toughness, Philadelphia,Pa

[4] **SEP 1315** Stahl-Eisen Prüfblätter Entwurf 09.86, Düsseldorf

[5] **DVM Merkblatt 001** 09.86 Berlin

[6] **Galileo Galilei** Discorsi e Dimonstrationi Matematische Intorno à due nuowe scienze, Leiden 1638, zitiert in Ostwalds Klassiker der exakten Wissenschaft Nr. 11 Leipzig 1980

[7] **Galileo Galilei** in Ref [6] auch zitiert in Jürgen Mittelstraß Neuzeit und Aufklärung Berlin, New York 1970

[8] **Haft, Fritjof,** EDV und Recht in Elektronische Datenverarbeitung im Recht Bd.1 Berlin 1970

[9] **EN 10 045-**1 [1990], EN 10 045-2 [1992] Charpy-V Impact Bending Test - Charpy-V Kerbschlagbiegeversuch, CEN, Brüssel.

List of Figures

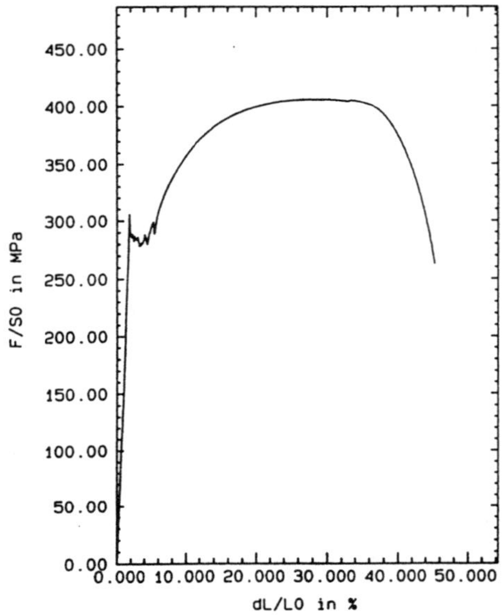

Prüfer: Adolfs
Werkstoff: Baustahl St37
Probennummer: 1
Bemerkung: 120 HV30
Probenform: Rundprobe
Probenquerschitt: 78.54 mm2
Dehngeschwindigkeit: 13.5 %/min
Prüftemperatur: 20° C
Elastizitätsmodul: 205623 MPa
Obere Streckgrenze: 306 MPa
Untere Streckgrenze: 278 MPa
Zugfestigkeit: 406 MPa
Bruchdehnung: 43.0 %
Brucheinschnürung: 32.8 %

Fig. 1 Record of a computerised tensile test

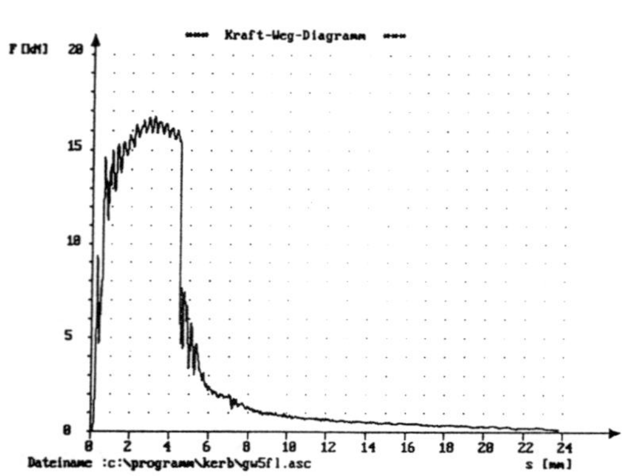

Name der Datei :c:\programm\kerb\gw5fl.asc
Name des Pruefers :Matheis
Kerbschlaghammer 300 J
Werkstoff :20 MnMoNi 5 5
Prueftemperatur [C] :RT
Probenentnahme :quer
Probenbezeichnung :GW5FL
Typ des Diagramms :E

* Kennzeichnende Werte der Kraft *
Fgy [kN] = 13.08476
Fm [kN] = 16.28255
Fu [kN] = 15.604
Fa [kN] = 4.704

* Kennzeichnende Werte der Arbeit *
Wm [J] = 35.96032
Wu [J] = 57.75718
Wa [J] = 57.75718
Wt [J] = 78.24133

* Kennzeichnende Werte des Weges *
sm [mm] = 3.166613
su [mm] = 4.525391
sa [mm] = 4.525391
st [mm] = 26.16262

Anteil des Verformungsbruches (%) 33.05717

Fig. 2 Record of an instrumented Charpy-V Impact Bending Test

Stress-strain computational simulation of main components of underground LHD machines

V. Ignatov,[a] K. Lebedenko,[a] S.P. Claridge[b]

[a]*Department of Mining Machines, Donetsk State Technical University, 340000, Donetsk, Ukraine*

[b]*Department of Mechanical and Manufacturing Engineering, University of Portsmouth, Portsmouth, Hampshire PO1 3DJ, UK*

Abstract

A stress-strain analysis has been undertaken of the basic structure of the acting unit of an underground Load-Haul-Dump (LHD) machine currently being used in the Ukraine, with the use of a personal computer and related software. The resulting stress-strain state of the bucket, boom and associated units under maximum load conditions is presented and discussed, particularly the stress distribution in the respective structures and the interaction of the various parts through the coupling and reaction points.

1 Introduction

Underground mining vehicles, such as Load-Haul-Dump (LHD) machines require the following attributes; compactness, manufacturing quality, long term strength, durability and limited mass. The specification of the LHD type PD-8 machine is given in Table 1 below.

Load Lifting Capacity	8.5 tonnes
Holding capacity of Bucket	4.0 m^3
Diesel Engine Power	140 kW
Overall Length	9360 mm
Overall Width	2500 mm
Overall Height	2350 mm
Bucket Dumping Height	1620 mm
Mass	23.4 tonnes

Table 1. Overall Specification of LHD PD-8 Machine

When analysing the performance of LHD-machine components, it is found that the boom and the bucket have the least reliability with numerous failures reported in the vicinity of the fixing pins attaching the bucket to the boom and the boom to the front semi-frame. These include cracking and fracturing of the metal-work lugs and hydrocylinders and cracking in the bucket and boom structural welds.

The current understanding of the stress-strain state within the structural members and connecting members is insufficient.

Much of the design process of this type of equipment is currently based on simple empirical relationships and experience and therefore a more systematic approach to the design is necessary. Studies of the LHD mining equipment have been undertaken and published in Poland (Ref. 1) but these have limited to general descriptions and the machine productivity and control.

A full experimental strain investigation has been completed on this equipment under real loading loading conditions, over a usual working cycle/period and this data helped establish the typical maximum stress-strain conditions at the strategic locations on the LHD machine.

However, this strain measurement data did not provide the complete picture and therefore it was decided to initiate an imitative simulation, utilising both CAD and finite element analysis, in order to obtain a better understanding of the stress-strain state and allow rapid resolution of the important strength to weight efficiency.

In 1994, the dynamics of the bucket and boom components of the PD-8 type LHD were analysed by Semenchenko and Ignatov et.al. (Ref 2.).

2 The Basic Structure of a LHD

The LHD machine belongs to a group of multimass spatial dynamic, pneumatic wheeled, mining machines. For the LHD to have manoeuvrability and compactness in underground conditions, the body consists of two basic frames (front and back) connected with a pin joint having vertical rotation axis. The acting unit comprises a bucket and boom, see Figure 1, which is connected to the front semi-frame with pin joints having horizontal rotation. The boom and bucket assembly is also pin jointed and operated with the use of a hydraulic jack. All the structures of the LHD are fabricated from flat sheet or roll-formed structural alloy steel and welded. The material specification is given as 10XSND or 09G2S in accordance with GOST 19281-78 (Yield stress 330-400MPa).

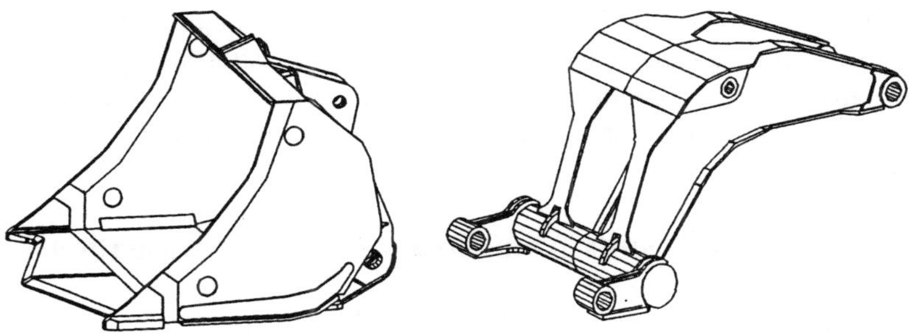

Figure 1 The bucket and boom of the LHD type PD-8 (by ACAD 11.0)

The LHD machines are complex assembled frame structures and consequently combine into unique kinematic chains. Each unique LHD kinematic chain configuration will cause different loading situations and consequently different stress distributions within the LHD structure.

3 Measurement Method

Taking into consideration the spatial and non-symmetrical nature of the loading on the LHD structure, it was possible to examine the two-dimensional stress state on the structural surfaces from the previous experimental strain analysis. Directional strain readings were taken using rosette strain gauges with an angle of 45 degrees and by evoking the rules of Mohr's Circle the principal strains at strategic locations on the surfaces were determined using:

$$\varepsilon_{1,2}=\frac{\varepsilon_0-\varepsilon_{90}}{2}\pm\frac{\cos45\sqrt{(\varepsilon_0-\varepsilon_{45})^2+(\varepsilon_{45}-\varepsilon_{90})^2}}{2} \tag{1}$$

The principal strain direction (angle) was determined using:

$$\varphi=\frac{1}{2}\arctan\frac{(2\varepsilon_{45}-\varepsilon_0-\varepsilon_{90})}{\varepsilon_0-\varepsilon_{90}} \tag{2}$$

For the two dimensional stress situation the maximum and minimum stresses were determined using:

$$\sigma_{12}=\frac{E}{1-\mu^2}(\varepsilon_{12}+\mu\varepsilon_{12}) \tag{3}$$

The maximum equivalent stress (σ_{ee})was evaluated in accordance with the maximum shear stress theory. Due to the relatively large size of the LHD structure strain gauge rosettes were mounted at strategic location using especially adapted resistance spot welding techniques. All gauges were connected to a temperature compensation block, conditioning amplifiers and data acquisition system.

4 Summary of Results from Strain Gauge Investigation

The strain measurement investigations of the working unit of the LHD machine were conducted at the testing area of the LKU Plant (Ukraine) and under real operating conditions in the mines of Tyrnyauz (Georgia) and Dokuchaevsk (Ukraine). Strain gauge readings were recorded through-out the whole working cycle of the equipment. This record was analysed to identify the most significant regimes of loading and apparent stress level.

A Digging out an obstacle, typically a large piece of rock being removed from a pile of mined rock, with the front edge (left or right) of the bucket. This causes a dangerous eccentric load at the left and right edge of the bucket. Identified as Loading AL and AR in Figure 2.

B A full or empty bucket impacting and 'clipping' to the boom body, as the unit is being prepared for the transportation position. This can be repeated a number of times by the operator within one loading cycle in an attempt to increase quantity of rock in the bucket. The average number of these 'clipping' actions per scoop is usually 2-3.

C The boom body turning from its transportation position to the upper position where the bucket is unloaded, when the hydraulic jacks are extended to a maximum.

D Rock mass unloading and the bucket impacting at the boom body. These impacts can be repeated several times within one working cycle of the machine. The average number of impacts being 2-3.

An example of the stress-time history, derived from one of the rosette strains of the boom during operation of a complete working cycle of a LHD machine in short face is shown in Figure 2a below. Figure 2b below shows a segment of one of the boom rosette strains in the region noted AL, identified as the time when loading is applied to the left edge of the bucket. The following stages occurred within this process:

a) the bucket hydraulic jack working
b) the boom hydraulic jack working
c) The successive working of the bucket and boom hydraulic jacks

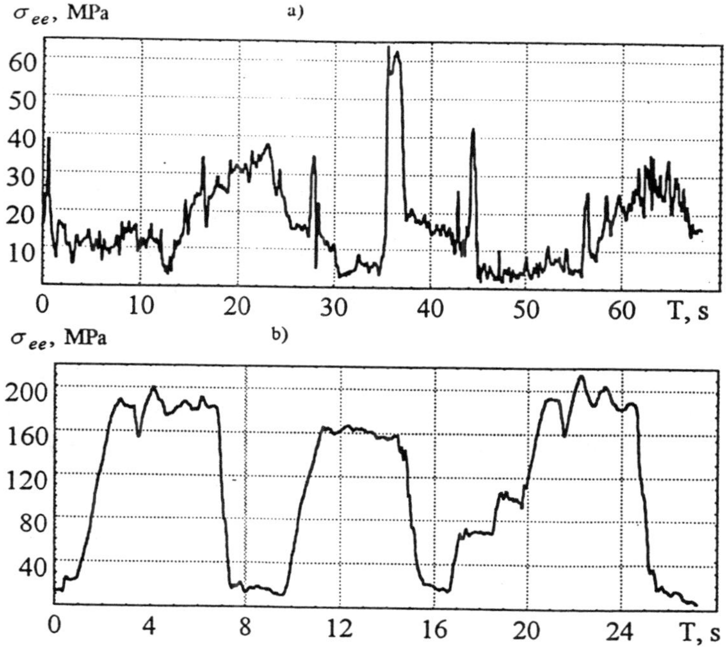

Figures 2 Equivalent stress of the boom during the whole working cycle (a) In the regime AL and (b) on the LHD

Analysis of the stress state within the whole working cycle at the real face shows that there is a transition of loading regimes and that the load is transmitted from one component to another through the pins and fixtures. In addition, the cyclic nature of the loading also has to be considered. In order to model the boom and bucket assembly using finite elements, it was decided to use the AL loading regime as this appeared to be the most damaging.

5 Design of the Finite Element Model

Using finite element software, models were created of both bucket and boom. 730 elements and 1194 quadrilateral elements with corner nodes, see Figure 3, were used for the bucket and the boom respectively.

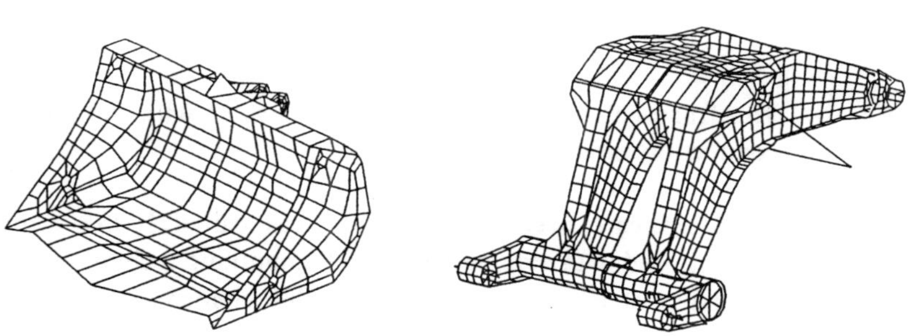

Figure 3 The 'plate' FE-model of bucket and boom (by RUSS)

It is possible to criticise the simple quadrilateral element in terms of accuracy. However, it was felt that in order to obtain a complete picture of the stress-strain state and to determine the maximum and minimum stress levels this type of modelling would be sufficient as an initial step.

For the model to be accurate, it was necessary to study the interaction of the assembled parts, through the joint and lugs. It was necessary to account for real stiffness of the structures and clearances at the connection points (Note the fork pin clearance can vary between 0.35 - 0.5 mm).

In modelling the connections between the bucket and boom, beam and truss finite elements were used. Beam element were used for modelling the cylindrical pin in bending, compression and even torsion and truss elements for only tension and compression. The truss elements were used in a manner which allowed the reactions in the vertical and horizontal plates of the pin assembly and taking into account any load redistribution as a consequence of clearance in the joint unit. A sectioned fork-pin unit, showing the elastic truss elements is shown in Figure 4.

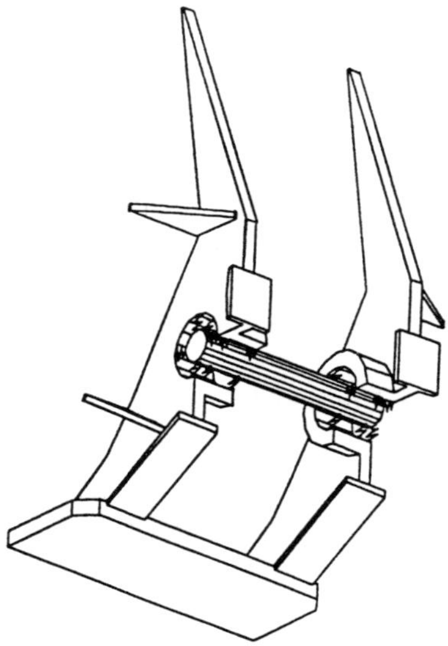

Figure 4. The fragment of the LHD bucket fork joint with pin and its connection model.

The finite element software used in the study was called RUSS (from Moscow) which when used with available personal computing technology has limitations. This version of RUSS was limited to 7000 elements.

Consequently, due to limited computing hardware and time, the FE models were simplified. For example the bucket was modelled by a 'plate' FE Model in the assembly with both a 'plate' and 'beam' FE models for the boom Additionally the beam model geometrical parameters were calculated on the basis of the structural peculiarities. The system hydraulic jacks by which the bucket and the boom are controlled were represented in the model by truss-elements.

6 Simulation Results

The stress-strain state for the working unit of the LHD was investigated under the different loading regimes. The results are presented in a conventional stress graph, where the numbers of the elements are on the x-axis and the equivalent stress magnitude on the y-axis. This allows easy determination of the maximum and minimum values in strategic locations around the structure as well as evaluating any middle stress quantities.

There are obviously differences between the equivalent external stress values, at the surface of the flat plates and the conventional internal stress values. The distribution of equivalent stress on external surface of the boom model, under the AL loading regime is presented below in Figure 5.

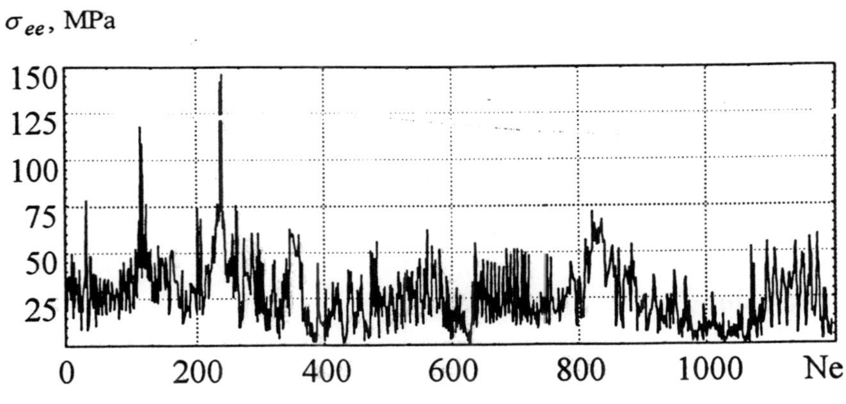

Figure 5. Equivalent stress through FE-model of boom under loading regime AL

The maximum equivalent stress in the lugs of the bucket hydraulic jack connection is 160 - 183 MPa, whereas, at the the bucket side walls and working blade it is 140 -158 MPa. The minimum stress is zero. These results show the high level of stress non-uniformity in this structure and confirms the necessity to reduce the stress in this structure in order to increase its working life.

Further analysis of the stress at the fork-pin joint truss-elements shows that there is a multi-component force vector in the joint units which can be resolved into 3 forces and 2 moments in the centre of the hinge joint (moreover these force moment vectors are perpendicular to the hinge axis at the horizontal and vertical plates). Thus, because of the flexibility of the bucket body, bucket hinge lugs and boom body, the force moments created in the hinge joint give additional loading to the various parts of the structure and can cause the mating elements of the joint to seize.

The influence of the torsional stiffness of the boom elements, particularly the boom front forks, was examined. It was thought that reduction in moment at the expense of the torsional elasticity would be beneficial. If, for example the force moment on the horizontal and vertical plates was reduced from 8-8.5 kNm to 3-4 kNm at the expense of a redesign of the front forks, the equivalent stress in the pin fork elements was significantly lowered.

The maximum equivalent stress for the LHD boom, based on the 'plate' model, see Figure 5, was 140 -146 MPa at a stress concentration location (where the boom foot locates on the LHD semi-frame).

The finite element results obtained from the personal computer were compared with the experimental strain results obtained earlier, for similar loading conditions. The measurements from ten strain gauge rosettes compared well with results from the computer analysis with differences of between 6 - 9% being noted.

For the specific case of the bucket loaded at its edge, the maximum computed strains and corresponding maximum displacements at a specific location on the bucket were found to be 5-18 mm. This also compared well with the observed displacement measurements of the machines in service.

7 Conclusions

The finite element model has successfully been used to model the stress-strain state of a LHD working unit, including the interconnecting hinge units.

The development of an finite element model which combines a 'plate' FE model (bucket and boom components) and 'beam and truss' FE models for the units of connection gives a precise description of the stress-state existing in the interconnecting hinge units.

Maximum and minimum stress for the LHD bucket and boom assembly were displayed and can be used to modify the design to reduce stress non-uniformity and improve the performance of the equipment in service.

Although not addressed in this study it is thought that strain rate is a significant factor in these LHD failures and this will be a focus of attention in future studies. It is also hoped that further work will include a computational fatigue analysis of strategic locations of the bucket and boom components based on the stress time histories and materials data.

References

1. Antoniak J, "Urzandzenia i Systemy Transportu Podziemnego w Kopalniach" - Katovice, Slack, p 777, 1990

2. Semenchenko A.K., Ignatov V.I., et al. "The LHD PD-8 Mathematical Model as the Three-dimensional Multimass Dynamic System", Izvestiya Vuzov, Gorny Zhurnal, 1994, 3, pp 15-19, 1994.

INDEX OF AUTHORS